T0324170

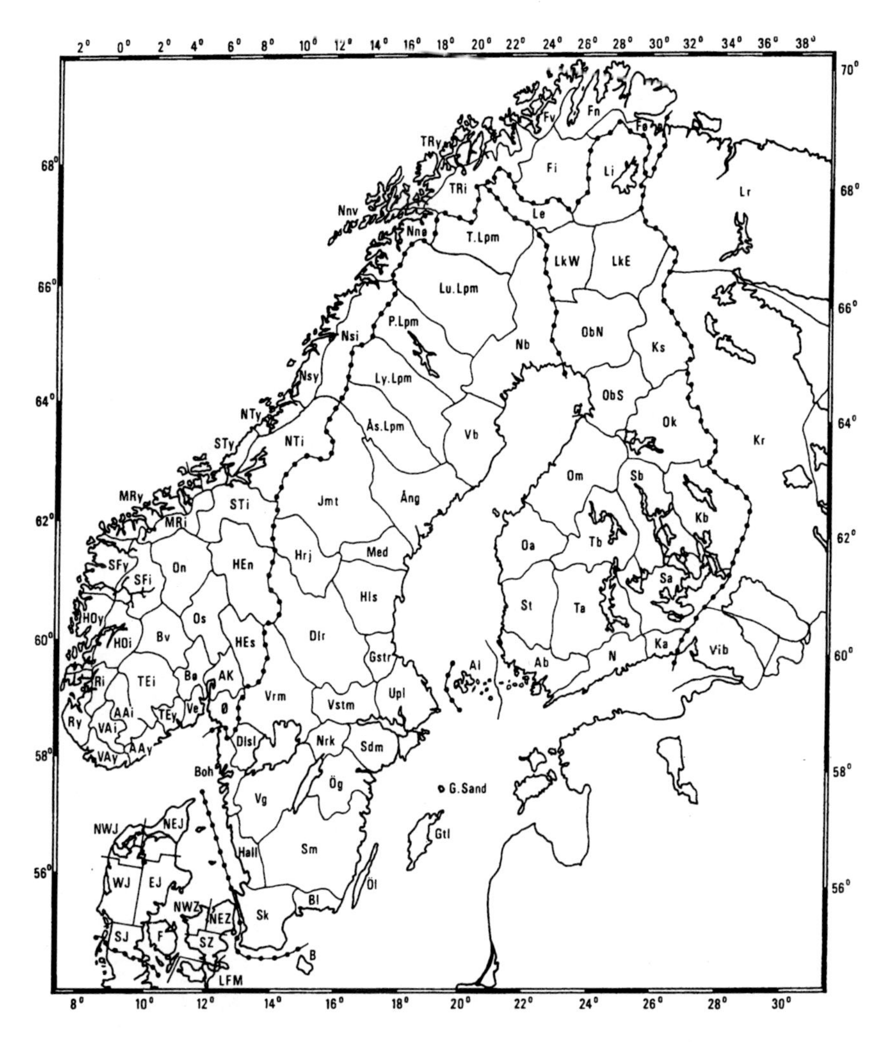

List of abbreviations for the provinces used throughout the text, on the map and in the following tables.

DENMARK

SJ	South Jutland	LFM	Lolland, Falster, Møn
EJ	East Jutland	SZ	South Zealand
WJ	West Jutland	NWZ	North West Zealand
NWJ	North West Jutland	NEZ	North East Zealand
NEJ	North East Jutland	B	Bornholm
F	Funen		

(Continued on back cover)

FAUNA ENTOMOLOGICA SCANDINAVICA
Volume 6 1977

The Elachistidae (Lepidoptera) of Fennoscandia and Denmark

by

E. Traugott-Olsen
&
E. Schmidt Nielsen

SCANDINAVIAN SCIENCE PRESS LTD.
Klampenborg . Denmark

Edited by
Leif Lyneborg

World list abbreviation
Fauna ent. scand.

Text composed by
The Setting Room Limited
Tunbridge Wells, Kent, England

Text printed by
Vinderup Bogtrykkeri A/S
7830 Vinderup, Denmark

Plates reproduced and printed by
Grafodan Offset A/S
3500 Værløse, Denmark

Publication date
15 October, 1977

ISBN 87-87491-14-1

Contents

Plate-figures are given on pp. 128–251, and a review of these can be found on p. 127.

Abstract

The Elachistidae (Lepidoptera, Gelechioidea) of Northern Europe are monographed. The seventy-seven species in six genera constitute the species hitherto recorded from Finland, Sweden, Norway, Denmark and Great Britain and nine species currently known from adjacent regions. All taxa are keyed and described. Both sexes are illustrated in full colour, the plates being accompanied by diagnostic characters. Wing venation, male and female genitalia of all species, and many leaf-mines are illustrated.

Twelve of the treated species, including *kebneella* sp.n., are arranged in a new genus *Biselachista* gen.n. Three additional W.Palaearctic and five Nearctic species are placed in new combination with *Biselachista* gen.n. Females of *Elachista parasella* Tr.-O., *E. littoricola* Le March. and *Biselachista abiskoella* (Bengts.) are described for the first time. One neotype and seventeen lectotypes are here designated.

The following new synonyms are established in Elachistidae:
Atachia Wocke, 1876 = *Elachista* Treitschke, 1833.
Elachista fuscogrisella Rebel, 1936 = *Elachista pigerella* (Herrich-Schäffer, 1854).
Elachista boursini Amsel, 1951 = *Elachista biatomella* (Stainton, 1848).
Elachista extensella Stainton, 1851 = *Elachista atricomella* Stainton, 1849.
Elachista zetterstedtii Wallengren, 1852 = *Elachista atricomella* Stainton, 1849.
Elachista longipennis Frey, 1885 = *Elachista atricomella* Stainton, 1849.
Elachista ranenensis Strand, 1919 = *Elachista pulchella* (Haworth, 1828).
Elachista albimarginella Hering, 1924 = *Elachista pulchella* (Haworth, 1828).
Elachista monosemiella Rössler, 1881 = *Elachista cerusella* (Hübner, 1796).
Elachista subcollutella Toll, 1936 = *Elachista subocellea* (Stephens, 1834).
Poeciloptilia pullella Herrich-Schäffer, 1855 = *Elachista pullicomella* Zeller, 1839.
Elachista laetella Rebel, 1930 = *Elachista subalbidella* Schläger, 1847.
Elachista mitterbergeri Rebel, 1906 = *Biselachista serricornis* (Stainton, 1854).
Elachista bistictella Tengström, 1848 = *Cosmiotes freyerella* (Hübner, 1825).
Elachista baltica Hering, 1891 = *Cosmiotes freyerella* (Hübner, 1825).
Elachista roesslerella Wocke, 1876 = *Cosmiotes consortella* (Stainton, 1851).
Elachista tyrrhaenica Amsel, 1951 = *Cosmiotes consortella* (Stainton, 1851).

The following synonyms are established in Heliozelinae (Incurvariidae):
Dyselachista Spuler, 1910 = *Heliozela* Herrich-Schäffer, 1853.
Aechmia saltatricella Fischer von Röslerstamm, 1841 = *Heliozela sericiella* (Haworth, 1828).

Introduction

During the last century, the Elachistidae have generally been recognized in various lists and catalogues of Lepidoptera from the Nordic countries, and several valuable taxonomic and faunistic papers have been published from this area during the last forty years. No work giving full treatment of all species in the family occurring in northern Europe has been published until now.

Colour illustrations, generally both of male and female, are presented for each species. Further, illustrations of the male and female genitalia and the venation of almost all species are provided; the larval mines of 25 species are also figured.

Keys, based on external characters and on male and female genitalia, are provided to genera and to the species of each genus. The keys are not intended to indicate phylogenetic interrelationships.

For each species, synonyms are chronologically listed with a bibliographical reference to the original description. Where possible, the original publication was examined; reprint pagination is only used to a minor extent. Printed dates of publication are accepted as correct, unless published evidence to the contrary exists. Infrasubspecific names *(International Code of Zoological Nomenclature,* Article 45) are not adopted. A synoptic redescription of the male of each species is given, including wing expanse (from tip to tip of forewing) and dealing mainly with such characters as cannot easily be deduced from the colour-plates. Females are only described where they differ from males. Descriptions of genitalia attach importance to diagnostic features. Individual variation is included in the description where possible, but is otherwise treated separately. *Differentia* are generally not given in the descriptions, as they are given in the keys and the short diagnosis with the colour-plates.

The general distribution of each species is very briefly outlined; a more detailed information on distribution in Fennoscandia and Denmark is given in the catalogue. Under the heading "Biology" notes on immature stages, food-plants, phenology and biotopes are given. Food-plants are also listed in a separate index. The nomenclature of plants is that of Lid (1974); no key is provided to larvae, but reference is given to Hering (1957).

All illustrations except for those of leaf mines and chaetotaxy were prepared by E. Traugott-Olsen. The text was prepared in collaboration; the introduction and final preparation of the manuscript was by E. Schmidt Nielsen.

During our work we have received much support and assistance. We are particularly grateful to Mr. I. Svensson, Österslöv; Dr. N. L. Wolff, Zoological Museum, Copenhagen; Dr. H. Krogerus, Helsinki; Dr. M. von Schantz, Helsinki and Mr. M. Opheim, Oslo, who, by the impetus given by their Scandinavian Microlepidoptera Meetings, initiated the modern study of Elachistidae in Scandinavia. They provided valuable information, discussion and loaned us material. Dr. K. Sattler and Dr. G. S. Robinson, British Museum (Natural History), London; Dr. J. D. Bradley, Commonwealth Institute of Entomology, London; Dr. H.-J. Hannemann, Museum für Naturkunde an der Humboldt-Universität zu Berlin; Mr. E. C. Pelham-Clinton, Royal Scottish Museum, Edinburgh; Dr. U. Parenti, Museo ed Istituto di Zoologia Sisematica, Turin; Dr. J. Klimesch, Linz, Austria, and Mr. O. Karsholt, Præstø, Denmark, loaned material and engaged in valuable discussion on some of the worst problems.

Further loans of material have been arranged by Dr. A. Lillehammer, Zool. Mus., Oslo; Dr. A. Löken, Zool. Mus., Bergen; Lic. P. Douwes, Mr. G. Samuelsson and Mr. R. Danielsson, Zool. Inst., Lund; Mr. B. Gustafsson, Naturhist. Riksmus., Stockholm; Dr. W. Hackman and M. Jalava, Zool. Mus., Helsinki; Dr. M. Geishardt, Museum Wiesbaden; Dr. J. Razowski, Institute of Systematic and Experimental Zoology, PAN, Kraków; Dr. A. Riedel, Instytut Zoologicny, PAN, Warszawa; Dr. F. Kasy, Naturhistorisches Museum, Wien; and R. W. Hodges, U.S. Department of Agriculture, Washington.

From Dr. H. Steuer, Bad Blankenburg, we borrowed many preserved larvae and his excellent mine herbarium. The lepidopterous section from the main part of the H.P.S. Sønderup herbarium, kept by Mr. N. U. Møller, Ålborg, was also kindly placed at our disposal.

Further material and help has been received from the late Dr. P. Benander, Höör; Mr. B. Å. Bengtsson, Löttorp; Mr. O. Buhl, Odense; Mr. M. Carlsson, Enskede; Mr. D. J. Carter, London; Dr. R. Gaedike, Eberswalde; Mr. H. Hendriksen, Allerød; Mr. P. L. Holst, Lemvig; Mr. E. Jäckh, Bidingen; Lic. J. E. Jelnes, Copenhagen; Mr. R. Johansson, Växjö; Mr. J. Jonasson, Göteborg; Mr. J. Kyrki, Oulu; Mr. K. Larsen, Copenhagen; Mr. J. Lundquist, Hillerød; Mr. K. R. Tuck, London; Mr. G. Pallesen, Århus; Mr. E. Palm, Føllenslev, the late Mr. E. Pyndt, Sakskøbing; and Mr. K. Pedersen, Sakskøbing.

We also thank Mrs. G. Lyneborg for drawings of leaf mines and chaetotaxy, Mr. G. Brovad and Mr. B. W. Rasmussen for photographical work, and Mrs. M. Bévort who typed the manuscript.

Special thanks are due to Dr. G. S. Robinson, British Museum (Natural History), London, who critically read the manuscript to correct the language and also proposed several important alterations, to Dr. N. P. Kristensen, Zoological Museum, Copenhagen for very valuable discussions and critical remarks during the work, and to the managing editor, Dr. L. Lyneborg, for much support.

We owe a debt of gratitude to everybody for their kind and valuable assistance.

Family Elachistidae

Elachistidae Bruand, 1850: 50.
Poeciloptilina sensu Herrich-Schäffer, 1857: 58.
Elachistinae Swinhoe & Cotes, 1889: 721.
Cycnodioidea Busck, 1909: 92.
Cycnodiidae Busck, 1911: 46.
Cycnodiadae Hampson, 1918: 387.
Aphelosetiadae Hampson, 1918: 387.
Elachistini Handlirsch, 1924: 886.
Aphelosetiidae Brues, Melander & Carpenter, 1954: 258.

Diagnosis

Small species. Head smooth-scaled, frons vertical, neck tufts raised above head. Antenna about three-quarters length of forewing, directed backwards in repose; scape with pecten, without eye-cap; flagellum rather smooth, often finely serrate, rarely with short cilia. Maxillary palpi very small; labial palpi rarely short and drooping, generally slender, porrect to recurved, diverging. Haustellum present. Hind tibia with long hairs above and beneath. Forewing narrow; hindwing lanceolate, much narrower than length of cilia (fringe). Forewing with all R veins to costa, hindwing with Rs to apex through axis of wing.

Male genitalia with spinose gnathos and digitate process from inner surface of base of valva.

Female genitalia with distinct antrum and often also with distinct colliculum.

Morphology of adult Elachistidae

Head (Text-fig. 1).

Smoothly scaled; three differently coloured scale-groups are often recognized: frontal, clypeal and vertical. Behind vertex two groups of raised scales (**neck tufts**) from **patagia. Chaetosemata** and **ocelli** absent. **Maxillary palpi** very short, one- or rarely two-segmented. **Haustellum** (proboscis) rarely reduced, generally shorter than labial palpi, segments weakly sclerotized, basal third with scales, sensillae simple, with a strong conical spine on a slightly enlarged semispherical socket without ribs.

Labial palpi three-segmented, about twice as long as width of head, diverging, curved, porrect to recurved, except in a few genera with secondarily reduced, short drooping palpi. First segment about one-quarter length of the long and curved second segment, third segment generally slender and acute, less than half as long as second. The scale covering of the first segment is smooth, while scales above and beneath second segment and (more rarely) third segment are raised; the underside of the second segment is often particularly strongly roughened.

Antennae (Text-figs. 2-6).

These are half as long as forewing or longer, never longer than forewing. **Scape** large, without eye-cap, but often with well developed pecten; reduction of pecten takes place from distal end. In many species the segments of **flagellum** are clothed with scales of varying colour, darkest in the middle of the segment and lighter towards junctions, giving the antenna an annulated appearance. Often the flagellum is simple, filiform (Text-fig. 4), but scales may be raised to a varying degree giving a serrate impression (Text-figs. 5, 6). Further, the flagellum is provided with fine hair-like scales, cilia; in a few species these are raised (Text-figs. 2, 6), in others they are appressed or curved.

Wings (Text-figs. 7-9, 38, 39).

Shape of **forewing** varies from short and broad to very narrow and elongate. **Hindwing** is always narrow to lanceolate, narrower than length of **cilia** (fringe). Several brachypterous Elachistidae are known from the Subantarctic islands, but in the Northern Hemisphere reduction of wings has not yet been observed in this family.

Terms applied to wings and wing pattern are indicated on text-figs. 38, 39. **Retinaculum** in forewing on subçosta in male and on subcosta and cubitus in female; **frenulum** with one bristle in male and with three bristles in female.

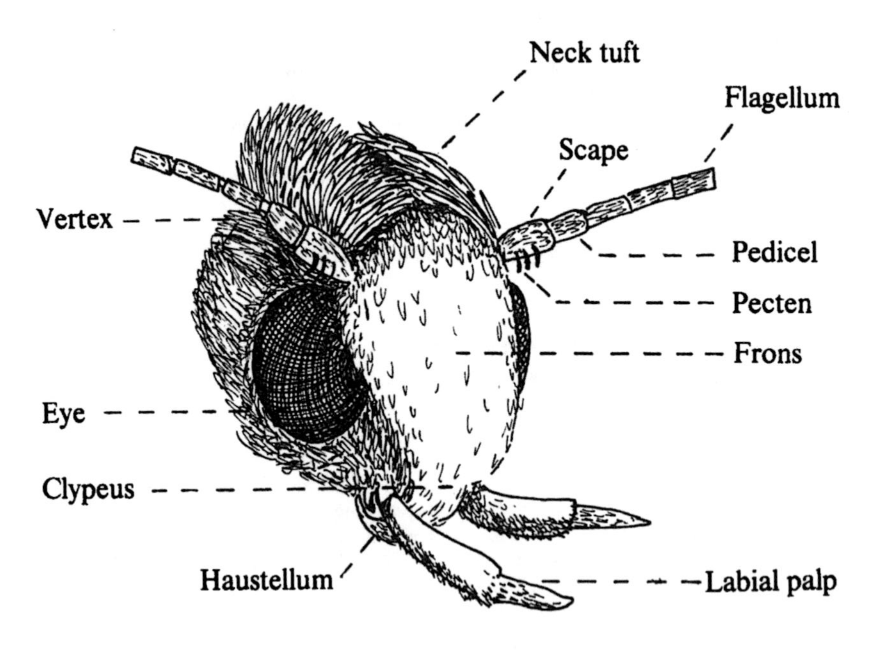

Text-fig. 1. Head of an *Elachista*.

Forewing with R five-branched, all branches to costa, R_1 arising from before middle, remote from R_2, which arises from well before the upper angle of distal cell, R_4 and R_5 stalked or coalescent, chorda only present in *Mendesia*, M primitively three-branched, generally two branched, single in *Cosmiotes*. M-stem in cell only distinct in *Mendesia*, but distal portion of this stem is often visible as a short spur on in-

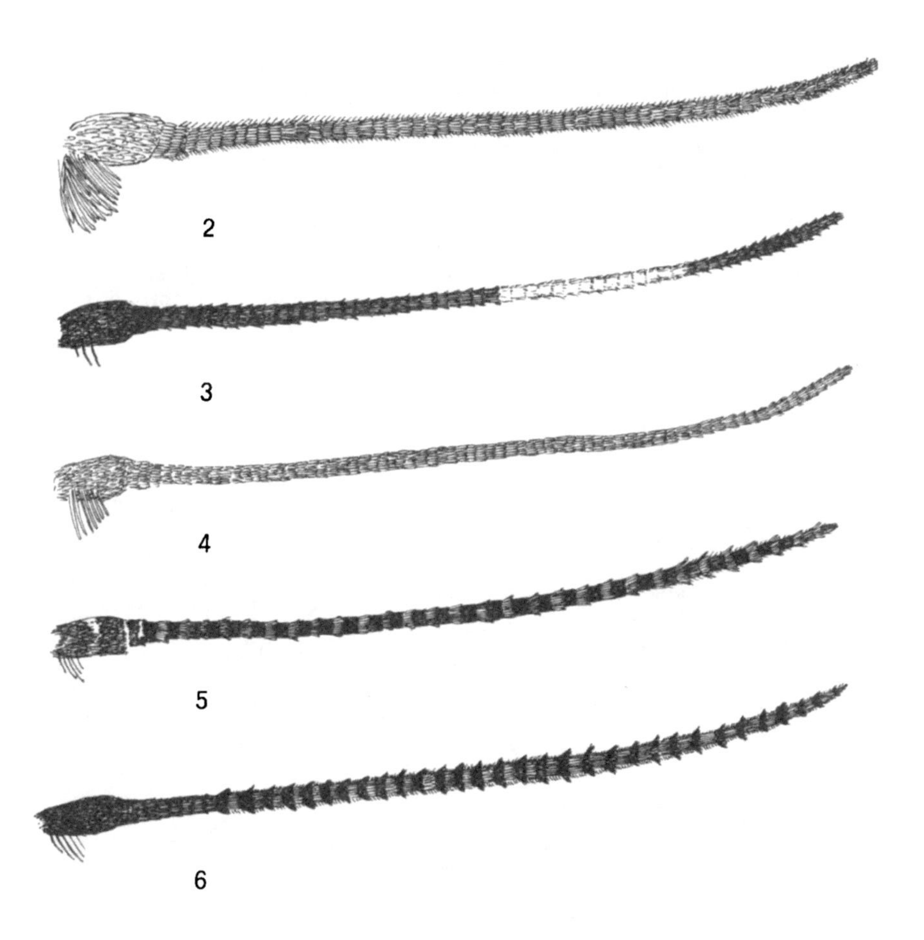

Text-figs. 2–6. Antennae of Elachistidae. – 2: *Mendesia farinella* (Thnbg.), ciliate type; 3: *Stephensia brunnichella* (L.); 4: *Elachista argentella* (Cl.), with appressed scales; 5: *Elachista* sp., sligthly serrate distally; 6: *Biselachista serricornis* (Stt.), strongly serrate and ciliate.

ner side of the outer margin of discal cell. CuA_2 always distinct. Only distal part of CuP (=Cu_2 auct.) sometimes distinct. Anal vein forked at base in most primitive species, otherwise simple.

Hindwing with $Sc + R_1$ remote from Rs, length varying from one-third of hindwing to well beyond middle. Rs extends along or very near the longitudinal axis to apex, often with a branch to costa. M_1 stalked with Rs, M_1 and M_2 stalked with CuA, coalescent or absent. Cell primitively closed, open in most species.

Individual variation in venation has not been extensively examined, but it is generally slight. In species with almost fused, shortly forked veins these veins may be totally fused in some specimens; the basal forking of 1A + 2A may be absent in species which generally have a long and distinct fork (e.g., *E. pigerella* (HS.) pl.-fig. 159 and fig. 1 in Steuer (1973)).

It is of special interest that the free branch from Rs to the costa of the hindwing can be absent or present in the same species (e.g., *E. argentella* (Cl.), pl.-figs. 191, 192 and *E. alpinella* Stt., pl.-fig. 168 and fig. 2 in Steuer (1973)). Much attention has been paid to this forking of Rs. Busck (1909, 1914) and Braun (1933) suggested the two-branched condition of Rs to be transitional between homoneurous and heteroneurous venation; this condition is now considered to be a specialization found in cases of reduction in heteroneurous venation; it is also observed in other narrow-winged families.

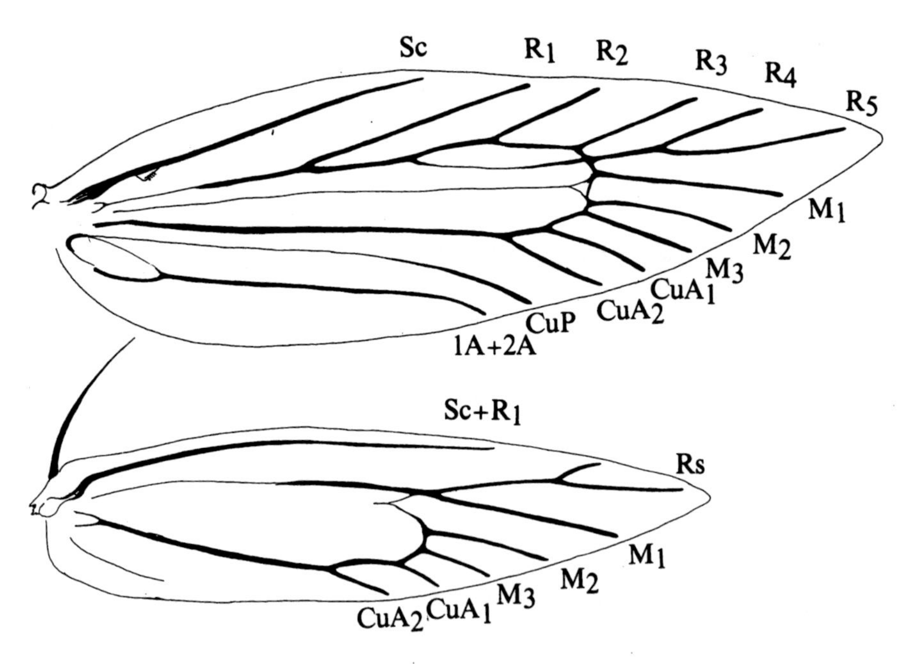

Text-fig. 7. Wing venation of *Mendesia farinella* (Thnbg.).

Legs (Text-fig. 10).

Fore tibia with a slender **epiphysis,** which is setose on the inner surface. Mid tibia with one pair of **apical spurs** and hind tibia with two pairs of spurs, **medial pair** before or well before the middle. Each segment of tarsus with a varying number of distal spines, of almost equal length to scales, only significantly longer in *Mendesia*.

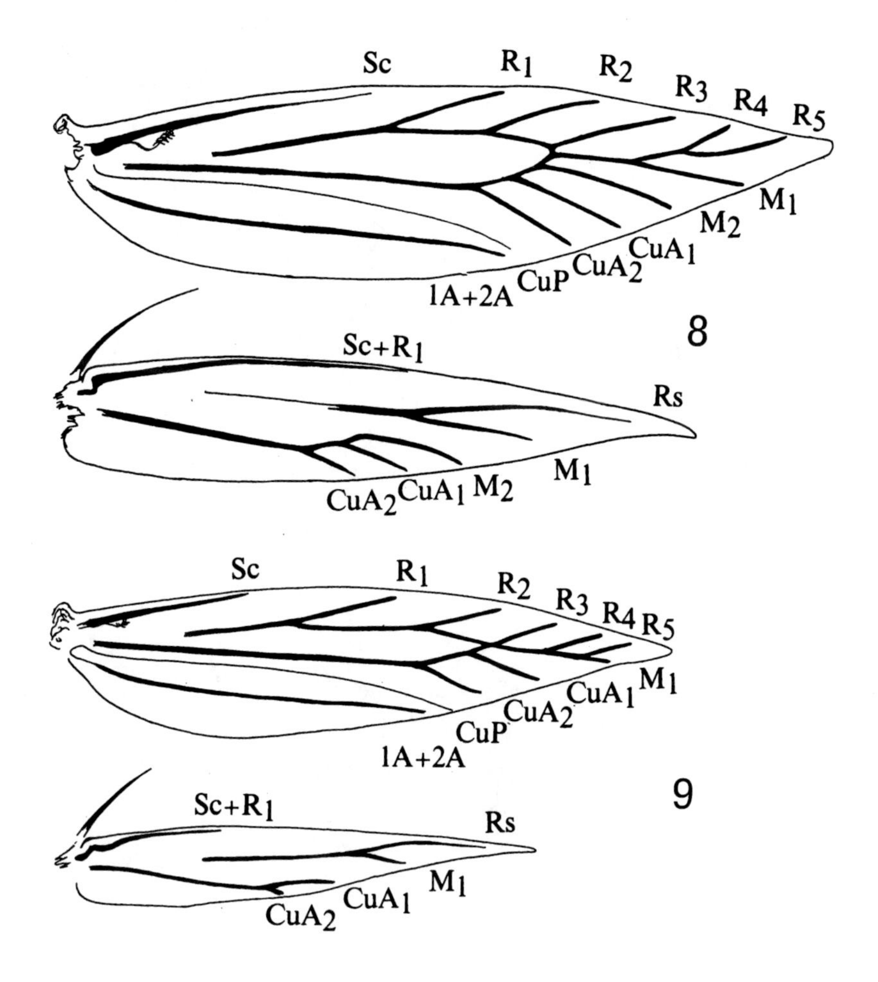

Text-figs. 8–9. Wing venation of 8: *Elachista alpinella* Stt.; 9: *Cosmiotes freyerella* (Hb.).

The terminology here applied to the genitalia follows that of Klots (1956) where possible. Recently, a review of the literature about the genitalia of the Lepidoptera and the origin of the various structures has been given by Matsuda (1976).

Male Genitalia (Text-figs. 11-17).

These belong to a rather generalized symmetrical gelechioid type, showing only few specializations. The **uncus** is generally very well developed, strongly reduced in some of the primitive genera and in the exclusively Nearctic genus *Hemiprosopa* belonging to the *Elachista*-group. In the genera with a large uncus, this is primitively very large, not or only slightly emarginate or cleft, showing specialization to deep indentation and separation into two distinct rounded or conical lobes. The lobes often have

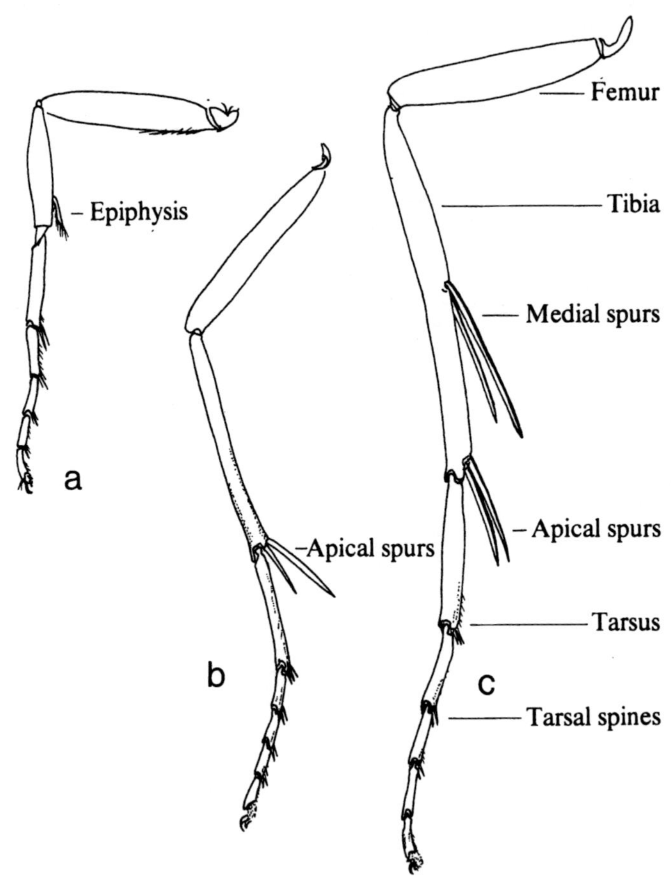

Text-fig. 10. Legs of *Elachista argentella* (Cl.) – a: fore leg; b: mid leg; c: hind leg.

strong, long setae (*bifasciella*-group), sparse short setae (*bedellella*-group), or thick or clavate setae (*Biselachista*); in *Cosmiotes* the lobes are claw-like, inwardly curved and often with bare tips. **Socii,** which arise from the base of the uncus, are small and hairy. The **tegumen** is always well developed, broad and truncate, the lateral margins and posterior margin often more strongly sclerotized and with numerous sockets. In several genera the anterior margin is very strongly sclerotized, often band- or arm-shaped.

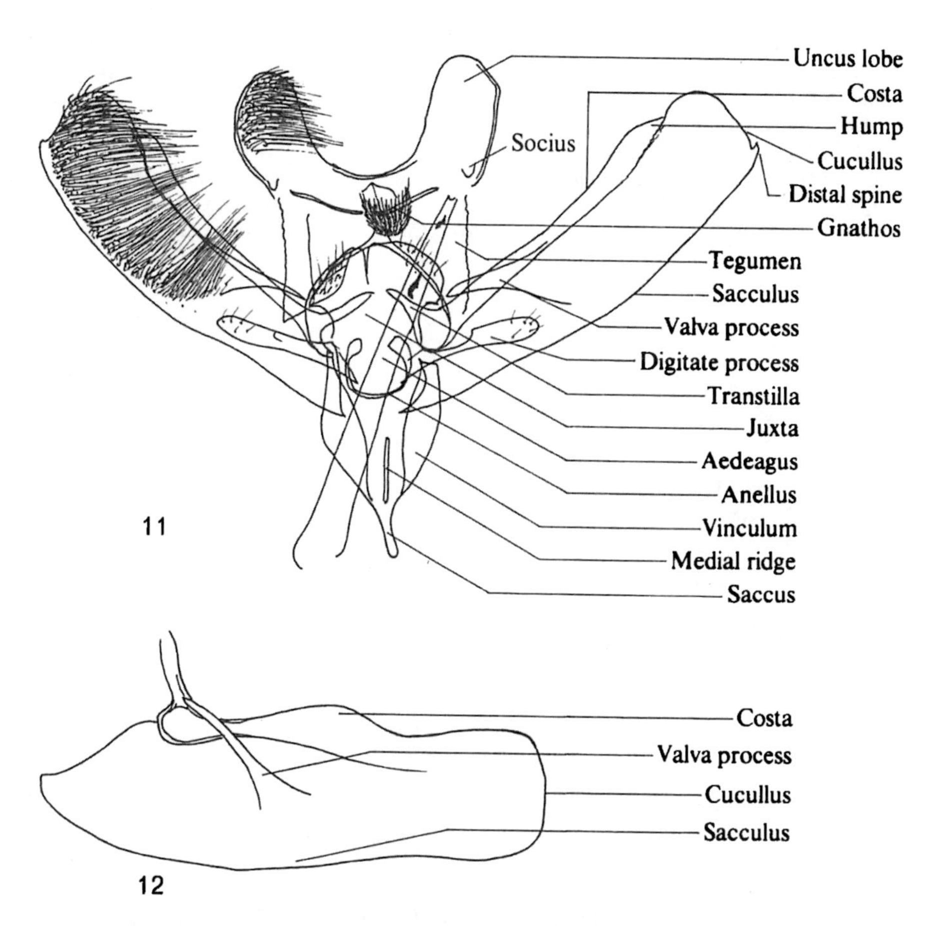

Text-figs. 11, 12. Morphology of male genitalia in Elachistidae. – 11: species of the *Elachista bifasciella*-group; 12: valva of species of the *Elachista bedellella*-group.

The **gnathos** is oval or ellipsoid, with transverse rows of comb-teeth, bifurcate (in *Biselachista*) or single (all other genera here treated), attached by two very strong arms to the caudal margin of the tegumen, just below the socii. The anal area is generally not distinctly sclerotized. The **vinculum** is a distinct, V- or U-shaped structure, often produced mid-ventrally into a short, often stout **saccus** and often provided with a medial sclerotized ridge, and dorso-laterally produced into two slender extensions articulating with the anterior extensions of the tegumen.

The **valva** is a merely simple, almost homogeneous structure, in which three not always distinctly demarcated regions are recognized: the dorso-proximal sclerotized margin is termed the **costa,** the ventral margin the **sacculus** and the distal, often enlarged portion, the **cucullus.** Costa is generally not setose, medially produced into a short, tapering **transtilla,** which only in *Cosmiotes* forms **labides**; distally, the costa is swollen into a **hump** (several *Elachista*) or produced into a thorn-shaped **spine** above cucullus (*Stephensia*); in many species the costa is straight to the hump, in others it is medially convex and provided with a sub-apical emargination before the cucullus (Text-fig. 12). The sacculus is often produced into a spine below the cucullus. The angles between the cucullus and costa and cucullus and sacculus are often of diagnostic importance. The distal two-thirds of the sacculus and cucullus are strongly setose. A process extends medially, almost from the middle of the valva, articulating with the lateral process of the juxta.

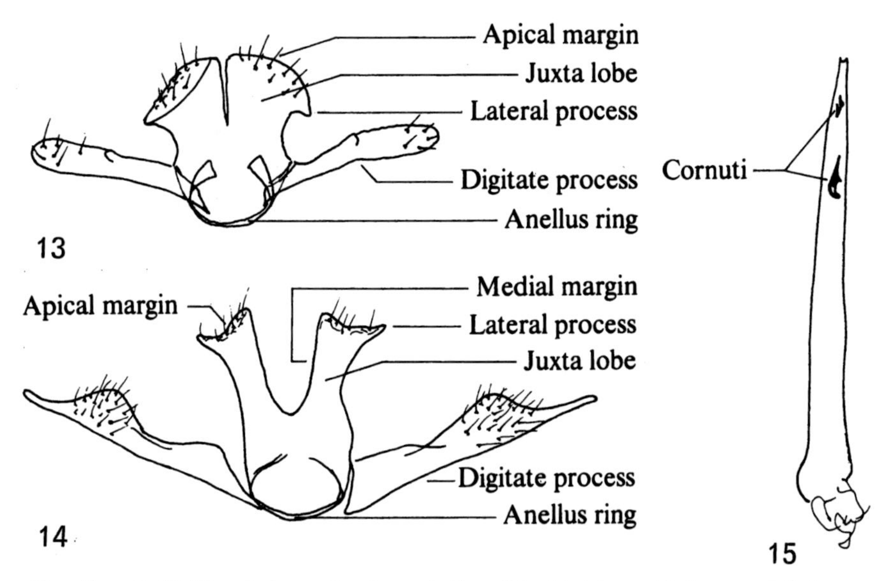

Text-figs. 13–15. Morphology of male genitalia of Elachistidae. – 13: juxta and digitate processes of species of the *Elachista bifasciella*-group; 14: juxta and digitate processes of species of the *Elachista bedellella*-group; 15: aedeagus.

The area between the tegumen, valvae and vinculum is closed by a membranous **diaphragma,** through which the **aedeagus** passes in a double cone, the **anellus;** below the aedeagus the diaphragma is sclerotized into a dorsally cleft shield-shaped structure, the **juxta,** the two lobes of which are here termed the **juxta lobes,** ("anellar lobes" of Dugdale (1971), "oberen Anellus Lappen" of Gaedike (1975)). The juxta lobes, formed by a medial slit, vary from simple rounded or conical structures to distally broadening lobes with a setose, outwardly oblique apical margin which is laterally produced into a short **lateral process** (Text-fig. 13); they may be elongate structures with the medial margin produced into a short process (Text-fig. 14). The juxta lobes are situated latero-ventrally to the aedeagus and basally fastened to the

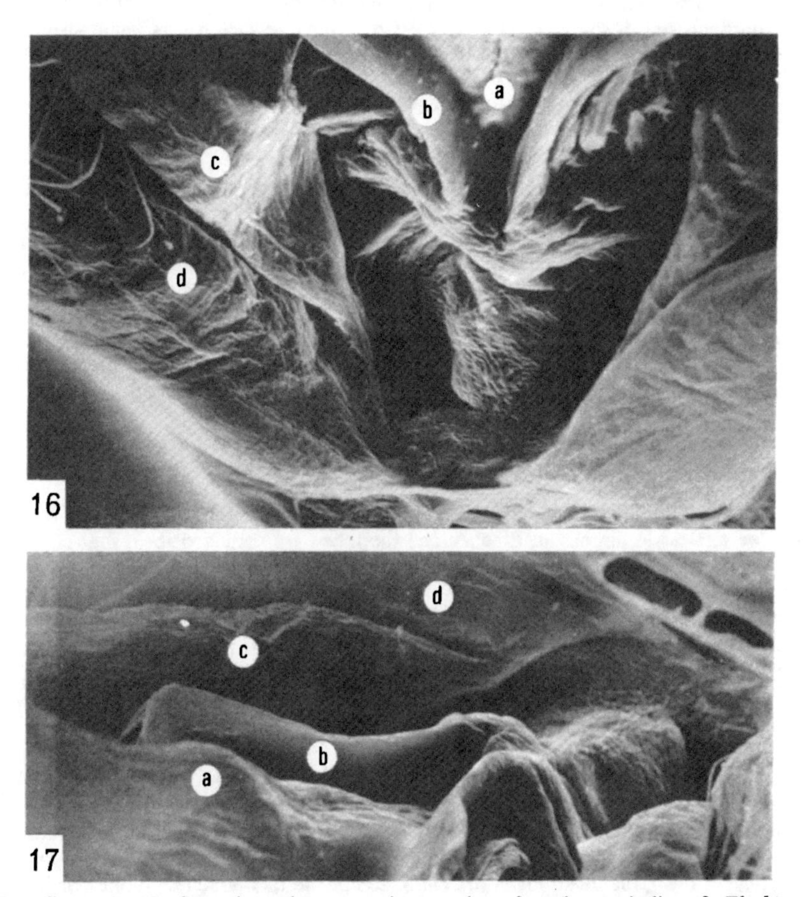

Text-figs. 16, 17. Scanning electron micrographs of male genitalia of *Elachista revinctella* Zell. – 16: ventral view; 17: ventro-lateral view; a: aedeagus; b: juxta; c: digitate process; d: valva.

inner base of the sacculus with the lateral processes articulating with the medial processes from the valva.

From between the juxta and the base of the valva a characteristic, setose, slender process arises, which is here termed the **digitate process.** Our use of this term is deliberate and designed to avoid the implications of homology in previously-applied terms. The true nature and origin of this process has long been a puzzle, and many terms have been used, e.g. "anellus lobes" (Metha, 1933), "sacculus" (Pierce & Metcalfe, 1935), "sacculus process" (Braun, 1948), "harpe" (Viette, 1954), "digitate process of the sacculus" (Bradley, 1963), "sacculus lobes" (Dugdale, 1971) and "shape" (Steuer, 1973). In preparations made for the light microscope it can appear fastened to the juxta-anellus complex or to the base of the valva, but in a scanning electron microscope examination of *Elachista revinctella* Zell. it was found strongly fastened to the inner margin of the base of the valva (Text-figs. 16, 17). Superficially similar processes arise from the mid-ventral part of the male genitalia in several Oecophoridae, but these apparently always arise from the juxta, and the "cuiller" (Hannemann, 1953) arises from the distal end of sacculus; the "clavus" (Hannemann, 1953) also appears different and much critical morphological examination will be necessary before homologies on the various processes in the male genitalia in the gelechioid families can be suggested.

The **aedeagus** (Text-fig. 15) is generally a simple tapering tube without a carina. The basal portion is bulbous, often divided into two or three lobes; the distal end may be chisel-shaped, square or indented to deeply cleft. The endophallus (**vesica**) is weakly sclerotized in a few species and often contains various sclerotized spines, **cornuti,** which are of high taxonomic value, even if both number and shape may vary. In *Biselachista* more strongly sclerotized, often band-shaped areas may be present in the wall of the aedeagus.

Female genitalia (Text-fig. 18)

The fused abdominal segments IX and X form a posterior pair of hairy lobes, **papillae anales;** in almost all Elachistidae the lobes are soft, and their shape varies from slender and triangular to short and broad. The lobes are entirely sclerotized only very rarely, but in the *bedellella*-group the anterior region from where the **apophyses posteriores** arise is sclerotized. The **apophyses anteriores** extend forward from the lateral parts of the sclerotized dorsal VIII; venter VIII is mainly membranous, but sclerotized to a varying extent. In the *bedellella*-subgroup a large distinct sternite is present; in other species a V- or U-shaped sclerotized band surrounds the copulatory orifice, the **ostium bursae** (e.g., in *E. argentella* (Cl.) and to a lesser extent in some species of the *unifasciella*-subgroup).

The ostium is situated mid-ventrally between the posterior margin of venter VIII and the middle of segment VII. The **bursa copulatrix** is always divided into two distinct parts, the **ductus bursae** and **corpus bursae.** The posterior part of the ductus is sclerotized to a varying degree, often in two areas. The **antrum** (termed "sterigma" by Dugdale (1971)), the most posterior sclerotized portion, is pouched, funnel- or bowl-shaped, and enlarged towards the ostium; the **ventral margin** of the antrum is mainly anteriorly curved, rarely straight or posteriorly curved; the dorsal margin is not distinct, while the lateral margins, **rims,** (termed "tails" by Traugott-Olsen (1974)), are distinct to a greater or lesser extent. The inner side of the antrum, the dorsal wall, ventral margin and rims may be spined in a pattern which varies between taxa.

The **colliculum,** the anterior sclerotized portion of the ductus bursae, is tube- or ring-shaped and varies much in length; in the *bifasciella*-group and *Biselachista* it is long and narrow, respectively fused with or separated from the antrum, while it is ring- or barrel-shaped in the *bedellella*-group; in some genera it is not distinct.

The membranous portion of the ductus is generally wider than the colliculum; near the inception of the **ductus seminalis** it contains a varying number of small sclerotized

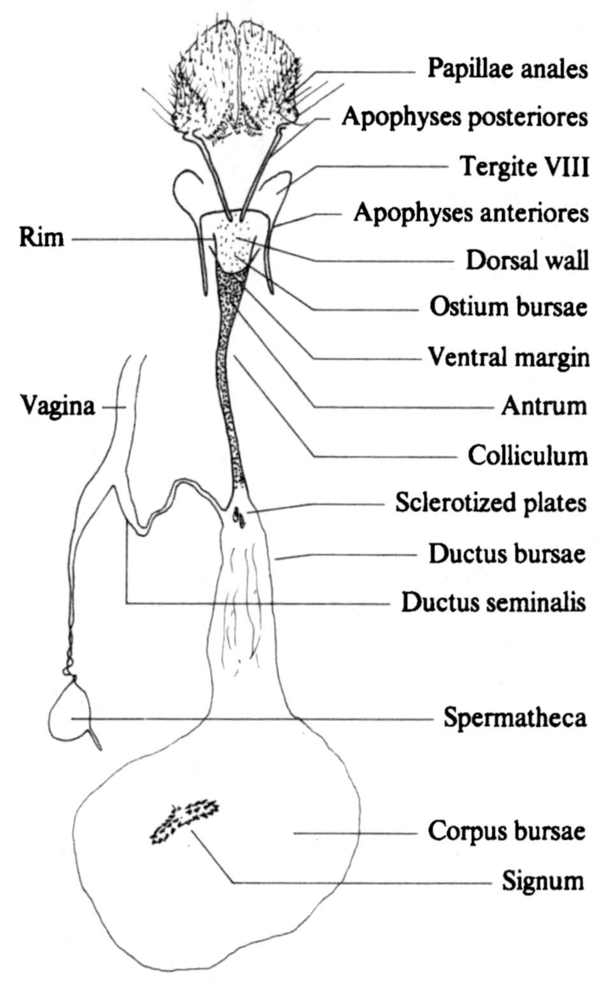

Text-fig. 18. Morphology of female genitalia of Elachistidae.

plates and the inner side is often minutely spined. The corpus bursae often has internal spines and further contains one to three sclerotized dentate plates of various shape, **signa,** which are absent in a few species.

The inception of the ductus seminalis on the ductus is anterior to the colliculum, often in a distinct pouch. In species with a long colliculum, the inception is almost in the middle of the ductus or even in the anterior part, while in species with a short ring-shaped colliculum the inception is in the posterior quarter.

The ductus seminalis leads to the **vagina,** as does the often long duct from the **spermatheca** (receptaculum seminalis); this region can be of some taxonomic importance in Elachistidae as part of the spermatheca, the **utriculus,** is sclerotized in some species, in others the spermathecal duct is distinct as a slightly sclerotized sinuous duct or it possesses a sclerotized ring.

A **spermatophore** may sometimes be present inside the corpus bursae (cf. Pl.-fig. 460), but it disintegrates very soon after having been deposited.

Individual variation in female genitalia is generally slight. Variation occurs particularly in the size of the spined patches on the dorsal wall of the antrum (e.g., in *E. ingvarella* Tr.-O., Pl.-figs. 429-436), in the degree of sclerotization along the ventral margin of the antrum (e.g., in *E. bifasciella* Tr., Pl.-figs. 422, 423 and in *E. ingvarella* Tr.-O., Pl.-figs. 429-436). Especially the number of sclerotized plates inside the ductus bursae may vary (e.g., in *E. ingvarella* Tr.-O., Pl.-figs. 429-436), or they may be present or absent in the same species (e.g., in *E. albifrontella* (Hb.) or *E. pomerana* Frey). A distinct signum may be present or absent in some species (e.g., *E. bifasciella* Tr., Pl.-figs. 422, 423, or *E. bedellella* (Sirc.), Pl.-figs. 466, 467).

Immature Stages

Eggs (Text-figs. 19, 20).

The egg is of the "flat" primitive type with the micropylar axis parallel to the substrate surface. It is generally ellipsoid or elongately oval, appressed to the surface, iridescent and strongly ridged. The ridges are mainly parallel to the longitudinal axis

19

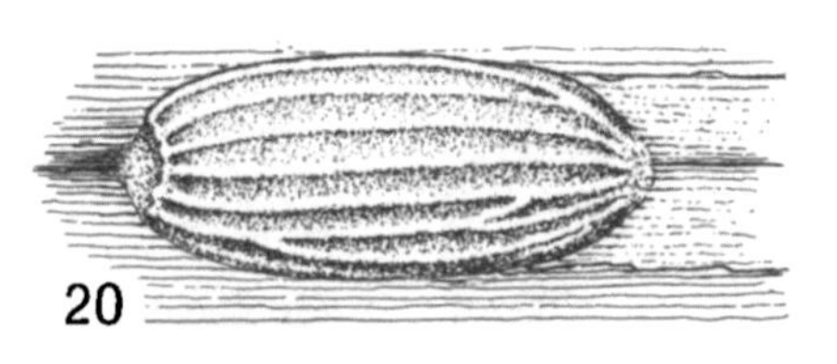

20

Text-figs. 19, 20. Eggs of Elachistidae. – 19: *Stephensia;* 20: *Elachista* (after Braun).

or more rarely irregularly branched, forming a rugose pattern. In *Stephensia* the egg is half inserted and semi-erect; this is probably a specialization at genus level as it is observed both in the Nearctic species (Braun, 1948) and in the Palaearctic species.

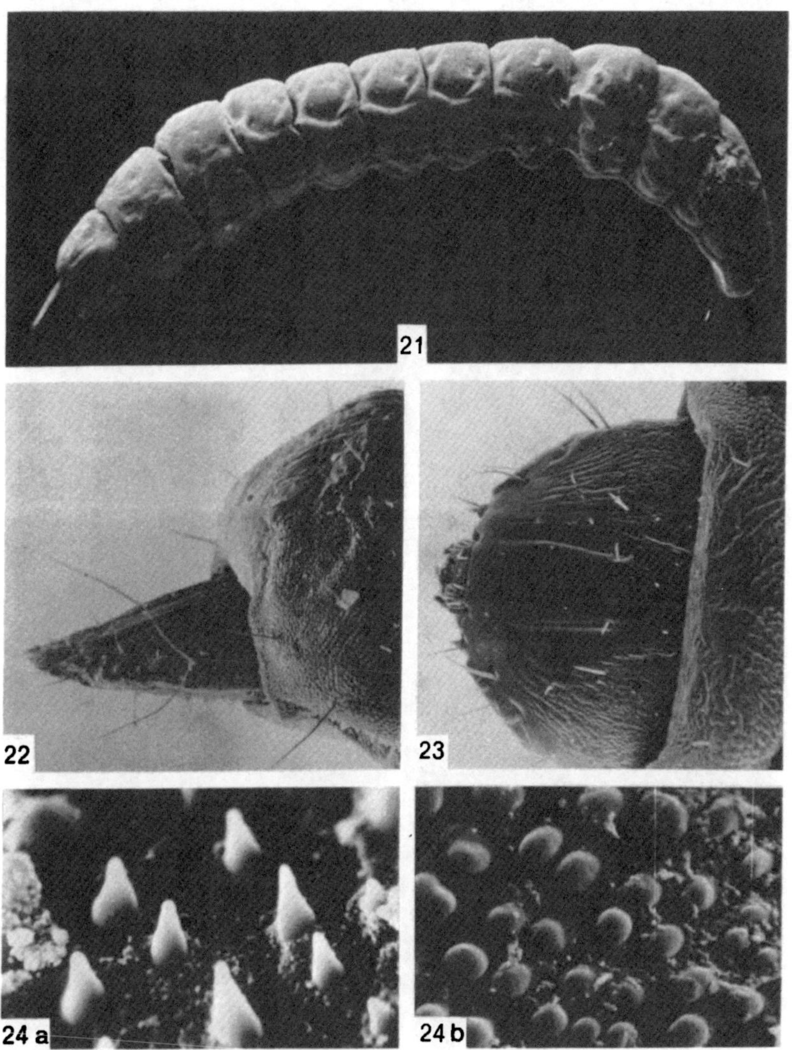

Text-figs. 21–24. Scanning electron micrographs of larva of *Elachista apicipunctella* Stt. – 21: larva; 22: head in lateral view; 23: head in dorsal view; 24a, b: integument of prothorax.

Larvae (Text-figs. 21-32).

The leaf-mining larva (Text-fig. 21) is small, less than 10 mm long, and the body is spindle shaped, with deep segmental constrictions. It is broadest at the pro- and mesothorax. The colour is generally whitish, greenish or yellowish, more rarely reddish or brownish; the ground colour turns darker if the larva is removed to outside the mine (Steuer, 1973). When feeding, the larva is more or less tinged with greenish by the green intestine; this greenish colour disappears during hibernation.

The surface sculpturing of the body integument varies widely. In the *bifasciella*-

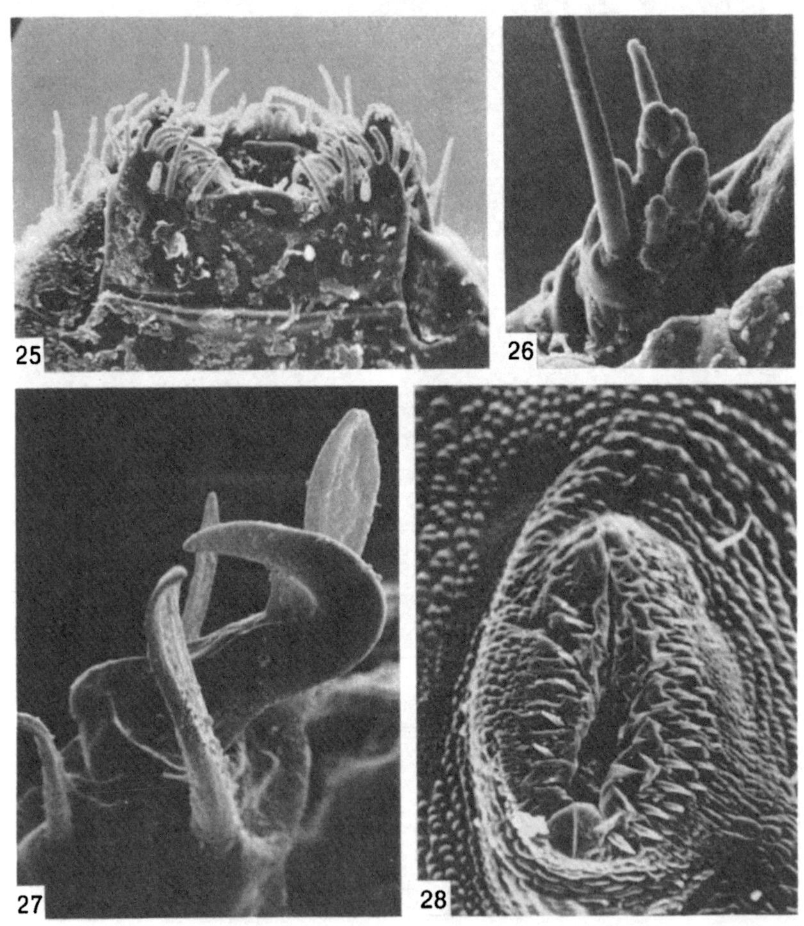

Text-figs. 25–28. Scanning electron micrographs of larva of *Elachista apicipunctella* Stt. – 25: oral opening in dorsal view; 26: antenna; 27: prothoracic leg; 28: abdominal proleg.

group the microsculpturing consists of variously shaped tubercles, and the form of these seems to be constant in each species. In *E. apicipunctella* Stt. the microtubercles are mainly flat-topped, interrupted by small regions with pyriform tubercles (Text-fig. 24). In all species of the *unifasciella*-subgroup, strongly sclerotized, oval or circular, flat tubercles are present; among these are scattered small circular blotches with more contiguous small tubercles and larger areas without sculpturing. Whether this sculpturing belongs to the "micro" or "macro" type as defined by Byers & Hinks (1973) is not clear.

The head is strongly sclerotized and flattened, narrower than the prothorax (Text-figs. 22, 23), prognathous, varying from elongate and narrow to short and broad. The genae often have longitudinal ridges and three dorsal and two or three ventral stemmata along the anterior margin. The antenna is short (Text-fig. 26), but not so reduced as in several other leaf mining families (Dethier, 1941; Hering, 1951). The labrum is large, the anterior margin with strong bristles (Text-fig. 25). The spinneret is atrophied and non-functional during the feeding period, and production of silk first starts at the time of pupation (Hering, 1951).

The prothoracic shield is longitudinally split into two major plates of varying distance apart; each plate may be further subdivided. The shape, size and colour of these plates is of high diagnostic value.

The thoracic legs are five-segmented and not reduced; the pretarsal setae are flattened, clavate, and as long as or longer than the claw (Text-fig. 27). Abdominal prolegs are also present, the crochets in a circle or penellipse and uniordinal or, rarely, partly biordinal (Text-fig. 28).

Little has yet been published about the chaetotaxy of Elachistidae and, for the present work, only a few species have been examined (*E. compsa* Tr.-O., *E. apicipuntella* Stt. and *E. gangabella* Zell.). Family characters cannot be outlined at present.

The chaetotaxy of *E. apicipuntella* Stt. is sketched in Text-figs. 29-32. Sockets on all segments are simple and low and no pinnaculae are present. On the prothorax, the very small seta just above and posterior to D2 is considered to be D1 and not MXD1, as D1 is also very small on all other segments; generally, D1 is anterior to D2. XD1 anterior to D2. XD2, SD1 and the small SD2 in an almost vertical row. The three prespiracular setae are anterio-ventrad of SD1; L1 and L3 approximate and only one SV seta present. On the mesothorax D1 is antero-dorsad of D2; SD2 is absent and the L setae are in an oblique row; no proprioreceptors are present. The D setae on abdominal segments I-VIII are similar to those of the mesothorax; SD1 is just above the spiracle, SD2 is absent, L3 is postero-ventrad of the spiracle and remote from L1 which is close to SV1; L2 is absent. On segment IX only four setae are present, aligned equidistantly in an almost vertical row. The absence of small proprioreceptors is confirmed by examination by scanning electron microscope.

Possible generic characters could be the absence of one SD and one L seta from the abdominal segments, as it appears from examination of the more primitive *"Scirtopoda" myosotivora* Müller-Rutz (Klimesch, 1939) that two SD and three L setae are present in that species (the generic position of this and some closely allied species will be published elsewhere by the authors). Braun (1948) also recognized three L setae on the abdominal segments of *Elachista albicapitella* Engel. Safe family chaetotaxy-characters can therefore only be deduced after examination of some of

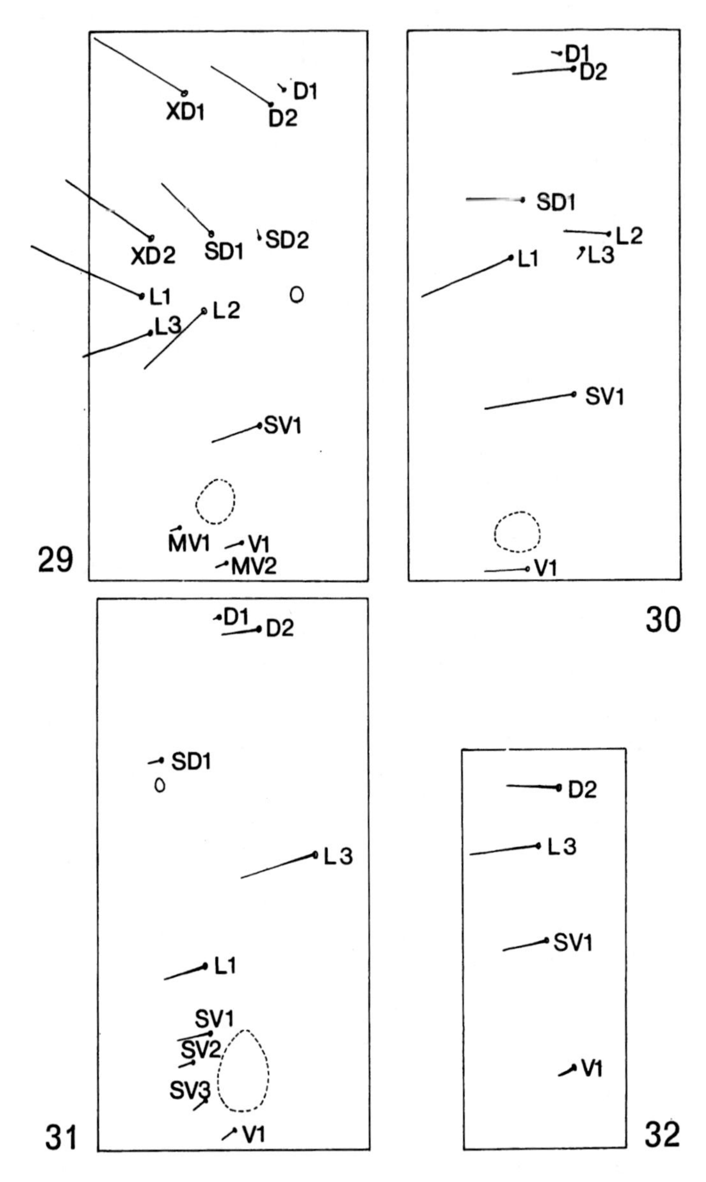

Text-figs. 29–32. Chaetotaxy of larva of *Elachista apicipunctella* Stt. – 29: prothorax; 30: mesothorax; 31: abdominal segment VI; 32: abdominal segment IX.

the generalized species. However, it must be emphasized that the presence or absence of setae in miners and borers is often inconstant, especially in Incurvariina, Nannolepidoptera and the tineoid families, while setal reduction is less common in internally feeding gelechioid families (MacKay, 1972; Nielsen, unpublished).

Pupae (Text-figs. 33-35).

The pupa is of gelechioid type, i.e., it is obtect, incomplete, and the fronto-clypeal suture is distinct and the abdominal segments are without dorsal spines (Mosher, 1916; Common, 1970; Hodges, *in press*).

The pupae of Elachistidae differ from this generalized type by the absence of the fronto-clypeal suture and maxillary palpi; likewise, the antennae are not separated posteriorly as is characteristic of Gelechioidea. The pupa is generally easily recognized by the presence of tubercles on the head and thorax and by being somewhat compressed and having longitudinal ridges. The pupae of Chrysopeleiinae lack maxillary palpi, but they further differ by presence of fronto-clypeal suture and absence of cremaster (Mosher, 1916).

The gelechioid pupa generally have segments 5-7 movable in male and 5-6 in females, but Mosher (1916) states that the abdominal segments of *Elachista praelineata* Braun are firmly soldered to each other, and that there are no movable

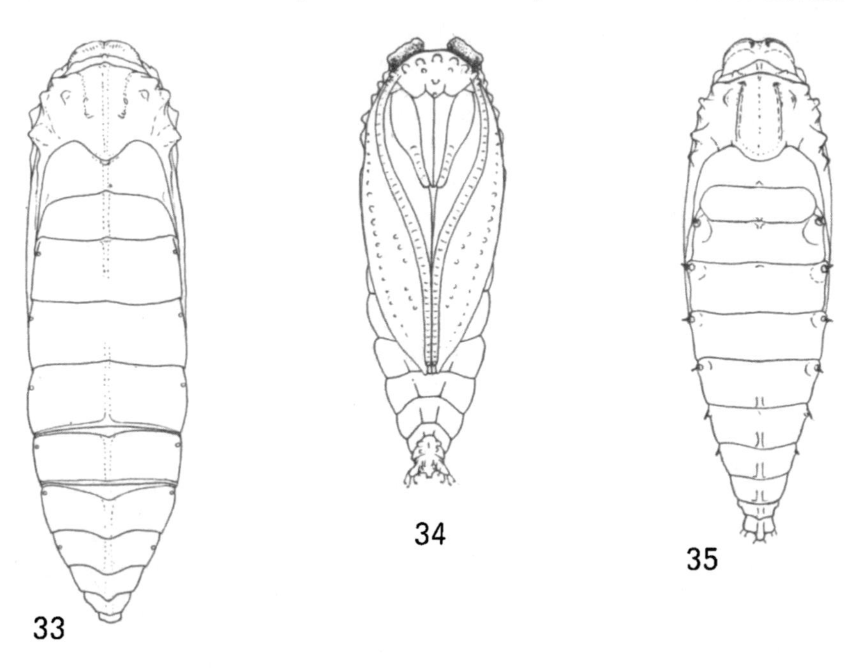

Text-figs. 33–35. Pupae of Elachistidae. – 33. *Elachista* in dorsal view; 34: *Elachista* in ventral view; 35: *Cosmiotes* in dorsal view (after Braun).

segments. The authors have been unable to confirm this fusion of abdominal segments.

The pupa is ovoid in dorsal view, widest in the thoracic region; only in the most generalized genera is the pupa round, in all others it ranges from somewhat depressed to strongly flattened.

In genera pupating under bark, the head has a very low dorso-ventral ridge, which is absent in genera with exposed pupae. In the case of primitive species, the moth emerges through a transverse slit in the basal, almost straight, portion of the generally λ-shaped ecdysis line. In specialized species the vertex and antero-dorsal portion of the thorax are also split. The head is, to varying degrees, set with tubercles, and *Cosmiotes* has two backwards-directed spines (Text-fig. 35). In *Biselachista*, a pair of flaps extends forwards from the upper margin of the vertex. A pair of dorso-lateral ridges and a medial ridge are generally present, produced to a varying degree or secondarily reduced, and running from the thorax to the posterior of the abdomen. The thorax often has two additional dorsal ridges between the medial and lateral ridges.

The wings, antennae and the dorsal and lateral parts of the thorax are often very rough owing to the presence of tubercles. Primitively, the wings are longer than the antenna, reaching segment VII, but in advanced taxa of almost equal length, reaching segment V or VI.

The abdomen is without tubercles, flattened, with the medial and lateral ridges generally prominent. The spiracles are situated on the lateral ridges or more ventrally, and are sometimes produced and bear a small spine, notably in *Cosmiotes*. In the Nearctic genus *Dicronoctetes* and the Melanesian *Eupneusta*, large prong-like processes extend from the spiracles. Larval prolegs are present in all species examined by the authors.

The cremaster on segment X is well developed in exposed pupae, forming one to four mainly ventral brushes of strong, often hooked or curved spines. It is reduced to a varying extent in specialized groups. The colour of the pupa is described under each species.

Bionomics

Adults of Elachistidae are small, inconspicuous, mainly nocturnal moths; a few species fly in sunshine during the day, but almost all species are active, flying low over herbage, at dusk and at dawn. Activity often starts in the late afternoon when adults hiding in the low vegetation move to the tips of the grass leaves. Adults are only moderately or not at all attracted by light and are best caught in a sweep net when flying freely over herbage or by sweeping vegetation. Various interception traps can be used with great advantage; the Malaise trap is particularly useful. During the day some species are often seen resting on tree-trunks. In repose (Text-fig. 36), the antennae are turned back along the costal margin of the forewing, the labial palpi are recurved, appressed to the head or directed forward, the fore legs are directed antero-laterally and the tornal area of the forewing is produced characteristically above the dorsum. When disturbed the moth runs extremely fast, interrupting its run by abrupt stops or by dropping to the ground and shamming death – "We think we have it, but yet we have it not" (Stainton, 1854a:249).

In southern Scandinavia most species are uni- or bivoltine, while almost all species are univoltine in northern Scandinavia; some of these species have three or more broods a year in C. and S. Europe. Univoltine species mainly fly in the early summer, and the number of species present as adults strongly decreases during and after July. Elachistidae overwinter as an egg or as a larva.

The generalized genera and a few species belonging to the *Elachista*-group of genera are miners in a few Dicotyledonous families: Boraginaceae, Lamiaceae, Asteraceae, Cistaceae, and Loniceracea, while all other species exclusively mine the Monocotyledonous families: Cyperaceae, Poaceae and Juncaceae.

Eggs are deposited singly on the lower leaves of the food-plants, freely attached to the surface or, more rarely, half-inserted into the tissue. When hatched, the young larva bores directly into the leaf, thus leaving the empty egg at the beginning of the mine. The mine is at first narrow, often thread-like, but later it gradually or abruptly widens into a blotch-mine. Changing of mines occur often, particularly frequently among species mining in narrow-leaved grasses and sedges, and in almost all species hibernating in the mine as a larva. Larvae of a few species end up as borers in stems and roots.

The mine is frequently difficult to identify, as the often narrow leaves of the food-plants do not permit a characteristic mine to develop. Most species are oligo- or polyphagous. In some species excrement ("frass") is deposited in a characteristic

Text-fig. 36. A specimen of *Elachista* in resting position.

way, in lines or piled up in a distinct part of the mine; in other species it is scattered haphazardly or simply gathered in the lowest part of the mine or, very rarely, it is extruded from the mine. A few species make a web-tube in the middle of the mine along the longitudinal axis, into which the larva retracts when disturbed.

Pupation takes place outside the mine (except for *E. quadripuntella* (Hb.) which pupates in the mine); primitively, pupation takes place under the epiderm or bark of the food-plant (Pl.-fig. 534) or under a dense cocoon attached to the concave side of a leaf; other species pupate under a flimsy cocoon (Pl.-fig. 535), but the principal mode of pupation is that in which the larva starts by making a slight web on the leaf surface, attaching the pupa to the web by a silken girdle and the cremaster (Pl.-fig. 536) just like a butterfly. The pupa is not extruded from the cocoon or girdle at ecdysis. Pupae and larvae of species living in damp areas have been reported as being able to withstand inundation, as the pupa from a larva which mined in a submerged leaf can float on the water surface (Gudmann, 1924).

Even if several species may be abundant, Elachistidae are generally not recognized as distinct pests in N. Europe. A few *Cosmiotes* species are reported to mine in wheat in Australia (Common, 1970), *Dicranoctetes saccharella* (Busck) is a pest on sugar-cane in Cuba (Busck, 1934: 170) and *Eupneusta solena* Bradley mines in sugar-cane in Papua New Guinea (Bradley, 1974). In spite of this, Elachistidae are undoubtedly important as miners in grasses and sedges in grassland ecosystems, and in other communities, e.g., in the herb layer in a Danish beech wood almost all leaves of *Melica uniflora* Retz. are mined by the larva of *Elachista revinctella* Zell.

A high rate of parasitism is reported for several species, and almost all parasites are hymenopterous. Thompson (1946) and Fulmek (1962) report the families Braconidae, Ichneumonidae and Eulophidae, including both larva and egg parasites, from Elachistidae. The only recorded dipterous parasite is *Actia nigroscutellata* Lundbeck (Tachinidae) (Mesnil, 1965: 825).

Rearing techniques for leaf-miners are outlined by Hering (1951) and, especially for Elachistidae, by Steuer (1973, 1976b).

Systematics and Classification

Position of the family Elachistidae

The family belongs to the large suborder Ditrysia, within which it always has been assigned to the large "tineoid complex". During the last part of the last century it was classed with the Cosmopterigidae, from which it was simultaneously separated by Walsingham (1909) and Busck (1909). Busck (1914) further considered Elachistidae (as Cygnodiidae) together with Coleophoridae, Lithocolletidae and Gracilariidae to be derived from Yponomeutoid stock, a viewpoint subsequently followed by Meyrick (1928) and Braun (1948).

In modern classifications the "tinoid complex" is generally divided into three super-families: Tineoidea, Yponomeutoidea and Gelechioidea. These divisions are recogniz-ed by Fracker (1915) and Mosher (1916), based on larval and pupal characters, respectively. Mosher further stressed a division of the superfamily Gelechioidea into two groups: Oecophoridae + Stenomidae + Cosmopterigidae + Elachistidae and Lavernidae + Scythrididae + Gelechiidae + Chrysopeleiidae, and added that Elachistidae should perhaps be upgraded to superfamily level. This was accepted by Forbes (1923), who included Elachistidae (as Cycnodiidae), Douglasiidae, and Heliodinidae in Cydnodioidea. Handlirsch (1924) recognized the subfamily Elachistinae and placed two tribes here, Elachistini and Cemiostomini. Despite this, several European authors until very recently followed Rebel's (1901) division of the family Elachistidae, into five groups of subfamily rank: Scythridinae, Momphinae, Heliozelinae, Coleophorinae and Elachistinae, a division rather similar to the one originally proposed by Bruand (1850) and adopted by Stainton (1854a).

In recent literature, Elachistidae are recognized as a distinct family of the superfamily Gelechioidea (Common, 1970; Brock, 1971; Hodges, *in press*). Members of this superfamily are recognized by the presence of scales on the base of the haustellum (proboscis), their upturned labial palpi, the absence of chaetosemata, the larva having three prespiracular setae on the prothorax with abdominal setae L1 and L2 approximate. The pupa lacks dorsal abdominal spines, and the labial palpi and prothoracic femora are not or only rarely visible.

Together with Oecophoridae, Coleophoridae, and Agonoxenidae, Elachistidae are among the earliest diverging lines from the most primitive Gelechioidea, having primitively retained the chorda in the cell, three-branched M and basally forked 1A + 2A in the forewing, possessing symmetrical male genitalia without a distinctly demarcated valva and having the gnathos present as a large spinose knob.

The family Elachistidae as here defined shows several specializations from the gelechioid ancestor described by Hodges (*in press*), and appears to be a probable monophyletic entity. Braun (1948) also included *Coleopoeta* Walsingham, 1907 in Elachistidae, although with great reservations; she noted the presence of strong dorsal abdominal spines and the atypical structure of the male genitalia (gnathos without spinose knob, juxta simple, digitate processes absent and valva short and square) and specializations in the larva, pupa and biology. As the inclusion of this genus would badly upset the Elachistidae as a monophyletic entity, *Coleopoeta* must be tentatively transferred to a monobasic taxon, which is closely related to the Elachistidae and Oecophoridae.

Even if very few papers descriptive of the morphology of the Gelechioidea are, as yet, available, the following probable autapomorphies or characters of uncertain status can be listed for the Elachistidae (numbers refer to Text-fig. 37):

1. Reduction in size (the ancestral gelechioid was probably much larger, e.g., like Oecophoridae and Ethmiidae (Hodges, *in press*)).
2. Absence of strong dorsal spines on the abdomen of adults.
3. Presence of a digitate process in the male genitalia; as mentioned above it cannot with certainty be homologized with the juxta process of Oecophoridae.
4. Absence of the fronto-clypeal suture, maxillary palpi; abdominal segments not movable in the pupa (Mosher, 1916).
5. General bionomy. Larvae are leaf-miners, not making a portable case and not tying leaves, pupating outside the mine.

Whether or not the presence of short cylindrical postero-distal processes from each flagellum segment and long clavate pretarsal setae on the thoracic legs is a family character needs further investigation.

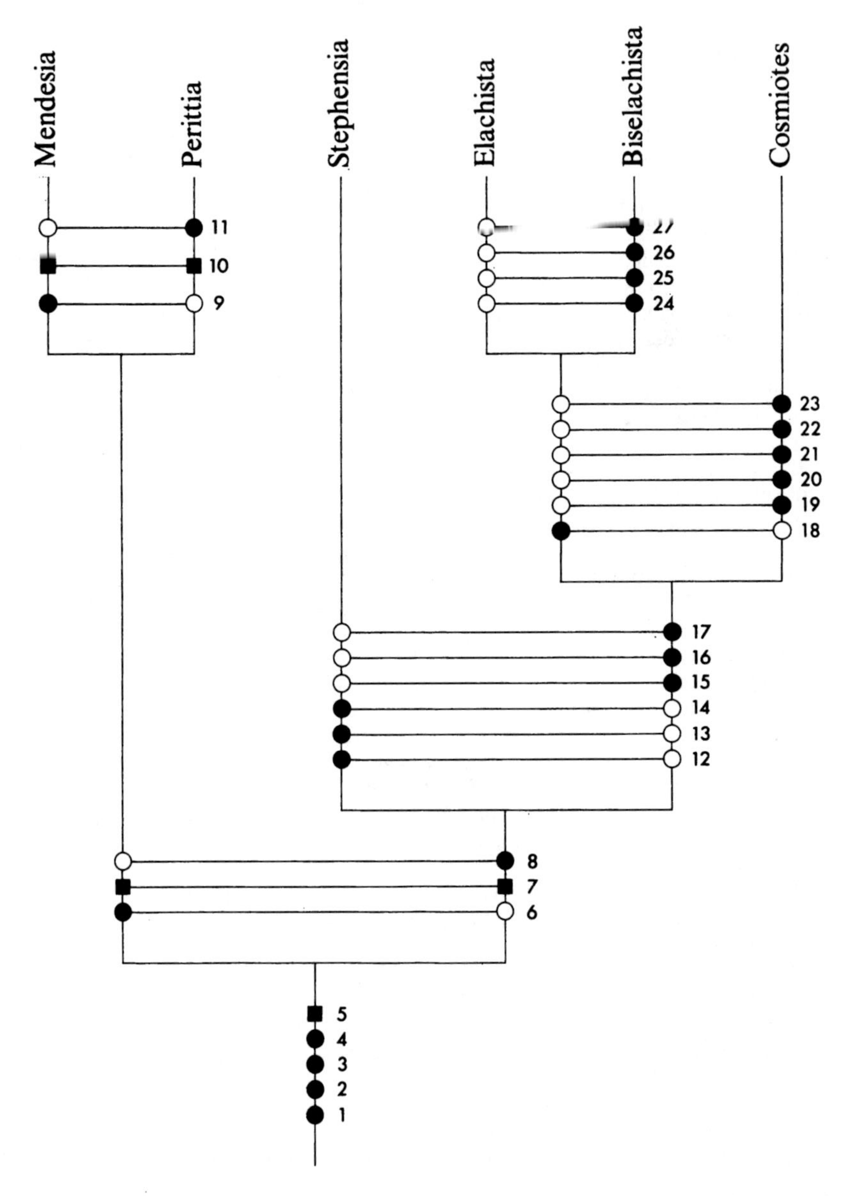

Text-fig. 37. Proposed phylogenetic classification of N. European genera of the family Elachistidae. Numbers refer to enumeration in text. Open circles: plesiomorphies; filled circles: apomorphies; filled square: interpretation uncertain.

Classification within Elachistidae

A tentative phylogenetic classification of the N. European genera of Elachistidae is presented in Text-fig. 37, based on the following characters:

6. A probable autapomorphy for *Mendesia* + *Perittia* is strong reduction of the uncus into two small setose lobes. In most primitive gelechioid families, the uncus is well developed.
7. In *Mendesia* + *Perittia* the valva is short and broad; in *Stephensia – Cosmiotes* it is long and slender; in Oecophoridae the valva is mainly intermediate.
8. The presence of a strong, sclerotized, prominent digitate process is a probable autapomorphy for *Stephensia – Cosmiotes*.
9. The spur from the base of Sc in the forewing of male is absent in *Mendesia*.
10. *Mendesia* species feed on Boraginaceae, *Perittia* on Loniceraceae.
11. M is two branched and the chorda is absent or very weak in the forewing of *Perittia*, respectively three-branched and present in *Mendesia*.
12. The costa of the valva has a thorn-shaped spine above the cucullus.
13. Tergite VIII has short anterior processes from the bases of the apophyses anteriores.
14. The egg is semi-erect, half-inserted into leaf tissue.
15. *Elachista-Cosmiotes* feed as leaf miners in Monocotyledons.
16. The sacculus is produced into a small spine below the cucullus, often secondarily reduced.
17. The antrum and colliculum are well-developed.
18. In *Elachista* + *Biselachista* the pupa is generally attached by a single girdle and the cremaster; in *Cosmiotes* pupation takes place in a dense cocoon.

19–23. Probable autapomorphies for *Cosmiotes* are the presence of a cilial spot (Text-fig. 39), claw-like uncus lobes, a transtilla produced into labides, the juxta ventrally large and duck-beak shaped, the pupa with produced spiracles, each with a spine.

24–27. Probable autapomorphies for *Biselachista* are the consistent wing-pattern (Text-fig. 38), divided gnathos, presence of coarse setae on the ventral side of the papillae anales, and a membranous zone between the distinct antrum and the long tube-shaped colliculum.

It appears from Text-fig. 37 that this tentative phylogenetic classification based on the genera represented in northern Europe is very similar to the relationships outlined by Braun (1948) for the Nearctic genera and based on genitalia and venation. The most primitive grade, *Mendesia* + *Perittia*, includes about 5 percent of the species. Both *Mendesia* and *Perittia* and some of the primitive genera in the *Stephensia-Cosmiotes* branch (in Scandinavia only represented by *Stephensia*) feed on Dicotyledons, while the remaining species mine in Monocotyledons and constitute approximately 95 percent of the total number of species of Elachistidae.

Remarks on the *Elachista*-group of genera

Braun (1948), in her revision of the N. American species, accepted a broad demarcation of the genus *Elachista*, including the bulk of the species in the family, a similar grouping as that accepted by authors during the last century, except that she recognized *Cosmiotes.*

Herrich-Schäffer (1855) divided the genus on external characters, mainly on wing pattern. A division very similar to this was used by Stainton (1849, 1854a, 1858a, b) and Frey (1859) and accepted with small alterations by subsequent authors.

Pierce & Metcalfe (1935) divided *Elachista* into three groups: A – with "uncus plain", B – "uncus with short stiff setae" and C – with "sacculus long, aedeagus slender". Group C is the genus *Cosmiotes,* while groups A and B, respectively, correspond to sections I and II in Braun (1948). Braun further recognized a third section, section III, including a few broad-winged species with characteristic female genitalia. One species in group A and one in group B of Pierce & Metcalfe (1935) belong to Braun's section III.

From *Elachista* sensu Braun (1948) *Biselachista* gen.n. is hereby separated, while the remaining species are kept in a single large genus, which from Text-fig. 37 appears to be potential paraphyletic, and which at present cannot be consistently subdivided at generic or subgeneric levels. In the present paper *Elachista* is divided into four species-group based on venation and genital morphology. The *gleichenella*-group corresponds to section III of Braun (1948), the *bifasciella*-group to section II (except for species now placed in *Biselachista* gen.n.) and the *bedellella*-group to section I; a fourth group, the rather heterogenous *tetragonella*-group is also recognized.

The linear sequence here presented differs in some aspects from what has been presented in recent lists (e.g., Krogerus et al. (1971); Bradley et al. (1972)). Within each group, the sequence applied is based on the colour of the head and the general wing pattern, the presence or absence of strong sexual dimorphism and especially on the extent of emargination of the uncus and the reduction in venation (e.g., number of R and M veins, development towards long-stalked or coalescent R_4–R_5 and R_5–M_1, presence or absence of fork from Rs to costa of hindwing, from closed cell in hindwing to open cell with inwardly produced lower angle to outwardly curved angle).

In the *bifasciella*-group the most generalized species are *E. alpinella* Stt. and *E. diederichsiella* Her.; from an ancestral form resembling these dark species almost without sexual dimorphism, two distinct lineages arise, one with pronounced sexual dimorphism, and one with a whitish or greyish ground colour (both are treated as subgroups), while the large subgroup of dark species consists of a dark-headed line (*poae-alpinella*) and a pale-headed line (*alpinella-apicipunctella*). The *bedellella*-group is divided into three distinct subgroups.

For each group or subgroup a short diagnosis is given in the text.

Nomenclature

No attempt has yet been made to review the total of less than fifty generic names applied to the family Elachistidae. The generic nomenclature adopted here largely confirms with that currently used in the W. Palaearctic and Nearctic faunas. Existing satisfactory subjective and objective synonyms stated elsewhere are generally listed. Two new synonyms of *Elachista* are published separately (Nielsen & Traugott-

Olsen,1977). The type-species of each genus listed is given, but without comments on its mode of fixation, as this is beyond the scope of this series.

At specific level, much revisional work has been done throughout the last three decades, mainly on single species or on smaller groups. Recently Parenti (1972, 1973, 1977) has begun a critical examination of species described from the W. Palaearctic region. During the preparation of the present work many types of N. European Elachistidae have been examined or re-examined but several species are treated in accordance with current usage; the latter is especially true for the many species described by Stainton from Britain, but the identity of these is generally fairly well known, both from Stainton's own publications (e.g., Stainton, 1858b, c) and from subsequent revisional works (e.g., Bradley, 1952, 1963) and is in agreement with the British check list (Bradley et al., 1972). All species described by Stainton from specimens received from J. Mann (Stainton, 1851) have been examined. If lectotypes have been designated these are cited in a note, under the valid name of the species. The results of the examination of some species unrelated to the N. European fauna are described elsewhere (Nielsen & Traugott-Olsen, *in press*).

All species described from Fennoscandia and Denmark and recognized as Elachistidae have been examined and the types fixed, if this has not previously been done. For four of the species originally placed in *Elachista* by Tengström (1848), no type-material is available (Jalava, *in litt.*); two of these, *E. trifasciella* Nyl. and *E. tristictella* Nyl., are here placed in synonymy in accordance with current usage, while the identities of *E. moniliella* Tgstr. and *E. salicis* Tgstr. are apparently unknown. The two species described in *Elachista* by Zetterstedt (1839) belong to the Gracillariidae and Nepticulidae.

Zoogeography

The family Elachistidae is known from all the major zoogeographic regions of the world, but the species are generally poorly known. Brief comments on the distribution of each genus are given after the description.

By far the largest number of species has been described from the northern hemisphere, mainly from the Nearctic region, and the European and Mediterranean parts of the Palaearctic region. Comparatively few species are known from the tropics and subtropics.

At present, a total of 68 species is known from the Nordic countries, 47 from Denmark, 63 from Sweden, 28 from Norway and 45 from Finland. In general, the highest number of species is reported from the southern parts of each country, 30–40 species from each district, while only 5–15 species are reported from the best-collected districts in the northern parts of Sweden and Finland. However, it must be strongly emphasized that the faunistics of the Elachistidae is very inadequately known, so the present distribution pattern is deficient.

In general, the species occurring in the area treated have their main distribution outside this, often being widely distributed in the W. Palaearctic region. Eight endemic species are known at present, all species recently described from northern Norway, Sweden and Finland, but all these must be supposed to be much more widespread, especially throughout the northern E. Palaearctic. Holarctic species have not yet been recognized, but the potential for overlap between the Palaearctic and Nearctic faunas is certainly high, and needs further examination.

Several species reach their northern distributional limit in Fennoscandia; many species especially do not go further than S. Norway, the Stockholm-area and the extreme south of Finland. One species, *Elachista kilmunella* Stt., has a discontinuous boreo-montane distribution; at least four species, *Elachista biatomella* (Stt.), *E. bifasciella* Tr., *E. rufocinerea* (Hw.) and *Biselachista scirpi* (Stt.) have an atlantic or atlanto-mediterranean distribution, while at least two species have an apparently discontinuous boreal distribution – *Elachista quadripunctella* (Hb.) and *E. littoricola* Le March.

Technical Remarks

Techniques for preparing the genitalia and wings of Lepidoptera have recently been described by Clarke (1941), Bentinck & Diakonoff (1967), Razowski (1973) and Reid (1976); the preparation of genitalia of "Microlepidoptera" in particular is extensively dealt with by Robinson (1976), to which paper reference should be made.

As the genitalia and wings of Elachistidae are among the smallest within the Lepidoptera, and as preparation and examination of genitalia is generally necessary for safe identification of specimens of this family, a few remarks on this topic are included.

The complete abdomen is boiled in 10% caustic potash (KOH) for 2–4 minutes, until the abdomen seems rather clear and sinks to the bottom, a sign that almost all lipids have been removed. Dissection and final arrangement is now done on a single slide, without transferring the genitalia from slide to slide, as this often disorders the setae and the fine structures. The authors use about 30% acetic acid (CH_3COOH), added to the slide with a very fine brush, as a dissection medium, as this dissolves lime soaps, has a low surface tension, evaporates slowly and permits easy descaling; debris and scales are removed from the abdomen and genitalia using fine brushes or snipe feathers, and needles are not used, as these easily harm or tear the genitalia. Alcohol and Euparal Essence are added with different brushes, while Euparal is added using a needle. It is very important to avoid disintegration of the male genitalia, and to have the genitalia totally "unfolded". The removal of the aedeagus from the genitalia cannot be recommended.

Wing venation preparations are made in a similar way. One pair of wings is removed from the insect by pressing the wings downwards at the base. When free, they are placed in a drop of 30% acetic acid on a slide and carefully pressed into the medium and descaled by gently rubbing with fine brushes or snipe feathers. All loose scales are removed by washing with absolute alcohol added using a fine brush. Wings are mounted on a slide, beneath a coverslip, using Euparal Essence and Euparal.

Key to North European genera of Elachistidae

1 EXTERNAL: Antenna strongly ciliate; hind tarsus strongly spined.
VENATION: Both forewing and hindwing with three M veins to termen.
MALE GENITALIA: Valva proximally very broad, gradually tapering distally.
FEMALE GENITALIA: Antrum almost balloon-shaped, ventral margin of antrum strongly posteriorly curved *Mendesia* Joannis (p. 37)

– EXTERNAL: Antenna not or only slightly ciliate; hind tarsus not strongly spined.
VENATION: Both forewing and hindwing with less than three M veins to termen.
MALE GENITALIA: Valva proximally very broad, abruptly tapering distally into a distinct cucullus, or with almost parallel margins.
FEMALE GENITALIA: Antrum funnel- or bowl-shaped, ventral margin of antrum straight or anteriorly curved, if posteriorly curved, produced into two strong lateral rims 2

2 (1) EXTERNAL: Antenna extending to just beyond middle of forewing; labial palpi short, drooping.
VENATION: Sc in forewing of male with a short spur from base of retinaculum.
MALE GENITALIA: Uncus with two very small setose lobes.
FEMALE GENITALIA: Papillae anales forming a short, broad triangle *Perittia* Stainton (p. 39)

– EXTERNAL: Antenna extending to well beyond middle of forewing; labial palpi long and slender, porrect to recurved, often strongly diverging, if short and drooping, antenna with few white segments beyond middle and with dark tip.
VENATION: Sc in forewing of male without a spur from base of retinaculum.
MALE GENITALIA: Uncus large, setose, generally divided into two lobes.
FEMALE GENITALIA: Papillae anales elongately triangular, elongately rounded, or short and broadly rounded 3

3 (2) EXTERNAL: Labial palpi short, drooping. Antenna with few white segments beyond middle, distal part dark. Hind tibia almost smooth.
MALE GENITALIA: Costa of valva distally produced into a short thorn-shaped process above cucullus (Pl.-fig. 234).
FEMALE GENITALIA: Tergite VIII with a short anterior process from base of apophyses anteriores (Pl.-fig. 399) *Stephensia* Stainton (p. 43)

– EXTERNAL: Labial palpi long and slender, porrect to recurved, rarely drooping. Antenna not with white segments beyond middle before dark distal part which is coloured as flagellum or lighter. Hind tibia with long setae above and beneath.
MALE GENITALIA: Costa of valva not produced into a process above cucullus.

FEMALE GENITALIA: Tergite VIII without an anterior process from base of apophyses anteriores 4

4 (3) EXTERNAL: Labial palpi very long and slender, strongly diverging. Forewing with cilial spot on termen at apex (Text-fig. 39).

VENATION: M_1 in forewing from R_5; only one M vein present in hindwing.

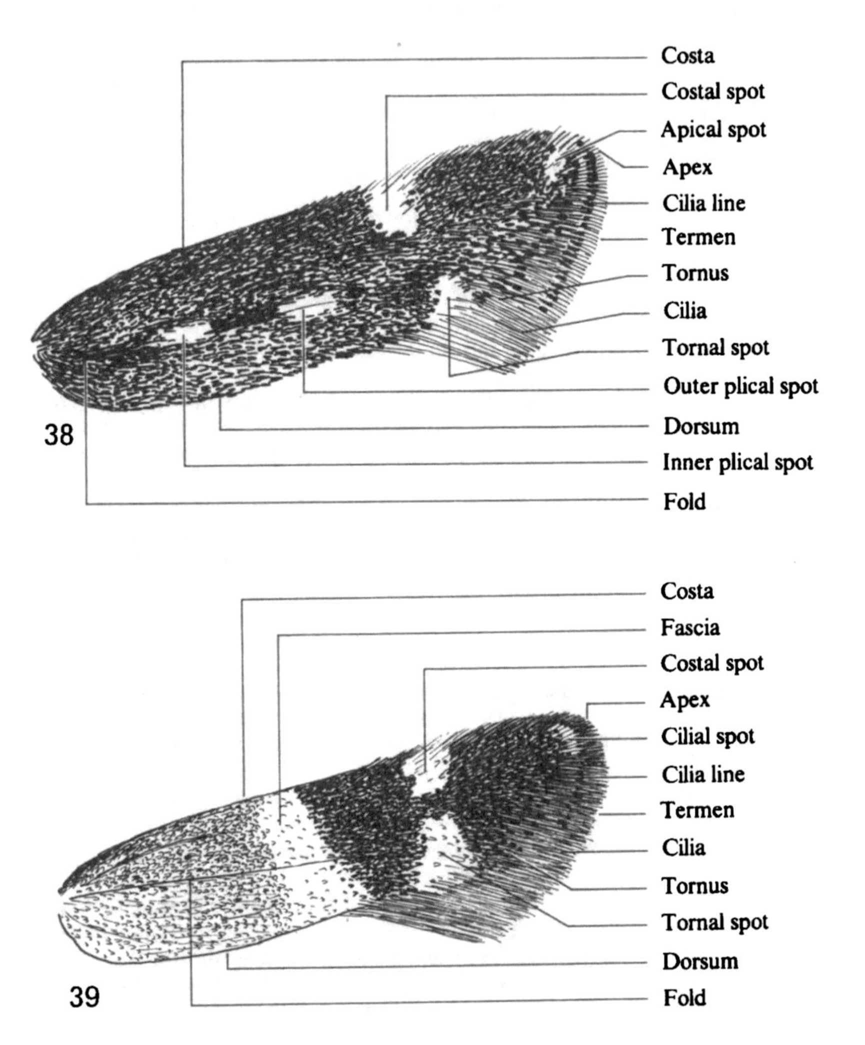

Text-figs. 38, 39. Forewings of Elachistidae. – 38: *Biselachista freyi* (Stgr.), male; 39: *Cosmiotes freyerella* (Hb.), female.

MALE GENITALIA: Uncus lobes pointed, claw-like. Transtilla medially produced into labides.
FEMALE GENITALIA: Ventral margin of antrum a short, distinct anterior sinus from posterior margin of venter VIII *Cosmiotes* Clemens (p. 269)

– EXTERNAL: Labial palpi long and slender, often diverging. Forewing without cilial spot.
VENATION: M_1 in forewing from stalk R_{4+5}; two M veins present in hindwing.
MALE GENITALIA: Uncus lobes not pointed, claw-like. Transtilla not produced into labides.
FEMALE GENITALIA: Ventral margin of antrum not a distinct sinus from posterior margin of venter VIII, generally in middle of anterior part of venter VIII or even more anterior 5

5 (4) EXTERNAL: Forewing with an apical streak from costa (Text-fig. 38) or with dark suffusion from middle of costa; spots not metallic, shining. Fold with dark suffusion or a black streak in the middle, often bound by lighter scales.
MALE GENITALIA: Uncus with thick or clavate setae. Gnathos bifurcate.
FEMALE GENITALIA: Antrum and colliculum distinct, separated by a short membranous zone. Papillae anales short, broadly rounded, with coarse, short, ventral setae *Biselachista* gen.n. (p. 252)

– EXTERNAL: Forewing without apical streak or, if present, a very small species without dark suffusion from middle of costa. Fold not distinctly darker-suffused or white species with a small black spot in fold.
MALE GENITALIA: Uncus with simple setae. Gnathos single.
FEMALE GENITALIA: Antrum and colliculum confluent, or, if separate, colliculum is only present as a short ring close to antrum. Papillae anales elongately triangular or rounded, generally with rather long setae
Elachista Treitschke (p. 46)

Genus *Mendesia* Joannis

Mendesia Joannis, 1902: 231.
Type-species: *Mendesia echiella* Joannis, 1902.
Triboloneura Walsingham, 1908: 54.
Type-species: *Elachista sepulchrella* Stainton, 1872.

Head (Text-fig. 40) short, smooth-scaled, neck tufts a single row of scales. Pecten of antenna prominent, made up of scales longer than width of scape, flagellum strongly ciliate in male, faintly ciliate in female. Labial palpi short, porrect to recurved, second segment rough-scaled on underside, third segment very short, pointed.

Forewing ground colour white, greyish white or light brownish, often with darker dots. Sexual dimorphism not pronounced.

Venation (Pl.-fig. 153). Forewing with R_4 and R_5 stalked, stem of R_{4+5} (chorda) forming accessory cell, M three-branched, stem visible in cell, CuP present distally, 1A and 2A distally fused, forked basally. Hindwing with Sc + R_1 very long, close to

Rs, Rs often with free branch to costa, cell closed, M three-branched, CuP hardly visible.

Mid tibia with two distal spurs of equal length, hind tibia with long hairs above and with only few long hairs beneath, middle spurs slightly beyond the middle, inner spur twice as long as outer, apical spurs equal long, situated at distal end; mid and hind tarsi with strong pointed spines.

Male genitalia. Uncus lobes very small, setose. Socii only indicated by a small group of setae. Gnathos elongate. Anterior margin of tegumen not reinforced. Valva very broad and short. Digitate process very small, membranous, setose. Juxta distinct, U-shaped. Vinculum rounded, without saccus. Aedeagus long and slender, without cornuti.

Female genitalia. Papillae anales elongate, triangular, with long setae. Sternite VIII narrow. Antrum weakly sclerotized, ventral margin posteriorly curved; colliculum not prominent. Ductus and corpus bursae without spines or signum.

Little is known about the immature stages. Joannis (1902) gave a short note on the larva and pupa of *M. echiella* Joannis, 1902. The larva mines in leaves of Boraginaceae; Joannis (1902) reports *Echium*, Walsingham (1907) *Symphytum* (this is probably a misidentification of *Echium* as *Symphytum* has not otherwise been reported from Tenerife) and Amsel (1935) mentions *Pondonosema*. Mendes (1909) reports that the pupa is freely attached to the surface of the food plant.

The genus includes very few species and is apparently restricted to the W. Palaearctic region including N. Africa, and the Middle East.

1. ***Mendesia farinella*** (Thunberg, 1794)
Text-figs. 2, 40; plate-figs. 1, 2, 153, 229–231, 396.

Tinea farinella Thunberg, 1794: 88.
Mendesia subargentella Dattin, 1932: 161.

Text-fig. 40. Head of *Mendesia farinella* (Thnbg.).

12–14 mm. Male (Pl.-fig. 1). Head, neck tufts, tegulae and thorax shining white; vertex and anterior part of tegulae faintly suffused with greyish brown. Scape of antenna white, upperside slightly suffused with brownish grey; flagellum greyish brown, slightly clothed with raised whitish scales, with many cilia, which are almost as long as width of flagellum. Labial palpi short, recurved, third segment short, pointed; palpi white, greyish beneath. Forewing ground colour white, basal third of costal margin suffused with greyish, otherwise without marks. Cilia white, no cilia line present. Hindwing grey-brown, distinctly lighter towards base; underside grey-brown. Cilia white, with base tinged yellowish brown. Abdomen whitish. Hind tarsus white.

Female (Pl.-fig. 2). Head and tegulae white. Antenna with short cilia.

Male genitalia (Pl.-fig. 229–231). Uncus lobes triangular, with a few short setae. Gnathos elongate. Tegumen with almost parallel sides. Valva broad proximally, costa almost straight; sacculus convex, rounded, tapering distally into a short, blunt, setose cucullus with a small dorsal thorn. Digitate process small, setose. Juxta U-shaped, lobes rounded, setose. Aedeagus long and slender, straight, base swollen, truncate, distal end pointed; without cornuti.

Female genitalia (Pl.-fig. 396). Apophyses slender, apophyses posteriores with proximal ventral ridge, longer than apophyses anteriores. Antrum balloon-shaped, weakly sclerotized, gradually tapering anteriorly into a short, slightly sclerotized colliculum with irregular folds. Corpus bursae without signum.

Distribution. In Denmark reported from N. Jutland and NEZ; from several districts in S. Sweden up to Dlr. and Upl.; not known from Norway; in Finland known from Al and Sa. – Otherwise from NW. USSR, N. and W. Europe and N. Africa.

Biology. Immature stages unknown. Hering (1957: 189) reports the larva from *Brachypodium,* but this is probably a misidentification of *E. argentella* (Cl.). Janmoulle (1945) considered the food-plant to be a grass, owing to the considerable similarity of *farinella* to the grass-feeding *argentella*, but later he (1947) changed this view. The food-plant is therefore still unknown, but as mentioned in the generic diagnosis, it most likely belongs to the Boraginaceae. Apparently univoltine; adults fly from late May to early July. The biotope is grassland, both inland and near the sea; in Denmark mainly found in dunes.

Genus *Perittia* Stainton

Perittia Stainton, 1854a: 177.
Type-species: *Aphelosetia obscurepunctella* Stainton, 1848.
Scirtopoda Wocke *in* Heinemann, [1876]: 465 (nom. praeocc.).
Type-species: *Tinagma herrichiella* Herrich-Schäffer, 1855.
Dyselachista Spuler *sensu* Lhomme (1951). See note.

Head (Text-fig. 41) smooth-scaled, neck tufts appressed to head or slightly raised. Antenna a little more than half as long as the forewing. Labial palpi short, drooping to slightly porrect, third segment pointed, almost as long as second segment. Haustellum short and thick, shorter than labial palpi.

Forewing ground colour greyish or brownish, often shining. Almost no sexual dimorphism.

Venation (Pl.-figs. 154, 155). In forewing on male Sc with a small spur towards costa from base of retinaculum, R_4 and R_5 stalked or coalescent, no chorda present. M_1 stalked with R_{4+5}, M_3 absent, CuP present distally, 1A + 2A forked at base. Hindwing with Rs + R_1 very long, close to Rs, which can have a free branch to costa, cell almost closed between M_1 and M_2.

Inner spur on mid tibia longer than outer spur, hind tibia with long hairs above and below, middle spurs before the middle, inner spur almost twice as long as outer spur, outer apical spur longer than inner apical spur. Spines on tarsi only slightly longer than scales.

Male genitalia. Uncus very small, setose. Socii only indicated by a group of setae. Gnathos elongate or rounded. Valva short and broad, costa straight, cucullus narrow, strongly sclerotized, setose. Digitate process very small or reduced, membranous, setose. Juxta deeply indented, lobes long, distally rounded or pointed. Vinculum short, rounded or truncate, anterior margin medially enlarged. Aedeagus with a rounded and truncate basal lobe, without cornuti.

Female genitalia. Papillae anales short, forming a broad triangle. Antrum sclerotized, colliculum not distinct, ductus bursae posteriorly with folds or spines. Corpus bursae set with spines inside, and with a single large signum.

The two species here dealt with both feed as leaf miners on Caprifoliaceae. Walsingham (1907) included three species from Tenerife feeding on Lamiaceae, but they do not belong in *Perittia* (Zimmerman & Bradley, 1950). *Perittia* is closely related to *Polymetis* Walsingham, 1907, and *Swezeyula* Zimmerman & Bradley, 1950, the larvae of which feed on Asteraceae and *Lonicera* respectively. Both the larvae of *Perittia* and *Swezeyula* (Swezey, 1950) quit the mine when mature and pupate in a cocoon under the epidermis or bark of the food-plant or in the ground. The pupa is

Text-fig. 41. Head of *Perittia herrichiella* (HS.).

smooth, without tubercles or longitudinal ridges. The immature stages of the species included in the very closely related genus *Onceroptila* Braun, 1948 are unknown.

Apparently *Perittia* contains few species, and at present the genus appears to be restricted to the W. Palaearctic, N. Africa and Asia Minor; several species are reported from the Mediterranean area. *Onceroptila* is restricted to the W. Nearctic and *Swezeyula* is reported from Hawaii and Japan. It must be noted that the generic relationships of several of the generalized species are still unclear.

Note. Lhomme (1951) included *herrichiella* HS. and *saltatricella* FR. in the genus *Dyselachista* Spuler, 1910: 424, type-species: *Aechmia saltatricella* Fischer von Röslerstamm, 1841: 249, by monotypy. Fischer von Röslerstamm (loc. cit.) based his description on specimens from Mann. A specimen labelled "Austria, Mann"; "Genital praeparat nr. A. 18.8.75,♂, E. Traugott-Olsen" in Naturhistorisches Museum, Vienna, is hereby designated as lectotype; the lectotype is conspecific with *Heliozela sericiella* (Haworth, 1828), **syn.n.** All specimens previously identified as *Dyselachista saltatricella* (FR.) and now studied by the authors are *H. sericiella* (Hw.). Spuler (1910), in the diagnosis of *Dyselachista,* described and figured the venation of a *Heliozela* Herrich-Schäffer, 1853; *Dyselachista* is hereby transferred to the Heliozelinae, being a junior subjective synonym of *Heliozela* HS., **syn.n.**

Key to species of *Perittia* based on external characters

1 Forewing shining bronzy grey-brown, with a distinct yellow-white spot on dorsum before tornus (Pl.-figs. 3, 4) 2. *herrichiella* (HS.)
– Forewing brownish grey, not shining, with a dark brown streak in the fold and a few white scales on tornus (Pl.-figs. 5, 6) 3. *obscurepunctella* (Stt.)

Key to species of *Perittia* based on male genitalia

1 Cucullus distinctly shorter than costa; juxta lobes distally blunt; aedeagus square-ended (Pl.-fig. 232) 2. *herrichiella* (HS.)
– Cucullus shorter or as long as costa; juxta lobes distally pointed; distal end of aedeagus pointed (Pl.-fig. 233) 3. *obscurepunctella* (Stt.)

Key to species of *Perittia* based on female genitalia

1 Antrum funnel-shaped, margins posteriorly almost parallel; posterior part of ductus bursae with fine internal spines (Pl.-fig. 397) 2. *herrichiella* (HS.)
– Antrum broad, with many longitudinal folds; posterior part of ductus bursae without spines, but with a longitudinally folded portion (Pl.-fig. 398) 3. *obscurepunctella* (Stt.)

2. ***Perittia herrichiella*** (Herrich-Schäffer, 1855)

Text-fig. 41; plate-figs. 3, 4, 154, 232, 397, 498-500.

Tinagma herrichiella Herrich-Schäffer, 1855: 260.

8–9 mm. Male (Pl.-fig. 3). Head, neck tufts and tegulae fuscous with a bronze sheen; thorax slightly darker. Antenna slightly serrate distally, colour similar to head, scape slightly lighter. Labial palpi short, drooping, upperside cream-coloured, underside bronzy violet. Forewing ground colour dark grey-brown with a brazen violet sheen; a square yellowish white spot which can be very small beyond middle of dorsum, mottled with dark brown-tipped scales beyond spot. Cilia grey-brown, paler beyond indistinct dark brown cilia line. Hindwing lighter than forewing, less shiny, cilia similar, base tinged light brownish. Abdomen grey, with a bluish or greenish reflection. Hind tarsus annulated light grey/blackish.

Female (Pl.-fig. 4). Very similar to male, forewing slightly darker, dorsal spot more distinct.

Male genitalia (Pl.-fig. 232). Uncus lobes very small, rounded, short setose. Valva short and broad, costa almost straight, sacculus strongly curved; cucullus distinct, narrow, with long setae. Juxta with narrow medial slit, lobes distally blunt, with only few setae. Aedeagus narrow, slightly curved, distally square-ended.

Female genitalia (Pl.-fig. 397). Apophyses posteriores more than twice as long as apophyses anteriores. Antrum with almost parallel sides, gradually tapering into ductus bursae, ventral margin anteriorly curved; a strongly sclerotized plate with a sinuous posterior margin across antrum. Posterior region of ductus bursae and corpus bursae with minute internal spines; signum rhomboid with few short teeth along margins.

Distribution. From few districts in E. Denmark; from several districts in S. Sweden up to Upl.; not known from Norway; in Finland from the most southern districts.

Biology. Larva red-brown. It mines in several species of *Lonicera, L. periclymenum* L., *L. xylosteum* L. and in introduced species (Buhr, 1942), also in *Symphoricarpos rivularis* Suksd. The mine (Pl.-figs. 498–500) is a large flat, transparent blotch, with scattered large blackish pieces of excrement inside (Hering, 1926). The egg is deposited near the base of the leaf, from where the larva first makes a narrow not widening mine more than 1 cm long with a central excrement line; the mine abruptly widens towards the mid-rib, forming the large blotch. The larva is present from late June to mid-August, when it quits the mine and bores into the stem or branch of the food-plant; in captivity, larvae can bore into cork or even furniture (Sønderup, 1924); the pupation site is just below the bark and is visible as a slight swelling on the surface. According to Lhomme (1951) pupation can also take place in the ground. Adults fly from early May to mid-June; univoltine. The moth can be seen flying around the food-plant in the sunshine in late afternoon.

Note. Superficially similar to *Heliozela* spp., which have scape without pecten and straight maxillary palpi.

3. ***Perittia obscurepunctella*** (Stainton, 1848)
Plate-figs. 5, 6, 155, 233, 398, 501, 534.

Aphelosetia oleella Fabricius *sensu* Stephens, 1834: 289.
Aphelosetia obscurepunctella Stainton, 1848: 2164.

8–10 mm. Male (Pl.-fig. 5) Head, neck tufts and tegulae brownish grey, thorax slightly darker. Antenna similarly coloured, distal half of flagellum finely serrate. Labial palpi short, drooping, whitish grey, darkest distally. Forewing ground colour brownish grey, mottled along margins with darker brownish; a short dark brown streak in middle of fold, a few whitish scales before and beyond this streak and at tornus. Cilia light brownish grey, mixed with dark brown along base and at apex; cilia line indistinct. Hindwing slightly lighter, darkest distally, cilia light greyish brown. Abdomen brownish. Hind tarsus brownish, segments lighter distally.

Female (Pl.-fig. 6). Very similar to male, outer white mark in fold often more distinct, tornal spot more prominent.

Male genitalia (Pl.-fig. 233). Uncus lobes two very small knobs, with short setae. Valva very short and broad, costa short, sacculus strongly curved to angular; cucullus sclerotized, curved, longer than costa, with long setae. Digitate process very small, setose, membranous. Juxta U-shaped, lobes long, tapering to a point, proximally setose, distally nude. Vinculum truncate, strong. Aedeagus basally bilobed, slightly curved, gradually tapering to distal pointed end.

Female genitalia (Pl.-fig. 398). Apophyses posteriores swollen posteriorly, less than twice as long as apophyses anteriores. Antrum broad, tapering anteriorly, with many longitudinal folds, especially along ventral margin. Ductus bursae with a slightly sclerotized, longitudinally folded, short posterior part. Corpus bursae with minute internal spines, signum very large, elongate, swollen in the middle, strongly dentate.

Distribution. From few districts in Denmark; in Sweden only from Sk.; not in Norway; from Finland known from few districts in the southern part. – Otherwise from W. USSR to Britain and C. Europe, but generally rare.

Biology. The larva is greenish grey, with dark dorsal line, head and prothoracic shield blackish (Meyrick, 1928). It mines in the leaves of *Lonicera periclymenum* L. and *L. xylosteum* L. The mine (Pl.-fig. 501) is a large, folded blotch, swollen on upperside, generally situated beyond the middle of the leaf; it is brownish and visible from both sides, containing scattered excrement and a characteristic round dark reddish-brown spot made of a slight web with excrement (Hering, 1926; Sønderup, 1949). According to Wörz (1957) the larva generally starts mining from the tip of the leaf. Pupation takes place in a cocoon in the bark on the food-plant or in a cork when kept in captivity (Pl.-fig. 534). Univoltine; larva from early June to early August. Hibernation as pupa, adults from mid-April to mid-June, mainly in the first part of May. Often seen resting on trunks during the day.

Genus *Stephensia* Stainton

Stephensia Stainton, 1858a: 269.

Type-species: *Phalaena (Tinea) brunnichella* Linnaeus, 1767.

Head (Text-fig. 42) smooth-scaled, neck tufts appressed. Antenna extending to well beyond the middle of forewing, pecten of a few fine hairs, flagellum finely ciliate ventro-medially, less so in female. Labial palpi short, drooping.

Costa of forewing slightly emarginate beyond the middle, ground colour reddish bronzy to blackish, with silvery or golden shining marks. Sexual dimorphism slight.

Venation (Pl.-fig. 156). Forewing with R_4 and R_5 short-forked or coalescent, no chorda present, M two-branched, M_1 from stalk R_{4+5}, only most distal portion of

CuP present, 1A + 2A coalescent from base. Hindwing with Sc + R_1 to middle of costa, Rs simple, cell open between M_1 and M_2.

Inner spur of mid tibia longer than outer spur; hind tibia smooth, scales and very few hairs appressed, middle spurs before the middle, outer spur shorter than inner spur, inner apical spur slightly shorter than outer. Spines few, only slightly longer than scales.

Male genitalia. Uncus lobes conical. Socii indicated by group of setae. Gnathos elongate. Anterior margin of tegumen not reinforced. Valva tapering distally, costa strongly sclerotized distally, with thorn-shaped process above cucullus. Digitate process long and strong, or bulbous. Juxta lobes almost conical. Vinculum rounded. Aedeagus without cornuti.

Female genitalia. Papillae anales elongate. Tergite VIII with anterior sclerotized process from base of apophyses anteriores. Signum present.

At present, only two species are included in *Stephensia,* the Nearctic *S. cunilae* Braun, 1932, and the W. Palaearctic *S. brunnichella* (L.). The larvae of both mine in leaves of Lamiaceae. The egg is half inserted into the tissue of the leaf; the larva first makes a narrow, fine mine towards the tip of the leaf where the mine is expanded into a large blotch. Pupation takes place under a fine web on a leaf.

4. ***Stephensia brunnichella*** (Linnaeus, 1767)
Text-figs. 3, 42; plate-figs. 7, 8, 156, 234–236, 399, 502.

Phalaena (Tinea) brunnichella Linnaeus, 1767: 898.
Microsetia stephensella Douglas, 1848: 21.

8–9 mm. Male (Pl.-fig. 7). Head, neck tufts, tegulae and thorax shining lead-grey, with a violet or bronzy sheen. Antenna blackish, except for five white segments beyond the middle, faintly ciliate, distal half slightly serrate. Labial palpi short, drooping, lead-

Text-fig. 42. Head of *Stephensia brunnichella* (L.).

grey. Forewing ground colour bronzy brown with bluish sheen and golden marks; a transverse fascia gradually widening towards dorsum close to base; an almost straight, distinct fascia of same width in the middle of the wing; an elongate, inwardly oblique streak beyond an elongate, triangular tornal spot before apex. Cilia brownish, no cilia line present. Hindwing lighter greyish, with similar cilia. Abdomen shining bluish grey. Hind tarsus blackish, distal end of segments with narrow whitish ring.

Female (Pl.-fig. 8). Antenna without cilia. Forewing more reddish brown, cilia violet-black.

Male genitalia (Pl.-figs. 234–236). Uncus with broad, V-shaped indentation, lobes conical, blunt-tipped, setose. Gnathos very long. Valva gradually tapering distally, cucullus not distinct, costa distally produced into a short, thorn-shaped process without setae above cucullus. Digitate process large, to middle of tegumen, without setae. Juxta lobes strongly sclerotized, conical. Vinculum rounded. Aedeagus long, straight, square-ended; without cornuti.

Female genitalia (Pl.-fig. 399). Apophyses posteriores longer than apophyses anteriores. Antrum weak, shallowly funnel-shaped, ventral margin broadly U-shaped, lateral rims broad, dorsal wall with small patch of minute spines and two distinct setae, antrum anteriorly tapering into a narrow, strongly folded short part of ductus bursae. Corpus bursae with scattered minute spines; signum elongate, medially thickened, strongly dentate.

Distribution. In Denmark only known from LFM and SZ; in Sweden from Sk. and Gtl.; not in Norway or Finland. 1. – Otherwise from Britain and widespread on the Continent from Belgium and France through Central and South Europe to Asia Minor.

Biology. The larva is greenish white, dorsal line dark green, head and prothoracic shield black. It mines the leaves of *Satureja vulgaris* (L.) Fritsch. The egg is deposited on the underside of the leaf near the mid-rib in the middle or proximal part. From here the larva first makes a very narrow mine with a fine central excrement line, the mine reaching the margin and distal end of the leaf, where the larva then makes a brownish blotch mine, visible from both sides and with tip and margins wrinkled over the upper side. Excrement large, blackish, scattered inside the mine (Pl.-fig. 502). The larva often changes leaf. Pupation takes place in a dense, whitish web outside the mine in a fold of the leaf; the pupa is short, rounded, light brown, without longitudinal folds and with a short, fine spine at each spiracle. Apparently bivoltine; larvae from autumn to April, hibernating in the mine, and again in July; adults fly during the day in May and August. The biotope is moist places in deciduous forests.

Note. In the Linnean collection at the Linnean Society, London, a single specimen is present under the name *brunnichella*. This has been designated as lectotype by Dr. J. D. Bradley, and is here cited for the first time. It is a female, labelled "N 115"; "Brunnichella"; "898"; "Lectotype"; "Genitalia slide prep. J. D. Bradley, 1977".

Genus *Elachista* Treitschke

Elachista Treitschke, 1833: 177.
  Type-species: *Elachista bifasciella* Treitschke, 1833 (see note).
Aphelosetia Stephens, 1834: 287.
  Type-species: *Tinea cygnipennella* Hübner, 1796,
    = [*Phalaena*] *argentella* Clerck, 1759.
Cycnodia Herrich-Schäffer, 1853: 46.
  Type-species: *Tinea cygnipennella* Hübner, 1796.
    = [*Phalaena*] *argentella* Clerck, 1759.
Poeciloptilia Hübner *sensu* Herrich-Schäffer, 1853: 55.
Phigalia Chambers, 1875: 107 (nom. praeocc.).
  Type-species: *Phigalia albella* Chambers, 1875.
Atachia Wocke *in* Heinemann, [1876]: 464.**Syn. n.**
  Type-species: *Symmoca pigerella* Herrich-Schäffer, 1854.
Neaera Chambers, 1880: 196 (nom. praeocc.).
  Type-species: *Laverna albella* Chambers, 1875 (nom. praeocc.).
    = *Elachista adempta* Braun, 1948.
Hecista Wallengren, 1881: 95.
  Type-species: [*Phalaena*] *argentella* Clerck, 1759.
Aphigalia Dyar, 1902: 544.
  Type-species: *Phigalia albella* Chambers, 1875.

Text-fig. 43. Head of *Elachista unifasciella* (Hw.).

Irenicodes Meyrick, 1919: 352.
Type-species: *Irenicodes eurychora* Meyrick, 1919.
Euproteodes Viette, 1954: 19.
Type-species: *Euproteodes galatheae* Viette, 1954.

Head (Text-fig. 43) smooth, neck tufts often raised, more rarely appressed. Scape of antenna with pecten of several to very few fine hairs proximally, rarely absent, flagellum about three-quarters of forewing, rarely ciliate, distal portion often slightly serrate. Labial palpi drooping to recurved, generally diverging, rather long and slender, rarely short.

Forewing ground colour varying from white to blackish, generally with lighter marks, which can have metallic sheen. Sexual dimorphism often pronounced.

Venation (Pl.-figs. 157-214). Forewing with R_4 and R_5 stalked or coalescent, M two-branched, M_1 free from near base R_{4+5} or from stalk R_{4+5}, M_2 free, stalked or coalescent with M_1 or CuA_1, CuP generally reduced, only most distal portion sometimes present, 1A + 2A rarely forked at base. Hindwing with Sc + R_1 to beyond middle of forewing or shorter, Rs simple or with free branch to costa, M two-branched, M_1 short or long-stalked with Rs, cell closed between Rs + M_1 and M_2 in a few generalized species, otherwise open.

Inner spur on mid tibia slightly longer than outer spur; hind tibia with a comb of long hairs above and only few or no hairs beneath, middle spurs in a varying position before the middle, inner spur longer than outer, apical spurs short, inner longer than outer. Spines generally slightly longer than scales.

Male genitalia. Uncus present, generally deeply indented, with many or few, long or short, setae. Socii small, short setose papillae. Gnathos single, comb-spined, elongate or rounded, on strong arms. Anterior margin of tegumen often strongly sclerotized. Valva of almost constant width, generally not or only slightly tapering, costa generally strong, medially produced into transtilla without labides, distally often forming a small hump before truncate or rounded cucullus, sacculus often with a small spine below cucullus; sacculus and cucullus with long setae. Digitate process present, rarely reduced. Juxta lobes variously shaped, apical margins generally setose, ventral part of juxta sometimes with a small tongue-shaped process. Vinculum often with a medial sclerotized ridge, saccus present or absent. Aedeagus often with bulbous base, with or without cornuti.

Female genitalia. Papillae anales broad or elongately triangular, generally soft, setose. Ostium bursae often near anterior margin of venter VIII; antrum and colliculum generally present, sclerotized to varying extent, confluent or separate; dorsal wall and inner side of antrum often spined. Ductus bursae and corpus bursae frequently with fine internal spines, ductus often with small sclerotized plates at inception of ductus seminalis, corpus generally with signum, which mainly consists of one dentate, elongate plate or one to three dentate bands or patches.

All species are miners in Poaceae, Cyperaceae and Juncaceae, and of these mainly the first two families. The life-history of many of the species is fairly well known. The egg is a flattened ovoid, appressed to the leaf surface, and not inserted into the tissues. The mine varies from very long and slender to a short blotch; the larva often changes mine. Pupation mainly takes place on a leaf, attached by a girdle and cremaster, or, rarely, in a fine cocoon. Hibernation generally occurs in the egg or larval stage.

The large genus *Elachista* is represented in almost all parts of the world, but mainly in the Holarctic Region. In the present paper 57 N. European species are dealt with and at least 50 more species occur in the W. Palaearctic area. Braun (1948) reported 48 species from N. America of which five must now be transferred to *Biselachista* gen.n. (see later). Only few species, from the Ethiopian, Neotropical and Oriental regions have, up to now, been placed in *Elachista*. Less than 30 species are known at present from the Pacific and Australian Region.

The genus is here divided into four major groups, of which two have been further divided into sub-groups. The basis for this is discussed in the introduction.

Notes. Treitschke (1833) originally placed 17 species in the genus *Elachista,* without designation of a type-species; only two of the original species are at present included in the Elachistidae.

Elachista canifoliella Treitschke, 1833 was subsequently designated as type-species by Boisduval (1836: 138), but under the *International Code of Zoological Nomenclature,* Article 69 (a) (iii) this is incorrect (viz. Sattler, 1973: 219). Later the following type-species were designated:

Tinea complanella Hübner, 1817 by Duponchel (1838: 25).
Tinea blancardella Fabricius, 1775 by Duponchel (1845: 221).
Phalaena (Tinea) rajella Linnaeus, 1758 by Walsingham (1907: 976).
Elachista bifasciella Treitschke, 1833 by Meyrick (1915: 210).

As the genus *Elachista* has, through the last and the present century, been in general use as later defined by Meyrick (1915: 210), this case ought to be referred to the International Commission on Zoological Nomenclature, which, in the interests of stability, should be asked to set aside all designations of type-species for *Elachista* made prior to that of Meyrick (1915).

Atachia Wocke is hereby placed in new synonymy with *Elachista* Treitschke. Wocke (1876) only included *S. pigerella* HS. in the genus, which was characterized by long labial palpi and basally forked 1A + 2A, but both these characters are also present in several species of *Elachista* and the latter feature may be absent in *pigerella* HS. Le Marchand (1952) placed *Atachia* close to *Mendesia* because of the forked Rs in the hindwing, but this character also appears in several *Elachista* species and in at least one *Cosmiotes*. Hering (1936) separated *Atachia* and *Elachista* on the well developed pecten in the former, but this is also present in some *Elachista,* and the character is generally very variable.

Key to species of *Elachista* based on external characters.

1	Forewing with metallic, shining fascia	2
–	Forewing without metallic, shining fascia	4
2 (1)	Fascia of forewing not reaching costa (Pl.-figs. 9, 10)	5. *regificella* Sirc.
–	Fascia of forewing reaching costa	3
3 (2)	Fascia sharply angled outwards; with a tiny distinct, metallic, distal spot between costal and tornal spots (Pl.-figs. 45, 46)	21. *nobilella* Zell.
–	Fascia almost straight; small metallic spot between costal and tornal spots often confluent with these, forming a sharply outward-bent fascia (Pl.-figs. 11, 12)	6. *gleichenella* (F.)

4 (1) Forewing plain-coloured, without cilia line 5
– Forewing with pattern and generally with cilia line present 10
5 (4) Forewing almost plain ochreous (Pl.-figs. 116, 117) 50. *subalbidella* Schl.
– Forewing pale grey, pale straw yellow, white, or grey brown to ochreous brown 6
6 (5) Forewing very broad, grey to ochreous brown (Pl.-figs. 13, 14) *pigerella* (HS.)
– Forewing not very broad, white, pale grey, or straw yellow 7
7 (6) Forewing very slender, pale grey (Pl.-fig. 73) or straw yellow (Pl.-fig. 74) *lastrella* Chrét.
– Forewing not very slender, white 8
8 (7) Large species, forewing more than 5 mm long (Pl.-fig. 77) 35. *argentella* (Cl.)
– Small species, forewing less than 5 mm long 9
9 (8) Head, tegulae, thorax and abdomen white. Cilia on forewing plain white. Hindwing very pale greyish (Pl.-figs. 86, 87) 39. *festucicolella* Zell.
– Head, tegulae, thorax and abdomen pale beige. Cilia at tornus of forewing faintly tinged with greyish or yellowish. Hindwing pale grey (Pl.-figs. 88, 89) *nitidulella* (HS.)
10 (4) Forewing ground colour dark, with one fascia in the middle, no costal and tornal spots present 11
– Colour and pattern of forewing different 17
11 (10) Fascia shining, yellow. Small species (Pl.-figs. 106, 107) 45. *chrysodesmella* Zell.
– Fascia not distinctly shining, yellowish to whitish. Medium-sized or larger species 12
12 (11) Cilia grey around apex and beyond cilia line (Pl.-figs. 114, 115) 49. *gangabella* Zell.
– Cilia more or less yellowish from apex to middle of termen beyond cilia line 13
13 (12) Fascia narrow or interrupted in the middle. Head and neck tufts pale ochreous white (Pl.-figs. 118, 119) 51. *revinctella* Zell.
– Fascia not narrow or interrupted in the middle. Head and neck tufts not both ochreous white 14
14 (13) Fascia straight, with almost parallel and darker margins (Pl.-figs. 110, 111) 47. *cingillella* (HS.)
– Fascia widening towards dorsum 15
15 (14) Head plain dark grey (Pl.-figs. 112, 113) 48. *unifasciella* (Hw.)
– Head pale ochreous on frons 16
16 (15) Outer margin of fascia distinct, strongly tinged with warm yellow (Pl.-figs. 120, 121) 52. *bisulcella* (Dup.)
– Outer margin of fascia blurred, not tinged with warm yellow (Pl.-figs. 108, 109) 46. *megerlella* (Hb.)
17 (10) Cilia of forewing white beyond distinct cilia line; ground colour varying from grey-brown to blackish brown (Pl.-figs. 27, 28, 98–105) 18
– Cilia of forewing not white beyond cilia line, or ground colour white or ochreous white 22
18 (17) Head and neck tufts bone white (Pl.-figs. 98, 99) *squamosella* (HS.)
– Head and neck tufts darker, not bone white 19
19 (18) Forewing ground colour greenish grey, tinged with ochreous, with a distinct

ochreous spot at base of dorsum (Pl.-figs. 104, 105) 44. *littoricola* Le March.
– Forewing ground colour without greenish, without an ochreous spot at base of dorsum 20

20 (19) Forewing beige-brown. Male (Pl.-fig. 100) with indistinct pattern, female (Pl.-fig. 101) with two pale fasciae 42. *bedellella* (Sirc.)
– Forewing ground colour grey or blackish grey, with one or two distinct whitish fasciae 21

21 (20) Forewing ground colour blackish grey; male (Pl.-fig. 102) with costal and tornal spots opposite, female (Pl.-fig. 103) with these spots almost confluent; cilia white all along termen 43. *pullicomella* Zell.
– Forewing grey at base, blackish beyond whitish fascia; costal and tornal spots never confluent; cilia only white at apex and one-third of termen (Pl.-figs. 27, 28) 12. *kilmunella* Stt.

22 (17) Forewing ground colour white, whitish or ochreous-whitish, without distinct pattern 23
– Forewing ground colour darker or paler greyish or brownish, with at least pale costal and tornal spots and generally a more or less prominent fascia before middle, or with blackish streaks in middle of wing 38

23 (22) Forewing wholly white with black dots 24
– Forewing whitish with yellow or brown scales, or cream-white with more or less dark-tipped scales, or plainly mottled with greyish or pale beige 27

24 (23) Forewing with only two black spots, without cilia line (Pl.-figs. 90, 91) 40. *dispilella* Zell.
– Forewing with two black spots and cilia line, or with more dots of black-tipped scales 25

25 (24) Forewing with only two black spots and black tipped scales forming a faint cilia line (Pl.-figs. 80, 81) 37. *triatomea* (Hw.)
– Forewing usually with two larger black spots and black tipped scales scattered over the wing 26

26 (25) Forewing with many distinct small dots, and with faint cilia line. Hindwing very pale greyish. Abdomen almost white (Pl.-figs. 92, 93) 41. *triseriatella* Stt.
– Forewing with many small, very fine dots, without cilia line. Hindwing and abdomen pale beige (Pl.-figs. 94, 95) *dispunctella* (Dup.)

27 (23) Forewing with two distinct white fasciae bound by yellowish or beige suffusion 28
– Forewing without distinct white fasciae 29

28 (27) White fasciae bound by yellowish suffusion (Pl.-figs. 84, 85) 38. *subocellea* (Stph.)
– White fasciae bound by beige suffusion, costal part of base suffused with dark grey (Pl.-figs. 82, 83) *collitella* (Dup.)

29 (27) Forewing almost white, with small black dots in lines, mainly along veins, not in the fold, scattered with beige (Pl.-fig. 78) or yellowish patches of scales (Pl.-fig. 79). Abdomen whitish 36. *pollinariella* Zell.
– Forewing white or cream-white, faintly suffused with pale beige or ochreous-red, or almost plainly mottled greyish to beige 30

30 (29) Forewing white with two black spots and few yellowish distal scales (Pl.-figs. 80, 81) 37. *triatomea* (Hw.)

– Forewing cream-white or mottled greyish to pale beige 31

31 (30) Forewing cream-white with a black spot in fold beyond middle (Pl.-figs. 75, 76) 34. *cerusella* (Hb.)

– Without distinct black spot in the fold beyond middle 32

32 (31) Forewing with a dark, distinct triangular streak from apex (Pl.-figs. 71, 72) 33. *rufocinerea* (Hw.)

– Forewing without darker triangular streak from apex 33

33 (32) Forewing ground colour cream-white, faintly suffused with beige (Pl.-fig. 69) or with yellowish patches (Pl.-fig. 70); scales without distinct darker tips 32. *anserinella* Zell.

– Forewing cream white and with scattered dark-tipped scales, or slightly mottled beige to greyish 34

34 (33) Forewing ground colour cream-white, faintly suffused with beige; very dark-tipped scales in fold, medially at dorsum and in distal part of wing (Pl.-figs. 96, 97) *rudectella* Stt.

– Forewing mottled beige to greyish, sometimes with faint fascia and costal and tornal spots 35

35 (34) Head pale beige (Pl.-fig. 67) 31. *pulchella* (Hw.)

– Head greyish 36

36 (35) Forewing with three dark streaks: one distally in the fold, one above this and one distal to the second (Pl.-figs. 53, 54) 24. *orstadii* Palm

– Forewing without such streaks 37

37 (36) Scales forming the cilia line with distinct white bases (Pl.-fig. 49) 23. *subnigrella* Dougl.

– Scales forming cilia line without white bases (Pl.-fig. 55) 25. *ingvarella* Tr-0.

38 (22) Forewing without fascia 39

– Forewing with a distinct fascia 43

39 (38) Forewing greyish or beige, with a black spot in the fold and one between this and apex (Pl.-figs. 19, 20) 9. *biatomella* (Stt.)

– Forewing without such black spots 40

40 (39) Large species. Forewing shining olive-brown with three (Pl.-fig. 15) or four (Pl.-fig. 16) white spots 7. *quadripunctella* (Hb.)

– Small species. Forewing not olive-brown 41

41 (40) Only costal and tornal spots present in forewing (Pl.-fig. 63) 29. *humilis* Zell.

– Forewing with three or four spots 42

42 (41) Forewing dark brown, with shining silver spots (Pl.-figs. 17, 18) 8. *tetragonella* (HS.)

– Forewing mottled greyish brown, with whitish grey spots (Pl.-figs. 21, 22) *martinii* Hofm.

43 (38) Head and neck tufts white (Pl.-figs. 41, 42) 19. *albifrontella* (Hb.)

– Head not totally white, neck tufts not white 44

44 (43) Head and neck tufts yellowish (Pl.-figs. 39, 40) 18. *luticomella* Zell.

– Head and neck tufts not yellowish 45

45 (44) Frons white, sometimes silky shining, vertex greyish brown 46

– Head greyish brown, or plain beige 50

46 (45) Costal and tornal spots confluent, forming a fascia (Pl.-figs. 43, 44) 20. *bifasciella* Tr.

– Costal and tornal spots not confluent to a fascia 47

47 (46) A third smaller distal spot between costal and tornal spots, all spots silky, shining (Pl.-figs. 47, 48) 22. *apicipunctella* Stt.

– No such third spot present, spots not shining 48

48 (47) Forewing very slender; costal spot almost square (Pl.-figs. 35, 36) 16. *compsa* Tr.-O.

– Forewing not very slender; costal spot more or less triangular 49

49 (48) Base of scales forming the cilia line white (Pl.-figs. 51, 52) *reuttiana* Frey

Base of scales forming the cilia line pale greyish (Pl.-figs. 37, 38) 17. *elegans* Frey

50 (45) Head and neck tufts beige 51

– Head and neck tufts dark brown or grey 59

51 (50) Forewing with distinct fascia, costal and tornal spots 52

– Forewing with blurred pattern 56

52 (51) Fascia not completely reaching costa (Pl.-fig. 32) 14. *alpinella* Stt.

– Fascia reaching costa 53

53 (52) Base of forewing whitish (Pl.-figs. 33, 34) 15. *diederichsiella* Her.

– Forewing from base to fascia pale beige or pale greyish 54

54 (53) Head paler beige than thorax (Pl.-fig. 67) 31. *pulchella* (Hw.)

– Head not paler than thorax 55

55 (54) Basal part of fold with dark-tipped scales (Pl.-figs. 56, 57) 25. *ingvarella* Tr.-O.

– Basal part of fold without dark-tipped scales (Pl.-figs. 53, 54) 24. *orstadii* Palm

56 (51) Head paler beige than thorax (Pl.-fig. 68) 31. *pulchella* (Hw.)

– Head not paler than thorax 57

57 (56) Forewing with three dark streaks: one distally in the fold, one above this, and one distal to the second (Pl.-fig. 53) 24. *orstadii* Palm

– Forewing without such streaks 58

58 (57) Base of scales forming the distinct cilia line white (Pl.-fig. 49) 23. *subnigrella* Dougl.

– Base of scales forming the cilia line not white (Pl.-fig. 55) 25. *ingvarella* Tr.-O.

59 (50) Fascia absent (Pl.-fig. 63) 29. *humilis* Zell.

– Fascia present 60

60 (59) Fascia very oblique to costa (Pl.-figs. 61, 62) 28. *pomerana* Frey

– Fascia not very oblique to costa 61

61 (60) Fascia most distinct at costa and dorsum, narrow or interrupted in the middle (Pl.-figs. 25, 26) 11. *atricomella* Stt.

– Fascia not narrow or interrupted in the middle 62

62 (61) Fascia bent outwards above fold and again below fold, forming a Z (Pl.-figs. 23, 24) 10. *poae* Stt.

– Fascia not Z-shaped 63

63 (62) Fascia distinct at dorsum, but not completely reaching costa (Pl.-figs. 30, 31) 14. *alpinella* Stt.

– Fascia complete or indistinct at costa and dorsum 64

64 (63) Fascia distinct in the middle, less distinct at costa and dorsum 65

– Fascia complete 67

65 (64) Small, distinct, light spot at base of dorsum (Pl.-figs. 65, 66) 30. *vonschantzi* Svens.

– Spots on base of dorsum absent 66
66 (65) Ground colour dark grey (Pl.-fig. 59) 26. *krogeri* Svens.
– Ground colour pale grey (Pl.-fig. 58) 27. *nielswolffi* Svens.
67 (64) Costal and tornal spots confluent 68
– Costal and tornal spots separate 69
68 (67) Ground colour cold bluish grey; costal and tornal spots confluent to form a sinuous fascia (Pl.-fig. 29) 13. *parasella* Tr.-O.
– Ground colour blackish grey. Costal and tornal spots confluent to form an outwardly V-shaped fascia (Pl.-fig. 60) 26. *krogeri* Svens.
69 (67) Bases of scales forming cilia line white (Pl.-fig. 50) 23. *subnigrella* Dougl.
– Bases of scales forming cilia line beige-grey (Pl.-fig. 64) 29. *humilis* Zell.

Key to species of *Elachista* based on male genitalia.

1 Uncus lobes with setae which are shorter than lobe is wide; costa of valva without distal hump; vinculum without a saccus 2
– Uncus lobes with setae which are longer than lobe is wide; costa of valva with a distal hump before cucullus; vinculum with a saccus 9
2 (1) Setae on uncus lobes numerous, almost as long as lobe is wide, or uncus lobes with scattered short setae; gnathos very large and uncus shorter than tegumen 3
– Setae on uncus lobes few, shorter or much shorter than lobe is wide; gnathos rather small or, if large, uncus longer or as long as tegumen 35
3 (2) Uncus with slight distal indentation or slight rounded emargination 4
– Uncus with deep V- or U-shaped distal indentation 5
4 (3) Uncus with slight, U-shaped, distal indentation; valva gradually tapering to the distinct cucullus, which is a prominen, almost circular lobe (Pl.-figs. 239, 240) 6. *gleichenella* (F.)
– Uncus with slight, rounded, distal emargination; valva gradually tapering distally, cucullus not distinct (Pl.-figs. 241, 242) *pigerella* (HS.)
5 (3) Uncus lobes almost conical, distally pointed, with scattered setae (Pl.-figs. 237, 238) 5. *regificella* Sirc.
– Uncus lobes widest in the middle or distally, distal end rounded, with dense, short setae 6
6 (5) Uncus lobe as broad as long; aedeagus with a distal row of small teeth (Pl.-figs. 243-245) 7. *quadripunctella* (Hb.)
– Uncus lobe distinctly longer than broad; aedeagus without cornuti or with several long distinct cornuti on a common plate 7
7 (6) Aedeagus with bulbous base, distal end blunt, with several long cornuti on a common plate (Pl.-figs. 250, 251) *martinii* Hofm.
– Aedeagus base only slightly enlarged, tip pointed, without cornuti 8
8 (7) Gnathos rounded; valva narrowest before cucullus; aedeagus strongly pointed (Pl.-figs. 246, 247) 8. *tetragonella* (HS.)
– Gnathos triangular; valva narrowest at base; aedeagus distally spear-ended (Pl.-figs. 248, 249) 9. *biatomella* (Stt.)

9 (1) Aedeagus without cornuti (Note: cornuti can be very small and are easily extruded during preparation) 10
– Aedeagus with cornuti 21
10 (9) Distal end of aedeagus chisel-shaped or pointed 11
– Distal end of aedeagus indented or deeply cleft (Note: one of the lobes can be almost membranous) 12
11 (10) Gnathos round, sacculus rounded towards cucullus; aedeagus distally chisel-shaped (Pl.-figs. 252, 253) 10. *poae* Stt.
– Gnathos transversely elongate; sacculus pointed below cucullus; aedeagus distally pointed (Pl.-figs. 284–286) 22. *apicipunctella* Stt.
12 (10) Vinculum with a medial ridge 13
– Vinculum without a medial ridge 15
13 (12) Uncus lobes rather small, very wide apart; distal end of aedeagus deeply cleft, one of the lobes almost membranous (Pl.-figs. 298, 299) 27. *nielswolffi* Svens.
– Uncus lobes large, rounded, close; distal end of aedeagus with shallow indentation 14
14 (13) Valva almost straight, cucullus almost right angled to sacculus; digitate process slender, distally swollen (Pl.-figs. 309, 310) 32. *anserinella* Zell.
– Valva strongly curved, cucullus rounded towards sacculus; digitate process rather short, club-shaped (Pl.-figs. 287–289) 23. *subnigrella* Dougl.
15 (12) Tegumen widening distally; gnathos elongately teardrop-shaped; digitate process slender (Pl.-figs. 278–280) 20. *bifasciella* Tr.
– Tegumen with parallel sides or widening proximally; gnathos rounded or transversely enlarged; digitate process club-shaped 16
16 (15) Cucullus rounded towards sacculus 17
– Sacculus produced distally below cucullus 18
17 (16) Digitate process short, club-shaped; vinculum tapering into a distinct narrow saccus (Pl.-figs. 281–283) 21. *nobilella* Zell.
– Digitate process slender, sides almost parallel; vinculum tapering into a very short, often not distinct saccus (Pl.-figs. 311–313) 33. *rufocinerea* (Hw.)
18 (16) Uncus lobes wide apart, incision widely U-shaped 19
– Uncus lobes close, incision V-shaped 20
19 (18) Uncus lobes very narrow; distal end of aedeagus slightly indented (Pl.-figs. 305, 306) 30. *vonschantzi* Svens.
– Uncus lobes rather broad; distal end of aedeagus deeply indented (Pl.-figs. 314, 315) *lastrella* Chrét.
20 (18) Uncus lobes rounded; gnathos transversely elongate; distal end of aedeagus slightly indented (Pl.-figs. 316, 317) 34. *cerusella* (Hb.)
– Uncus lobes tapering distally; gnathos rounded; distal end of aedeagus deeply indented (Pl.-figs. 292, 293) 24. *orstadii* Palm
21 (9) Vinculum with a medial ridge 22
– Vinculum without a medial ridge 28
22 (21) Aedeagus with one small cornutus (Pl.-figs. 259, 260) 12. *kilmunella* Stt.
– Aedeagus with more than one cornutus (Note: in some species the cornuti can be very small) 23
23 (22) Gnathos indented on underside; aedeagus deeply cleft, one lobe weakly sclerotized; one cornutus a cluster of 3–4 small teeth (Pl.-figs. 296, 297)

26. *krogeri* Svens.

– Gnathos not indented on underside; aedeagus not or only slightly indented distally, one of the cornuti often with more than one point, but not divided into distinct teeth 24

24 (23) Distal end of aedeagus slightly indented, cornuti small 25

– Aedeagus square-ended or distally pointed, cornuti large 26

25 (24) Digitate process club-shaped; medial ridge on vinculum very strong, saccus short and broad; aedeagus gradually tapering from base to distal end (Pl.-figs. 302–304) 29. *humilis* Zell.

– Digitate process with almost parallel sides; medial ridge on vinculum often weak, saccus short and narrow; aedeagus strongly tapering distally (Pl.-figs. 300, 301) 28. *pomerana* Frey

26 (24) Aedeagus with 3–4 long, slender, pointed cornuti (Pl.-figs. 269, 270) 16. *compsa* Tr.-O.

– Aedeagus with two spoon-shaped cornuti 27

27 (26) Uncus lobes very wide apart, incision widely U-shaped; aedeagus pointed (Pl.-figs. 266–268) 15. *diederichsiella* Her.

– Uncus lobes close, incision V-shaped, very deep; aedeagus square-ended (Pl.-figs. 290, 291) *reuttiana* Frey

28 (21) Uncus lobes close, incision V- or narrowly U-shaped; aedeagus with one cornutus 29

– Uncus lobes wide apart, incision widely U-shaped; aedeagus with two cornuti 30

29 (28) Distal hump on costa of valva short, distinct; aedeagus almost straight (Pl.-figs. 276, 277) 19. *albifrontella* (Hb.)

— Distal hump on costa of valva not distinct; aedeagus bent near base (Pl.-figs. 273–275) 18. *luticomella* Zell.

30 (28) Aedeagus with two small cornuti of almost equal length 31

– Aedeagus with two rather large cornuti of unequal length 33

31 (30) Cucullus rounded towards sacculus; dorsal margin of digitate process crenellated; aedeagus square-ended (Pl.-figs. 261, £62) 13. *parasella* Tr.-O.

– Sacculus produced into a small spine below cucullus; dorsal margin of digitate process smooth; aedeagus distally tapering or distal end indented 32

32 (31) Saccus narrow and pointed; aedeagus distally tapering (Pl.-figs. 271, 272) 17. *elegans* Frey

– Saccus not tapering; distal end of aedeagus indented (Pl.-figs. 294, 295) 25. *ingvarella* Tr.-O.

33 (30) Tegumen widening distally; distal end of aedeagus indented (Pl.-figs. 254–258) 11. *atricomella* Stt.

– Tegumen with parallel sides or widening proximally; aedeagus square-ended 34

34 (33) Valva widest in the middle; cucullus at an obtuse angle or rounded to sacculus (Pl.-figs. 263–265) 14. *alpinella* Stt.

– Valva widest distally, cucullus at almost a right angle to sacculus (Pl.-figs. 307, 308) 31. *pulchella* (Hw.)

35 (2) Uncus longer than or as long as tegumen 36

– Uncus distinctly shorter than tegumen 37

36 (35) Uncus very slightly indented; aedeagus square-ended (Pl.-figs. 318, 319) 35. *argentella* (Cl.)

– Uncus deeply indented; aedeagus with two blunt distal lobes (Pl.-figs. 320-323) 36. *pollinariella* Zell.

37 (35) Valva enlarged at base, beyond this long and narrow with parallel margins, cucullus not produced 38

– Valva widest at base, in the middle or distally, costa convex before a sub-apical emargination, sacculus more or less concave, cucullus produced 41

38 (37) Aedeagus with a row of pointed cornuti 39

– Aedeagus without cornuti 40

39 (38) Gnathos transversely enlarged; ventral part of anellus with tongue-shaped process; cornuti long, length increasing distally (Pl.-figs. 332, 333) *nitidulella* (HS.)

– Gnathos rounded; ventral part of anellus without tongue-shaped process; cornuti short, of almost equal length (Pl.-figs. 334, 335) 40. *dispilella* Zell.

40 (38) Gnathos long, slender; aedeagus coniform, distally tapering into a pointed, curved apex (Pl.-figs. 336, 337) 41. *triseriatella* Stt.

– Gnathos teardrop-shaped; aedeagus very long and slender (Pl.-figs. 338, 339) *dispunctella* (Dup.)

41 (37) Aedeagus with cornuti 42

– Aedeagus without cornuti 48

42 (41) Cornutus an elongate conglomerate with multiple teeth (Pl.-figs. 326, 327) *collitella* (Dup.)

– Aedeagus with one distinct cornutus or 7–9 long distinct spines 43

43 (42) Aedeagus with 7–9 long cornuti of increasing length towards distal end (Pl.-figs. 330, 331) 39. *festucicolella* Zell.

– Aedeagus with one cornutus 44

44 (43) Aedeagus extremely short and thick, three times as long as broad at base, with one prominent spoon-shaped cornutus and several sclerotized plates along distal margin (Pl.-figs. 340, 341) *rudectella* Stt.

– Aedeagus more than four times as long as broad at base, cornutus not spoon-shaped, distal margin without sclerotized plates 45

45 (44) Digitate process small and short; apical margin of juxta lobes produced into an extremely long, tapering process; aedeagus almost bottle-shaped (Pl.-figs. 344-346) 42. *bedellella* (Sirc.)

– Digitate process large and club-shaped; apical margin of juxta lobes rounded or truncate; aedeagus gradually tapering distally 46

46 (45) Apical margin of juxta lobes produced into a rounded lobe; cornutus elongate, slender, on a plate (Pl.-figs. 356, 357) 48. *unifasciella* (Hw.)

– Apical margin of juxta lobes truncate; cornutus thorn-shaped 47

47 (46) Cornutus small, not serrate (Pl.-figs. 324, 325) 37. *triatomea* (Hw.)

– Cornutus very large, concave margin serrate, base with fine teeth (Pl.-figs. 328, 329) 38. *subocellea* (Stph.)

48 (41) Uncus lobes short, rounded; gnathos almost circular; digitate process prominent, enlarged in the middle, distally pointed; juxta lobes long and slender; aedeagus with a deep distal slit in one side 49

– Uncus lobes rather long, often tapering distally; gnathos elongate or rounded;

digitate process slender, club-shaped, swollen near distal end or absent; juxta lobes rather short; aedeagus without a distal slit 51

49 (48) Cucullus rounded towards costa; distal end of digitate process curved (Pl.-figs. 360, 361) 50. *subalbidella* Schl.

– Cucullus truncate; distal end of digitate process straight 50

50 (49) Costa of valva with a deep subapical emargination; apical margin of juxta lobes rounded medially (Pl.-figs. 364, 365) 52. *bisulcella* (Dup.)

– Costa of valva with a slight subapical emargination; apical margin of juxta lobes produced into a short medial and a short lateral process (Pl.-figs. 362, 363) 51. *revinctella* Zell.

51 (48) Digitate process very small or absent; ventral part of anellus with a broad tongue-shaped process 52

– Digitate process of normal size, slender or club-shaped; ventral part of anellus with or without a narrow tongue-shaped process 53

52 (51) Digitate process absent; aedeagus square-ended (Pl.-figs. 349, 350) 44. *littoricola* Le March.

– Digitate process very small; aedeagus tapering to a pointed distal end (Pl.-figs. 347, 348) 43. *pullicomella* Zell.

53 (51) Cucullus rounded; digitate process very slender; apical margin of juxta lobes rounded medially 54

– Cucullus truncate; digitate process club-shaped or enlarged dorsally near distal end; apical margin of juxta lobes produced into a medial process 55

54 (53) Gnathos elongate; ventral part of anellus with a narrow tongue-shaped process (Pl.-figs. 351, 352) 45. *chrysodesmella* Zell.

– Gnathos rounded; ventral part of anellus without tongue-shaped process (Pl.-figs. 358, 359) 49. *gangabella* Zell.

55 (53) Uncus lobes distally pointed; valva widest in the middle, subapical emargination very weak (Pl.-figs. 342, 343) *squamosella* (HS.)

– Uncus lobes distally rounded; valva widest in the middle, sub-apical emargination distinct 56

56 (55) Digitate process strong, slightly swollen distally; apical margin of juxta lobes produced into a very long, tapering medial process (Pl.-figs. 353, 354) 46. *megerlella* (Hb.)

– Dorsal margin of digitate process enlarged beyond the middle, ventral margin straight; apical margin of juxta lobes produced into a small medial knob (Pl.-fig. 355) 47. *cingillella* (HS.)

Key to species of *Elachista* based on female genitalia.

(Note: Female of *E. nielswolffi* Svens. is unknown.)

1 Dorsal VIII fringed along posterior margin with long medial setae 2

– Dorsal VIII not fringed with long setae 3

2 (1) Antrum plus colliculum longer than apophyses posteriores, colliculum with prominent internal teeth (Pl.-fig. 400) 5. *regificella* Sirc.

– Antrum plus colliculum shorter than apophyses posteriores; colliculum anteriorly dentate (Pl.-fig. 401) 6. *gleichenella* (F.)

3 (1) Antrum and colliculum both sclerotized, confluent, antrum mainly funnel-shaped, gradually tapering into colliculum or pouched; signum single if present 4

– Antrum and colliculum generally not both sclerotized and confluent; antrum large, membranous or small and weakly sclerotized, ventral margin generally distinct; colliculum a short, often weakly sclerotized ring close to ostium. Signa absent, single, double or triple 32

4 (3) Corpus bursae without signum 5

– Corpus bursae with signum 9

5 (4) Antrum pouched, constricted towards colliculum; colliculum of equal length to apophyses posteriores 6

– Antrum funnel-shaped, gradually tapering into colliculum; colliculum shorter or longer than apophyses posteriores 7

6 (5) Apophyses posteriores longer than anteriores; ventral margin of antrum U-shaped; ductus and corpus bursae strongly folded (Pl.-fig. 407) 10. *poae* Stt.

– Apophyses of equal length; ventral margin of antrum anteriorly curved; anterior region of ductus bursae with longitudinal folds (Pl.-fig. 414) 14. *alpinella* Stt.

7 (5) Ventral margin of antrum almost straight; colliculum shorter than apophyses posteriores (Pl.-fig. 448) 33. *rufocinerea* (Hw.)

– Ventral margin of antrum anteriorly curved; colliculum longer than apophyses posteriores 8

8 (7) Ventral margin of antrum deeply U-shaped, without spines; dorsal wall of antrum with fine spines (Pl.-figs. 422, 423) 20. *bifasciella* Tr.

– Ventral margin of antrum widely U-shaped, with coarse spines; dorsal wall of antrum with spines behind ventral margin (Pl.-fig. 447) 32. *anserinella* Zell.

9 (4) Ventral margin of antrum posteriorly curved; colliculum four times as long as apophyses posteriores (Pl.-figs. 408, 409) 11. *atricomella* Stt.

– Ventral margin of antrum straight or anteriorly curved; colliculum less than three times as long as apophyses posteriores 10

10 (9) Antrum bowl-shaped or pouched, very constricted towards colliculum 11

– Antrum more or less funnel-shaped, gradually tapering into colliculum 15

11 (10) Apophyses slender, posteriores longer than anteriores; dorsal wall of antrum with elongate patch of spines; signum very small (Pl.-figs. 450, 451) 34. *cerusella* (Hb.)

– Apophyses of almost equal length; spines on dorsal wall not in an elongate patch; signum of normal size 12

12 (11) Antrum bowl-shaped, short; colliculum with several dentate sclerotized plates or signum with very long teeth distally 13

– Antrum pouched, margins almost parallel; colliculum as long as or longer than apophyses anteriores; ductus bursae with a few oval plates or without teeth and plates 14

13 (12) Colliculum half as long as apophyses posteriores; signum with teeth of almost equal length (Pl.-fig. 415) 15. *diederichsiella* Her.

– Colliculum longer than apophyses posteriores; signum with extremely long distal teeth (Pl.-fig. 445) 30. *vonschantzi* Svens.

14 (12) Colliculum of almost the same length as apophyses posteriores (Pl.-fig. 419)

18. *luticomella* Zell.
– Colliculum twice as long as apophyses posteriores (Pl.-fig. 443)
28. *pomerana* Frey
15 (10) Ventral margin of antrum almost straight or slightly anteriorly curved 16
– Ventral margin of antrum deeply anteriorly curved 18
16 (15) Colliculum very narrow, shorter than apophyses posteriores (Pl.-fig. 449)
lastrella Chrét.
– Colliculum broad, as long as or longer than apophyses posteriores 17
17 (16) Dorsal wall of antrum with spines both before and behind ventral margin; colliculum twice as long as apophyses posteriores (Pl.-fig. 425)
22. *apicipunctella* Stt.
– Dorsal wall of antrum mainly with spines anterior to ventral margin; colliculum of almost equal length to apophyses posteriores (Pl.-fig. 426)
23. *subnigrella* Dougl.
18 (15) Antrum widely V-shaped, ventral margin widely U-shaped, almost to anterior margin of antrum 19
– Antrum funnel-shaped, ventral margin anteriorly curved or U-shaped, rarely to beyond middle of antrum 20
19 (18) Ventral margin of antrum strongly reinforced; ductus bursae with 5–7 teeth (Pl.-figs. 410–412) 12. *kilmunella* Stt.
– Ventral margin of antrum not reinforced; ductus bursae with a single small plate with a tooth (Pl.-fig. 413) 13. *parasella* Tr.-O.
20 (18) Papillae anales flat, leaf-like, sclerotized, with very short setae (Pl.-fig. 424)
21. *nobilella* Zell.
– Papillae anales soft, with setae of varying length 21
21 (20) Membranous part of ductus bursae with one to several sclerotized anterior plates or teeth inside 22
– Membranous part of ductus bursae without sclerotized plates or teeth 28
22 (21) Ventral margin of antrum anteriorly U-shaped, reaching middle of antrum 23
– Ventral margin of antrum anteriorly curved, reaching about one-third of antrum 25
23 (22) Apophyses posteriores longer than anteriores; antrum strongly spined on inner side; colliculum less than twice as long as apophyses posteriores (Pl.-fig. 446)
31. *pulchella* (Hw.)
– Apophyses of almost equal length; antrum without or almost without internal spines; colliculum about twice as long as apophyses posteriores 24
24 (23) Anterior portion of ductus bursae with many longitudinal folds (Pl.-fig. 427)
reuttiana Frey
– Anterior portion of ductus bursae without longitudinal folds (Pl.-figs. 430–441)
25. *ingvarella* Tr.-O.
25 (22) Antrum gradually tapering from ventral margin to colliculum 26
– Antrum slightly pouched before tapering into colliculum 27
26 (25) Dorsal wall of antrum strongly spined (Pl.-figs. 420, 421)
19. *albifrontella* (Hb.)
– Dorsal wall of antrum without spines (Pl.-fig. 416) 16. *compsa* Tr.-O.
27 (25) Lateral rims of antrum very strong, dorsal wall with fine spines (Pl.-fig. 442)
26. *krogeri* Svens.

– Lateral rims of antrum very short, dorsal wall sparsely spined (Pl.-figs. 417, 418) 17. *elegans* Frey

28 (21) Dorsal wall of antrum without spines (Pl.-fig. 416) 16. *compsa* Tr.-O.

– Dorsal wall of antrum with spines 29

29 (28) Rims of antrum short, ventral margin reaching to one-third of antrum, colliculum very narrow (Pl.-fig. 420) 19. *albifrontella* (Hb.)

– Rims of antrum long, ventral margin reaching to middle of antrum, colliculum rather broad, widening anteriorly 30

30 (29) Signum very small and weak (Pl.-figs. 422, 423) 20. *bifasciella* Tr.

– Signum normal sized, bent 31

31 (30) Ventral margin of antrum and rims very strong, dorsal wall with fine spines, antrum smoothly curved (Pl.-fig. 444) 29. *humilis* Zell.

– Ventral margin of antrum and rims not reinforced, dorsal wall with strong or coarse spines, antrum gradually tapering (Pl.-fig. 428) 24. *orstadii* Palm

32 (3) Signum a weak plate with two prominent, curved teeth and often a few small additional teeth 33

– Signum an oval, dentate plate, two or three dentate patches or totally absent 35

33 (32) Colliculum present (Pl.-fig. 405) 9. *biatomella* (Stt.)

– Colliculum absent (Pl.-fig. 404) 8. *tetragonella* (HS.)

35 (32) Corpus bursae without signum 36

– Corpus bursae with one to three oval patches, or band-shaped signa 39

36 (35) Spermatheca sclerotized; venter VIII membranous 37

– Spermatheca not sclerotized; venter VIII sclerotized 38

37 (36) Middle region of ductus bursae with internal spines (Pl.-fig. 457) *collitella* (Dup.)

– Anterior two-thirds of ductus bursae and posterior part of corpus bursae with internal spines (Pl.-fig. 458) 38. *subocellea* (Stph.)

38 (36) Anterior margin of sternite VIII curved anteriorly; colliculum a very short ring (Pl.-figs. 466, 467) 42. *bedellella* (Sirc.)

– Anterior margin of sternite VIII emarginate; colliculum absent (Pl.-fig. 468) 43. *pullicomella* Zell.

39 (35) Corpus bursae with a single elongate or oval dentate signum 40

– Corpus bursae with more than one signum 50

40 (39) Anterior margin of papillae anales not sclerotized; antrum large, membranous or indistinct; dorsal wall spined or with two prominent lateral patches of spines 41

– Anterior margin of papillae anales sclerotized; only ventral margin of antrum generally distinct, sclerotized; dorsal wall almost without spines or with two lateral patches of spines 43

41 (40) Antrum not distinct; two elongate patches of spines postero-laterally to the short colliculum (Pl.-fig. 402) *pigerella* (HS.)

– Antrum large, membranous, dorsal wall with spines 42

42 (41) Colliculum as long as apophyses posteriores; ductus bursae with many transverse anterior folds (Pl.-fig. 406) *martinii* Hofm.

– Colliculum very short; ductus bursae without transverse anterior folds (Pl.-fig. 403) 7. *quadripunctella* (Hb.)

43 (40) Venter VIII sclerotized, ostium bursae situated in the middle or anterior part of

sternite VIII 44

– Venter VIII membranous or with a U-shaped sclerotized medial area; ostium bursae situated at anterior margin of segment VIII 45

44 (43) Ostium bursae circular, in the middle of sternite VIII; no colliculum present (Pl.-fig. 469) 44. *littoricola* Le March.

– Ostium bursae rounded, in anterior part of sternite VIII; colliculum very short (Pl.-figs. 466, 467) 42. *bedellella* (Sirc.)

45 (43) Papillae anales large, sclerotized, sparsely setose 46

– Papillae anales soft, except anterior margin, strongly setose 47

46 (45) Venter VIII with U-shaped sclerotized area surrounding ostium bursae; ductus and corpus bursae with internal spined patches (Pl.-fig. 452) 35. *argentella* (Cl.)

– Tergite VIII with medial, H-shaped, sclerotized area, venter VIII membranous; ductus and corpus bursae without internal spines (Pl.-figs. 453–455) 36. *pollinariella* Zell.

47 (45) Ductus bursae without transverse folds and spines (Pl.-fig. 462) 41. *triseriatella* Stt.

– Ductus bursae with transverse folds and spines in anterior portion 48

48 (47) Tergite VIII with straight margins; colliculum wider than long (Pl.-fig. 459) 39. *festucicolella* Zell.

– Tergite VIII 8-shaped; colliculum longer than wide 49

49 (48) Two spined patches lateral to ostium; colliculum barrel-shaped (Pl.-fig. 461) 40. *dispilella* Zell.

– A patch set with fine spines behind ostium; colliculum almost as wide as long (Pl.-fig. 460) *nitidulella* (HS.)

50 (39) Corpus bursae with three distinct signa (Pl.-fig. 475) 48. *unifasciella* (Hw.)

– Corpus bursae with two signa 51

51 (50) Signa band-shaped 52

– Signa patch-shaped 54

52 (51) Antrum large, funnel-shaped; ductus bursae with a medial spined patch (Pl.-fig. 464) *rudectella* Stt.

– Only ventral margin of antrum distinct; ductus bursae without internal spines 53

53 (52) Ventral margin of antrum almost straight; two spined patches lateral to colliculum (Pl.-fig. 476) 49. *gangabella* Zell.

– Ventral margin of antrum almost circular, strong; no spined patches lateral to colliculum (Pl.-fig. 463) *dispunctella* (Dup.)

54 (51) Each signum with less than ten teeth 55

– Each signum with more than ten teeth 57

55 (54) Antrum membranous, indistinct; signa very small, each with 2–3 teeth (Pl.-fig. 470) 45. *chrysodesmella* Zell.

– Antrum at least slightly sclerotized, ventral margin distinct; signa with 2–8 teeth 56

56 (55) Ventral margin of antrum with spines (Pl.-figs. 471, 472) 46. *megerlella* (Hb.)

– Ventral margin of antrum without spines (Pl.-figs. 473, 474) 47. *cingillella* (HS.)

57 (54) Signa two distinct, sclerotized plates with long, slender teeth (Pl.-fig. 477)

50. *subalbidella* Schl.

– Signa weakly sclerotized, with short, almost triangular teeth 58

58 (57) Anterior half of ductus bursae with internal spines (Pl.-fig. 456) 37. triatomea (Hw.)

– Ductus bursae without internal spines 59

59 (56) Venter VIII sclerotized; ventral margin of antrum widely V-shaped; colliculum almost membranous (Pl.-fig. 465) *squamosella* (HS.)

– Venter VIII membranous; ventral margin widely U shaped, colliculum short, sclerotized 60

60 (59) Ventral margin of antrum with spines on inner side; teeth on signa rather distant (Pl.-fig. 478) 51. *revinctella* Zell.

– Ventral margin of antrum without spines on inner side; teeth on signa rather close (Pl.-fig. 479) 52. *bisulcella* (Dup.)

The *gleichenella*-group

Head pale or silvery, thorax and tegulae with bronzy or golden sheen; forewing blackish brown with silvery or golden marks, often with a bluish reflection; hindwing broad.

Venation (Pl.-figs. 157, 158). Forewing with R_2 and R_3 wide apart, R_4 and R_5 stalked, M_1 from stalk R_{4+5} or coalescent with R_5, M_2 free, base of Cu_2 opposite base of R_2 or between R_2 and R_3, CuP reduced, 1A + 2A simple. Hindwing cell open.

Male genitalia. Uncus slightly or deeply indented, lobes with short, stiff setae. Gnathos large. Anterior margin of tegumen strongly sclerotized. Digitate process slender. Juxta lobes narrow. Vinculum short, with or without medial ridge, without saccus. Base of aedeagus bilobed.

Female genitalia. Characteristic group of long setae present on posterior portion of dorsal VIII. Ostium bursae near anterior margin of venter VIII. Signum prominent.

Larvae mine in Cyperaceae and Juncaceae; pupation takes place outside the mine in a fine cocoon on a leaf or between loosely spun leaves on the ground.

5. ***Elachista regificella*** Sircom, 1849
Plate-figs. 9, 10, 157, 237, 238, 400, 503, 504.

Elachista magnificella Duponchel *sensu* Zeller, 1847: 891.
Elachista magnificella Tengström, [1848] 1847: 147, *nec* Duponchel, 1843.
Elachista regificella Sircom, 1849: XLII.
Poeciloptilia geminatella Herrich-Schäffer, 1855: 301, 309.

8–9 mm. Male (Pl.-fig. 9). Head coppery brown; tegulae and thorax plain blackish brown, shining. Antenna dark brown, darkest towards base, distal third serrate. Labial palpi slender, descending and only slightly recurved, upperside whitish, underside brownish, distal part of second segment often with a brownish ring. Forewing ground colour dark bronzy brown, with distinct golden shining marks; a broad fascia from costa to dorsum near base, a fascia passing only two-thirds across the wing, not reaching costa, from middle of dorsum; a prominent, square spot at costa, beyond a triangular tornal spot. Cilia of same colour as wing, less shiny, no cilia line present.

Hindwing rather broad, brown, with cilia of same colour. Abdomen blackish-brown, anal tuft lighter. Hind tarsus ringed blackish brown/greyish white.

Female (Pl.-fig. 10). Distal third of antenna silky white. Inner fascia of forewing often more narrow, especially towards dorsum; costal and dorsal spots enlarged, the latter almost confluent.

Male genitalia (Pl.-figs. 237, 238). Uncus lobes conical, with short setae. Gnathos prominent, broad, elliptical to square, dorsal margin slightly incurved. Valva widest at base, costa strong, cucullus rounded. Digitate process narrow, elongate, with parallel setose margins. Juxta lobes slender, lateral processes long. Vinculum short, triangular, with medial sclerotized ridge. Basal portion of aedeagus gradually narrowing; cornuti absent, but distal half with several sclerotized folds.

Female genitalia (Pl.-fig. 400). Apophyses posteriores about three times as long as apophyses anteriores. Ostium bursae broad, at anterior part of sternite VIII; antrum short, bowl-shaped, ventral margin and dorsal wall spined; colliculum tubular posteriorly, widening anteriorly, longer than apophyses posteriores, anterior part with prominent internal teeth. Corpus bursae with two patches of minute internal spines; signum narrow, elongate, angulate, dentate, dentation most prominent on concave margin.

Distribution. From all parts of Denmark; in S. Sweden up to Upl.; in Norway only from the Oslo-area; in Finland from SW. districts. – Widespread in Europe from USSR to Italy and Britain.

Biology. The young larva is lemon-yellow, greenish when full-grown. Head pale brownish, mouthparts dark brown (Frey, 1859). It starts mining in the upperside of the leaves of *Luzula pilosa* (L.)., *L. luzuloides* (Lam.) Dandy & Wilm., *L. sylvatica* (Huds.) Gaud. and probably other *Luzula* species in August-September, making a slender, *Stigmella*-like mine, later widening, with longitudinal folds on the epidermis like a *Phyllonorycter* mine (Pl.-fig. 503, 504). In late spring the mine is almost as wide as the leaf, whitish or brownish. A single larva often mines more than one leaf, and probably this is the reason for the very different mines obtained during the spring. Pupation takes place on a leaf (Martini, 1916) or on the ground (Frey, 1859). The adults fly from early June to mid-August, but the species is probably univoltine. The biotope is shady places where the food-plant grow, often under conifers and along wood edges.

6. ***Elachista gleichenella*** (Fabricius, 1781)
Plate-figs. 11, 12, 158, 239, 240, 401, 505, 506.

Tinea gleichenella Fabricius, 1781: 303.
Tinea gleichella Fabricius, 1794: 323.
Oecophora magnificella Duponchel *in* Godart, [1843] 1842: 321.
Elachista trifasciella Nylander *in* Tengström, [1848] 1847: 148.
Poeciloptilia fractella Herrich-Schäffer, 1855: 301, 309.

8–9 mm. Male (Pl.-fig. 11). Head and thorax shining, head silvery or lead-grey, thorax more bronzy to blackish or violet. Antenna blackish, silver greyish towards apex, dark ringed. Labial palpi descending, upperside light grey, shining, underside brownish grey. Forewing ground colour velvet blackish, with shining silvery or golden marks; a golden basal spot not reaching costa; a slightly outward bent fascia from before middle of costa to middle of dorsum; a prominent irregular spot on costa and a triangular tornal spot beyond middle; often a small distal dot between these spots. Cilia grey-brown, cilia line blackish. Hindwing light grey-brown, cilia similar. Abdomen grey-brown with reflecting scales. Hind tarsus with broad blackish and more narrow whitish rings.

Female (Pl.-fig. 12). Golden or silvery marks on forewing more prominent; golden basal area reaching costa and tegulae, fascia more distinct and distal spots confluent, forming a distinct outwards-pointed V-shaped fascia.

Male genitalia (Pl.-figs. 239, 240). Uncus with shallow indentation, uncus lobes with a row of few short setae along outer margin. Gnathos prominent, rounded. Costa of valva strong, valva narrowing to the projecting, semi-circular cucullus. Digitate process short, club shaped, with fine setae. Juxta lobes strongly sclerotized, slender, S-shaped. Vinculum very short. Aedeagus short, characteristically bent just below distal end, containing one big cornutus and a small distal patch of teeth.

Female genitalia (Pl.-fig. 401). Apophyses posteriores about three times as long as apophyses anteriores. Anterior margin of tergite VIII anteriorly curved. Ostium bursae a transverse slit near anterior margin of sternite VIII. Antrum plus colliculum short, half as long as apophyses posteriores, anteriorly dentate. Posterior portion of ductus bursae with longitudinal folds; corpus bursae with minute internal spines, signum prominent, elongate, oval, finely denate.

Distribution. From all parts of Denmark; from S. Sweden up to Med.; from SW. Norway and SW. Finland. – Widespread in E., C. and W. Europe.

Biology. Larva whitish grey, with a faint purple tinge, head dark brown, prothoracic shield purplish black, posteriorly black (Stainton, 1858c). The larva mines from September-October to April-May, hibernating in the mine, in many species of sedges, *Carex flacca* Schreb., *C. echnata* Murr., *C. ornithopoda* Willd., *C. montana* L., *C. paniculata* L., and *C. pilosa* Scop. and a few others *Carex*-species; furthermore, Frey (1859) mentions *Deschampsia caespitosa* (L.) Beauv. and *Luzula pilosa* (L.) Willd. The autumn mine is long and narrow, starting from the tip of the leaf on the upperside, first whitish green later brownish; in the spring the larva starts a new mine, which is as broad as the blade (Pl.-figs. 505, 506) (Wörz, 1957). Pupation takes place on the ground between loosely spun leaves; the pupa is rather short, light brown, wing-ribs strongly serrate. The adult fly from mid-June to mid-August. The habitat of this species varies from dark coniferous forests to open, sunny calcareous grassland.

Notes. Fabricius (1781: 303) based his description of *gleichenella* on a specimen in the Yeats collection; nothing is yet known about this collection.

Tinea gleichella Fabricius, 1794 is an unjustified emendation of *T. gleichenella* Fabricius, 1781, and therefore the specimens of *Elachista albifrontella* (Hb.) present under that label in the Thunberg collection (Benander, 1961) are only misidentifications and of no nomenclatorical significance, as Thunberg (1794: 30) refers this species to Fabricius (1794: 323).

Tinea metallella [Denis & Schiffermüller], 1775: 144 was by Fabricius (1794: 323) considered conspecific with *gleichenella,* but as the original description fits several species, this name has not been adopted here.

The *tetragonella*-group

From very large (13 mm) to very small (5 mm) species. Brownish or greyish, with or without marks.

Venation (Pl.-figs. 159–163). Varying from very generalized to strongly reduced. Forewing with R_4 and R_5 stalked or coalescent, M_1 free or from stalk R_{4+5}, 1A + 2A forked at base or simple. Hindwing with Sc + R_1 long, close to Rs, Rs sometimes with free branch to costa, cell closed or open.

Male genitalia. Uncus slightly or deeply indented, lobes with short, stiff setae on distal portion. Anterior margin of tegumen weakly sclerotized. Vinculum short, without medial ridge or saccus. Base of aedeagus broad, cornuti often absent.

Female genitalia. Papillae anales short, triangular, with short setae. Tergite VIII weakly sclerotized. Antrum large, membranous, dorsal wall with fine spines, colliculum weak or absent. Signum weak, dentate or with two prominent curved teeth. Larvae of all species mine in *Carex* or *Luzula*; pupation takes place inside the mine or outside in a loose web.

Elachista pigerella (Herrich-Schäffer, 1854) **comb.n.**
Plate-figs. 13, 14, 159, 241, 242, 402.

Symmoca pigerella Herrich-Schäffer, 1854: 111.
Elachista fuscochreella Frey, 1856: 304.
Elachista muehligiella Frey, 1856: 304.
Elachista fuscogrisella Rebel, 1936: 100. **Syn.n.**

11–13 mm. Male (Pl.-fig. 13). Head, tegulae and thorax ochreous brown to dark brown, finely mottled by yellowish white scale-tips, frons more yellowish grey. Antenna unicolorous dark brown, scape and pecten slightly yellowish, scales on distal half raised. Labial palpi slightly curved, short, third segment about one-quarter length of second, palpi ochreous grey on upperside, underside more brownish. Forewing ground colour brown, varying from ochreous brown to grey-brown, mottled with ochreous white by pale margins of the dark brown scales; without marks. Cilia blackish grey, cilia line very indistinct. Hindwing broad, lighter than forewing, cilia similar. Abdomen dark brown, anal tufts lighter. Hind tarsus with broad blackish rings and narrow distal whitish rings.

Female (Pl.-fig. 14) Slightly smaller, similar to male.

Male genitalia (Pl.-figs. 241, 242). Uncus long, very slightly indented, with short, stiff lateral setae. Gnathos rounded. Tegumen elongate, narrow. Valva long, narrowing distally, costa strong. Digitate process with parallel margins. Juxta lobes truncate, tapering laterally into pointed processes. Vinculum short, rounded. Aedeagus bulbous at base, curved, without cornuti.

Female genitalia (Pl.-fig. 402). Apophyses posteriores swollen one-fifth from base and distally, about twice as long as apophyses anteriores. Posterior margin of tergite VIII deeply emarginate, anterior margin slightly emarginate. Ostium bursae indistinct, antrum and margins not sclerotized, two patches of minute spines posterolaterally to colliculum; colliculum short. Signum oval, dentate.

Distribution. Not known from Scandinavia or Britain. Apparently rare and very local, known from a few localities in Germany, Switzerland, Italy and France, but will probably be found to be more widespread now when the immature stages and biology are known.

Biology. According to Steuer (1973) the larva is very slender and long, first yellowish, later paler, head and prothoracic shield black, the latter narrow, tapering anteriorly. It mines in the leaves of *Carex flacca* Schreb., starting near the tip and mines downwards in a very long, narrow mine, filled with excrement; when it reaches the base of the leaf it continues into the next leaf. Pupation on a leaf, attached by a single girdle and the cremaster. The adults appear from mid-July and fly to late August. Nothing is known about the hibernation, but this probably takes place as an egg or young larva. The biotope is moist places where the food-plant grows, often on calcareous ground.

Note. The lectotype of *E. fuscogrisella* Rbl. designated by Gaedike (1975) is a greyish brown specimen of *E. pigerella* (HS.).

7. ***Elachista quadripunctella*** (Hübner, 1825)
Plate-figs. 15, 16, 160, 243–245, 403.

Tinea quadrella [Denis & Schiffermüller] *sensu* Hübner, [1805], pl. 42, fig. 293.
Chionodes quadripunctella Hübner, [1825]: 420.

10–12 mm. Male (Pl.-fig. 15). Head, tegulae and thorax olive-brown, with metallic sheen. Antenna unicolorous brown, distal half slightly serrate. Labial palpi long, slender, slightly recurved; upperside greyish white, distal part of second and third segments brown beneath. Forewing shining olive-brown, suffused with silver-grey at base; with three distinct small silky white spots, one at two-thirds of costa, one at tornus opposite costal spot and one in the fold just before the middle of the wing. Cilia brown, no cilia line present. Hindwing rather broad, brownish, cilia grey-brown. Abdomen brown. Hind tarsus with broad black and narrow whitish rings.

Female (Pl.-fig. 16). Forewing with four white spots, three as in male and a fourth on costa before middle.

Specimens with marks reduced or absent are mentioned and named by Caradja (1920: 154).

Male genitalia (Pl.-figs. 243–245). Uncus with deep V-shaped indentation, lobes with short, stiff, lateral setae. Gnathos oval. Valva broadest at base, narrowing distally, angle between cucullus and costa sharp. Digitate process short, sides almost parallel. Juxta lobes with medial margin produced, apical margin bent and setose, with narrow lateral process. Vinculum triangular. Aedeagus short, sinuate, rounded distally, a few small triangular teeth in a row below distal end.

Female genitalia (Pl.-fig. 403). Apophyses posteriores four times as long as apophyses anteriores. Antrum smooth, funnel-shaped, membranous, tapering into colliculum, ventral margin U-shaped, dorsal wall with distinct spines; colliculum short, with longitudinal folds. Signum elongate, with parallel sides, dentate.

Distribution. In Scandinavia only known from Sweden, Vg., Halleberg. – Widespread in Europe from the Baltic to Italy, France and Belgium; not in Britain.

Biology. Larva dull whitish yellow, head pale brown, prothoracic shield weakly sclerotized, pale greyish brown (Stainton, 1858c). It is known from *Luzula luzuloides* (Lam.) Dandy & Wilm., *L. pilosa* (L.) Willd. and *L. sylvatica* (Huds.) Gaud. and a few other *Luzula* species, the leaves of which it mines from late September to late spring, hibernating in the mine which is long, broad, greenish to whitish or brownish, the loosened epidermis often forming longitudinal folds similar to those known from *Phyllonorycter*-mines. Pupation takes place inside the mine. The adults are reported to fly from late May to late July, generally considered univoltine, but Wörz (1957) reports two generations. The Swedish specimens were caught in the middle of July in dark, wet shady places with *Luzula* in a stand of old coniferous trees (Jonasson, *in litt.*)

Note. The specific name *Elachista quadrella* (Hb.) has been used for this species, but as *Chionodes quadrella* Hübner, 1805 is a misidentification of *Tinea quadrella* [Denis & Schiffermüller], 1775: 135, the former name is not available.

8. ***Elachista tetragonella*** (Herrich-Schäffer, 1855)
Plate-figs. 17, 18, 161, 246, 247, 404.

Poeciloptilia tetragonella Herrich-Schäffer, 1855: 301, 308.

6–7 mm. Male (Pl.-fig. 17). Head, tegulae and thorax dark brown, with violet or greenish sheen, face often with slight silvery sheen. Antenna dark brown, medial region slightly serrate. Labial palpi short, drooping, dark brown, with a few greyish scales scattered on upperside. Forewing ground colour dark brown to blackish brown with four distinct silvery white spots, often with a bluish reflection: one at costa beyond middle, one small sub-apical spot neither reaching costa nor termen, one tornal and one before the middle at the fold. Cilia grey-brown, cilia line inconspicuous, brownish. Hindwing grey, cilia similar. Abdomen dark brown. Hind tarsus with broad dark brown rings and narrow distal yellowish white rings.

Female (Pl.-fig. 18). Similar to male, but silvery white spots on forewing more prominent, and yellowish white rings on hind tarsus more narrow.

Male genitalia (Pl.-figs. 246, 247). Uncus with deep U-shaped indentation forming two elongate lobes, which are twice as long as wide, with short setae on distal halves. Gnathos oblong, rounded. Tegumen narrow. Valva rather narrow, widest before middle; costa strong, cucullus rounded. Digitate process slender, with fine setae. Juxta lobes apically truncate, produced laterally into short tapering processes, apical margin setose. Vinculum short, rounded. Aedeagus straight, tapering into strongly pointed distal end; no cornuti present.

Female genitalia (Pl.-fig. 404). Apophyses slender, posteriores longer than anteriores. Antrum smooth, funnel-shaped, membranous, finely spined inside; colliculum absent. Signum a weak plate with two prominent teeth, each often with a small heel.

Distribution. Not in Denmark; in Sweden only from Ög., Vg. and Dlsl.; not in Norway or Finland. – Not in Britain; scattered records from C. Europe, including France.

Biology. Larva greenish yellow, with broad grey dorso-lateral lines. Head yellowish brown, prothoracic shield yellowish, weakly sclerotized. It mines the leaves of *Carex montana* L. from April to the end of May. According to Stainton (1858c) it starts mining from the tip of the leaf, making a long, whitish, conspicuous blotch, which can be puckered; excrement is irregularly scattered in the mine. Martini (1916) states that the larva mines both upwards and downwards. Pupation in late May between loosely spun leaves. Adults fly from late May to early July, mainly in the first days of June.

Note. Specimens from Finland previously reported as *tetragonella* are *B. ornithopodella* (Frey).

9. ***Elachista biatomella*** (Stainton, 1848)
Plate-figs. 19, 20, 162, 248, 249, 405.

Aphelosetia biatomella Stainton, 1848: 2165.
Elachista boursini Amsel, 1951: 142. **Syn.n.**

7–9 mm. Male (Pl.-fig. 19). Head dull white to ochreous grey, often scattered with brownish; tegulae and thorax whitish grey, more or less brown anteriorly. Scape of antenna greyish white to brownish, flagellum brownish grey, scales on distal half slightly raised. Labial palpi drooping, third segment very short, upperside whitish, underside with a brownish streak. Forewing ground colour varying from whitish grey to brownish grey, suffused with brownish along costa and dorsum; a whitish streak before termen; the fold with whitish streak with distinct black elongate spot in the middle; a second elongate black spot placed in a more or less distinct white streak beyond middle, between fold-spot and apex; sometimes a third dark suffusion close to costa between these spots. Cilia grey; cilia line blackish, sometimes blurred; tornal area scattered with blackish tipped scales. Hindwing light brownish grey, cilia grey. Abdomen grey, anal tuft ochreous. Hind tarsus grey with whitish rings.

Female (Pl.-fig. 20). Very similar to male; ground colour of forewing less brownish and spots often less distinct.

Male genitalia (Pl.-figs. 248, 249). Uncus with very deep U-shaped indentation, proximal part of incision strongly sclerotized; uncus lobes long, narrowest in the middle, distally clavate, with short setae. Gnathos broad. Tegumen short, broad at base. Valva widest before middle; basal part of costa strong, cucullus rounded. Digitate process short, slender, with fine setae. Juxta lobes elongate conical, pointed, setose. Vinculum elongate, rounded. Aedeagus long, slender, pointed, without cornuti.

Female genitalia (Pl.-fig. 405). Apophyses of almost equal length. Antrum bowl-shaped, constricted anteriorly, membranous, ventral margin curved anteriorly, dorsal wall and antrum with interior spines; colliculum narrow, as long as apophyses posteriores. Signum a weakly sclerotized plate with two prominent curved teeth, both with a slender additional tooth from base.

Distribution. In Denmark along the West Coast of Jutland, and in NEJ and NEZ; in Sweden from Hall.; not in Norway and Finland. – Well known from C. Europe, Italy and the western part of Europe, including Britain.

Biology. This was described by Stainton (1858c). Larva yellowish, head dark brown, prothoracic shield light brown: in the leaves of *Carex flacca* Schreb. The egg is deposited in the angle of the leaf, from which the larva first mines upwards making a slender, narrow mine with a distinct excrement track; when the mine is 2–3 cm long the larva turns, gradually widening the mine, which no longer remains flat, but looks like a whitish blister, slightly puckered, the excrement track no longer distinct. Pupation is in the angle of one of the lower leaves of the food plant. Bivoltine; larvae of the first brood hibernate in the mine and feed again during the spring, larvae of the second brood present in July. Adults fly from late May throughout June and again in August. The biotope is open, exposed areas, like heaths and open grassland.

Note. The identity of *E. boursini* Amsel is based on the lectotype designated and figured by Parenti (1973). *E. disemiella* Zeller (1847: 89) has previously been placed in synonymy with *biatomella*, but is a distinct species belonging to the *argentella*-subgroup.

Elachista martinii Hofmann, 1898
Plate-figs. 21, 22, 163, 250, 251, 406.

Elachista martinii Hofmann, [1898] 1896: 143.

5–6 mm. Male (Pl.-fig. 21). Face light grey; neck tufts, tegulae and thorax dark grey. Antenna grey, with slightly lighter annulation, finely serrate almost from base. Labial palpi drooping to porrect, third segment short, upperside light grey, underside dark grey. Forewing ground colour mottled grey by scales with light bases and dark tips; with four whitish grey spots: one on costa beyond middle, one at apex, one before middle of dorsum to above fold and a tornal spot between costal and apical spots. Cilia grey along termen, lighter around apex and tornus; cilia line dark grey. Hindwing grey, cilia similar. Abdomen dark grey. Hind tarsus diffusely annulated dark grey/light grey.

Female (Pl.-fig. 22). Head, tegulae and thorax suffused with whitish grey. Forewing dark grey at base, blackish distally; spots white, more distinct and larger than in male, especially dorsal spot is prominent and square.

Male genitalia (Pl.-figs. 250, 251). Uncus very deeply indented, lobes elongate, clavate, distally rounded and with short setae. Gnathos elongate. Tegumen rather narrow. Valva widest in the middle; basal half of costa strong, cucullus rounded. Digitate process short, with fine setae. Juxta lobes short, truncate, apical margin slightly bent, tapering laterally. Vinculum short, rounded. Aedeagus short, basal portion wide, gradually narrowing, with four distinct cornuti of increasing length on a common plate.

Female genitalia (Pl.-fig. 406). Apophyses of almost equal length. Antrum funnel-shaped, membranous, margins curved slightly outwards, ventral margin deeply U-shaped, dorsal wall with minute spines behind ventral margin; colliculum weakly sclerotized, confluent with antrum, as long as apophyses posteriores. Ductus bursae strongly transversely folded just before corpus bursae, which has fine internal spines; signum a small dentate plate.

Distribution. Not from Scandinavia or Britain. – Widespread in southern Europe from the Balkans to France and up to the southern part of Germany.

Biology. Immature stages and biology were included in the original description by Hofmann (1896). The larva is slender and yellow, head dark brown or almost black, prothoracic shield light brown, anterior portion blackish. It is apparently monophagous on *Carex humilis* Leyss, in the leaves of which it starts mining upwards during the autumn, making a narrow mine with excrement deposited in irregular heaps. After hibernation inside the mine, it bores into the base of a new leaf, making a mine as wide as the blade with excrement scattered irregularly; if still not full-grown having reached the tip of the leaf, it turns or changes over to a new leaf. The pupa is light brown, dorsal and lateral ridges prominent, yellowish, lateral ridges with a fine dark brown line. Univoltine; larvae present from September to late May – early June, adults from first part of June to late July.

The *bifasciella*-group

Medium sized species. Dark grey, dark brown or blackish; often mottled species with a more or less conspicuous fascia across forewing before middle and a costal and a tornal spot beyond, or whitish ochreous or greyish species with or without brownish suffusion.

Venation (Pl.-figs. 164–190). Forewing with R_4 and R_5 stalked or coalescent, M_1 from stalk R_{4+5}, M_2 free or stalked with M_1, CuP reduced, 1A + 2A simple. Hindwing with cell open.

Male genitalia. Uncus lobes generally large, always with dense, long setae. Anterior margin of tegumen strongly sclerotized. Costa of valva strong, almost straight, distally with a hump before cucullus, sacculus often with a small distal spine below cucullus. Digitate process generally club-shaped. Juxta lobes with more or less oblique apical margins. Vinculum often produced into a short saccus and frequently with a medial ridge. Aedeagus base bulbous, with or without cornuti.

Female genitalia. Papillae anales elongate, triangular, anterior margin not sclerotized. Venter VIII membranous, ostium at anterior margin of segment VIII. The generally funnel-shaped antrum and tube-shaped colliculum sclerotized and confluent. Ductus bursae often with small sclerotized plates at inception of ductus seminalis. Signum a single dentate plate, if present.

The *bifasciella*-subgroup

Dark species, sometimes with white face. Sexual dimorphism not pronounced. Venation (Pl.-figs. 164–176). Forewing with R_4 and R_5 on a common stalk and M_1 from stalk R_{4+5}, M_2 free in most primitive species, in more advanced species R_3, R_4, R_5 and M_1 on common stalk or R_4, R_5, M_1 and M_2 on common stalk, M_1 and M_2 sometimes coalescent. Hindwing with Rs sometimes with free branch to costa, M_2 to or coalescent with CuA_1.

10. ***Elachista poae*** Stainton, 1855
Plate-figs. 23, 24, 164, 252, 253, 407, 507, 535.

Elachista poae Stainton, 1855: 47.
Elachista poella Morris, 1870: 214.

9–12 mm. Male (Pl.-fig. 23). Head dark grey-brown; tegulae and thorax almost black. Antenna uniformly dark grey-brown. Labial palpi only slightly curved, third segment a little shorter than second segment, palpi with upperside greyish white, underside grey-brown. Forewing ground colour grey-brown, darkest just beyond fascia; the angulated, often rather indistinct greyish white fascia from two-fifths of costa goes obliquely outwards, but is sharply bent just before and just after the fold, reaching dorsum in a rather obtuse angle, forming a Z; an indistinct, dull white mark at four-fifths of costa beyond a similar tornal spot. Cilia dark grey; cilia line black, distinct. Hindwing brown, with greyish cilia. Abdomen dark grey. Hind tarsus slightly ringed with greyish.

Female (Pl.-fig. 24). A little larger than male. Antenna ringed near base. Upperside of labial palpi almost white. Forewing pattern more distinct, Z-shaped fascia prominent and costal spot white, triangular, costal and tornal spots connected by a narrow, distinct, white fascia, sharply bent outwards in the middle.

Male genitalia (Pl.-figs. 252, 253). Uncus with U-shaped indentation, lobes rather large and elongate, outer margins bent beyond middle. Gnathos round. Valva slender, wider distally, costa with distinct hump. Digitate process club-shaped, upper margin slightly irregular. Juxta lobes medially produced, apical margins oblique, setose. Vinculum with short projecting saccus, without medial ridge. Aedeagus straight, distal end chisel-shaped, without cornuti.

Female genitalia (Pl.-fig. 407). Apophyses posteriores longer than apophyses anteriores. Antrum pouched, ventral margin U-shaped, rims short, dorsal wall with two small, minutely spined, elongate patches; colliculum tube-shaped, short, as long as apophyses posteriores. Anterior portion of ductus bursae and corpus bursae strongly folded longitudinally. Signum absent.

Distribution. From the eastern parts of Denmark; in Sweden only known from few districts in the southern part; not in Norway; in Finland only from Ab and Ta. – Known from C. and NW. Europe.

Biology. The larva is whitish yellow, dorsally greenish, head pale brown, prothoracic shield divided into four small plates, pale brown. It mines in the leaves of the aquatic grass *Glyceria*; *G. maxima* (Hartm.) Holmb. is frequently reported. The egg is deposited on the tip of the leaf, from where the larva mines downwards in a narrow, slender mine, forming a pale green track, which later turns reddish-brown, but is still difficult to see (Pl.-fig. 507); when it reaches the water it may turn and mine upwards (Stainton, 1858c) or continue downwards and later change to an adjacent submerged leaf through the close leaf sheaths, without having contact with the water during the transfer (Hering, 1951: 203).

The mine of *poae* is very similar to the primary mine of *Schoenobius forficella* (Thunberg, 1794) (Pyralidae), but the mine of *poae* is much flatter, looks greenish, and excrement is not placed in a distinct line but irregularly scattered, while the mine of *S. forficella* is more distinct, whitish and transparent with distinct excrement in the middle of the mine (Hering, 1925). Pupation takes place on the concave side of an unmined leaf about two feet above water, under a slight cocoon made up of transverse threads from one side of the leaf to the other (Pl.-fig. 535), which partly fold the leaf; when the leaf grows, this cavity flattens. The pupa is shining black, and easily observed by predatory birds and the collector. Emergence takes place in the morning between 8 and 10 a.m. (Buxton, 1914). Bivoltine; the larvae feed during April-May and in July-August, the adults fly from late May to late June and again in late August. Generally considered rather rare, but locally common in moist places along ponds, lakes and slow-flowing streams.

11. ***Elachista atricomella*** Stainton, 1849
Plate-figs. 25, 26, 165, 254-258, 408, 409, 508.

? *Microsetia obsoletella* Stephens, 1834: 264.
Microsetia exiguella Fabricius *sensu* Stephens, 1834: 264.
Elachista atricomella Stainton, 1849: 25.
Elachista alienella Stainton, 1851: 9.
Elachista extensella Stainton, 1851: 26. **Syn.n.**
Elachista zetterstedtii Wallengren, 1853: 219. **Syn.n.**
Elachista holdenella Stainton, 1854a: 252.
Elachista helvetica Frey, 1856: 289.
Elachista longipennis Frey, 1885: 102. **Syn.n.**

11–13 mm. Male (Pl.-fig. 25). Head mottled brown; neck tufts brown; tegulae and thorax darker brown, slightly mottled. Antenna plain brownish. Labial palpi slightly curved, upperside whitish, underside dark brown, third segment with a dark ring at base. Forewing ground colour blackish brown, slightly tinged darker along fold and dorsum; a white, slightly oblique fascia from one-fourth of costa, often interrupted before the fold; a white, triangular, distinct spot at four-fifths of costa, beyond a blurred brownish white tornal spot. Cilia dark grey, paler around tornus; cilia line distinct, black. Hindwing dark grey-brown, cilia lighter. Abdomen dark brown. Hind tarsus ringed blackish/white.

Female (Pl.-fig. 26). Rings on labial palpi more distinct. Scape of antenna and proximal part of flagellum blackish, distal part ringed black/grey. Forewing darker; fascia distinct, white, narrow and not interrupted on fold, and often separated from dorsum by a narrow streak of ground colour; costal and tornal spots more distinct, whitish.

Male genitalia (Pl.-figs. 254-258). Uncus with broad, U-shaped indentation, lobes large, distally rounded. Tegumen short, narrowest at base, a sclerotized band with many sockets along lateral and posterior margins. Valva slender and curved, costa strong with smooth hump, margins almost parallel. Digitate process club-shaped, widest before apex, margins smooth. Medial margins of juxta lobes distally pointed, apical margins sinuate, setose. Vinculum tapering into a short saccus, medial ridge absent. Aedeagus slightly bent before middle, with two cornuti, a long spoon-shaped one in the middle and a short conical or divided one more distally.

Female genitalia (Pl.-figs. 408, 409). Apophyses posteriores swollen at base, slightly longer than apophyses anteriores. Antrum large, lateral margins almost parallel, pouched towards colliculum, ventral margin curved posteriorly, with two short lateral rims, antrum with minute internal spines; colliculum tapering from antrum into a narrow tube, about four times as long as apophyses posteriores. Corpus bursae with minute spines, signum narrow, bent, with coarse teeth.

Distribution. From all parts of Denmark; in southern Sweden up to Upl.; not in Norway or Finland. – Otherwise reported from almost all countries on the Continent and Britain.

Biology. Larva pale yellowish, tinged greenish, head and prothoracic shield yellowish brown, the latter made up of two narrow elongate plates. According to Stainton (1858c) the larva mines downwards in the leaf of *Dactylis glomerata* L.

from March to May after hibernation, in a slender, almost linear, whitish mine with a very fine line of greyish excrement (Pl.-fig. 508). Each single larva may mine several leaves, but finally mines down into the stem and can even continue into the roots; when mature, it quits the mine and pupates outside on the stem, attached by a single girdle and the cremaster. Pupa light brown. *Carex, Milium* and *Melica* are also reported as food-plants (Hering, 1957; Wörz, 1957). Both larva and mine are difficult to separate from *E. luticomella* Zell. The adults fly from early June to mid-August, but probably univoltine. The biotope is wood edges, gardens, hedges and other calm places. Often seen resting on trunks and often caught at light.

Notes. *E. extensella* Stt. A specimen in fine condition in the Stainton collection in the British Museum (Nat. Hist.) labelled "234", "Extensella", "B.M. ♂ Genitalia Slide No. 20387" is hereby designated as lectotype.

Wallengren (1852) considered *E. zetterstedtii* close to *E. atricomella* Stt. and *E. bistictella* Tgstr., but later (1875: 76) synonymized it with *E. humulis* Zell.; from the specimens under that label in the Wallengren collection, Zool. Inst. Lund, Sweden, a lectotype has been designated by the present authors, which is conspecific with *E. atricomella* Stt.; it bears the following labels "7/6", "70", Genital præparat A.22.11.76 ♂ E. Traugott-Olsen".

E. longipennis Frey. The lectotype designated by J. D. Bradley and published first by Parenti (1977) is conspecific with *E. atricomella* Stt.

12. ***Elachista kilmunella*** Stainton, 1849

Plate-figs. 27, 28, 166, 259, 260, 410–412.

Elachista kilmunella Stainton, 1849: 25.
Elachista stagnalis Frey, 1859: 216.

9–11 mm. Male (Pl.-fig. 27). Head grey, tinged ochreous white; tegulae grey, tinged brownish white, especially along posterior margin; thorax dark grey. Antenna plain dark grey-brown. Labial palpi slender, porrect, slightly curved; upperside greyish white, underside grey-brown. Ground colour of forewing dark grey, slightly tinged with bluish grey; from two-fifths of costa a distinct, almost straight whitish fascia, narrowing towards or not reaching costa and dorsum, widest over fold; costal spot at four-fifths, white, opposite tornal spot which is often small, but the size of the two spots is very variable and often they are only slightly separated. Cilia light brownish grey, white from apex to middle of termen and grey around tornus beyond distinct brownish cilia line. Hindwing grey-brown, cilia similar. Abdomen grey-brown, anal tuft ochreous grey. Hind tarsus shining grey, slightly ringed whitish, outer side with black spots.

Female (Pl.-fig. 28). Forewing dark grey, especially beyond fascia which is wide and straight; costal and tornal spots more prominent.

Male genitalia (Pl.-figs. 259, 260). Uncus with deep, U-shaped indentation, lobes elongate, slender. Tegumen short, sides almost parallel. Upper margin of gnathos almost straight, otherwise rounded. Valva with almost parallel sides, hump of costa smoothly rounded, cucullus rounded dorsally, sacculus produced into a short spine below cucullus. Digitate process wide in the middle, tapering towards apex. Juxta lobes narrow at base, medially produced, apical margins setose, irregular. Vinculum with a medial sclerotized ridge; saccus distinct, short. Aedeagus long, almost straight, slender, tip square-ended, with one small pointed cornutus.

Female genitalia (Pl.-figs. 410–412). Apophyses of almost equal length, both pairs medially curved. Antrum wide, V-shaped, ostium very large, ventral margin almost to base of antrum, which is finely spined inside, lateral rims long, strong, dorsal wall with medial patch of minute spines; colliculum narrow, long, more than twice as long as apophyses posteriores. Ductus bursae with 5–7 small triangular teeth at inception of ductus seminalis, anterior to this weakly sclerotized. Signum oval, with short teeth.

Distribution. Not in Denmark; in Sweden from almost all districts, except the isles in the Baltic; from all parts of Norway except the SW. part; from almost all districts in Finland. – Elsewhere from NW. USSR and the mountains in C. Europe, in N. England and Scotland.

Biology. Little is known about the early stages, but the larva is presumed to mine in *Carex* and probably also *Eriophorum* (Krogerus, *in litt.*) Pupa almost plain yellowish brown, thorax with two rows of dorsal tubercles, with scattered lateral tubercles, attached by a single girdle. The adults fly through June and July; in boggy places.

13. ***Elachista parasella*** Traugott-Olsen, 1974
Plate-figs. 29, 167, 261, 262, 413.

Elachista parasella Traugott-Olsen, 1974: 267.

8–12 mm. Male (Pl.-fig. 29). Head grey, face suffused with ochreous grey; tegulae and thorax grey, posterior part of tegulae whitish and anterior part of thorax and tegulae dark grey. Antenna plain grey. Labial palpi porrect, third segment very short, pointed; palpi light grey, proximal parts of segments dull-coloured. Forewing ground colour grey, suffused with bluish grey caused by scales with light bases and dark bluish grey tips; a slightly curved, uninterrupted, whitish fascia from costa to dorsum before middle; a triangular whitish costal spot beyond middle, opposite a rather narrow, similar tornal spot, the two often confluent to a narrow outwardly angled fascia. Base of cilia pale grey, distal part greyish, cilia line faint. Hindwing grey, cilia similar, tinged brownish. Abdomen grey. Hind tarsus slightly tinged greyish/whitish, with dark grey spots on outer side.

Female. Worn, not in condition suitable for description.

Male genitalia (Pl.-figs. 261, 262). Uncus with deep, U-shaped indentation, lobes elongate, elliptically rounded. Gnathos round. Tegumen long and narrow. Valva slender, costa strong, hump not distinct, cucullus rounded, sacculus without distinct distal spine below cucullus, but often more strongly sclerotized distally. Digitate process widest at two-thirds, dorsal margin irregular, crenate, ventral margin straight. Juxta lobes narrow at bases, apical margins irregular, setose, outwardly oblique. Vinculum triangular, gradually produced into a short saccus, without medial ridge. Aedeagus long and slender, straight, basal portion only slightly bulbous, somewhat thicker in the middle, containing two very small spoon-shaped cornuti.

Female genitalia (Pl.-fig. 413). Apophyses slender, posteriores longer than anteriores. Antrum V-shaped, ventral margin U-shaped, rims long, fine, ventral margin and dorsal wall with fine spines; colliculum narrow and long, more than twice as long as apophyses posteriores, a small plate with a single small tooth at anterior end. Signum elongate, sinuous, with medial triangular projection, coarsely dentate. Anterior margin of sternites strongly sclerotized.

Distribution. Not in Denmark; from mountain areas in Sweden: Jmt. and P. Lpm.; in Norway known from STi, TRi and from Fv, and from several districts in northern Finland. – Otherwise not reported.

Biology. Nothing is known about the immature stages. The adults are known from late May to the middle of July.

Note. The description of the female genitalia is based on a single female from Suecia, Jmt., Enafors, UTM 33V UL71, 3–6.7.1972, I. Svensson leg. which is in worn condition. It is classed as *parasella* owing to its size and general appearance and to the fact that a male was caught in the same locality a few days earlier in the same year; all other known specimens of *parasella* are males.

14. ***Elachista alpinella*** Stainton, 1854

Plate-figs. 30–32, 168, 263–265, 414.

Elachista alpinella Stainton, 1854a: 254.
Poeciloptilia latipennella Herrich-Schäffer, 1855: 301, 308.
Elachista monticola Wocke *in* Heinemann, [1876]: 495.
? Elachista turfosella Frey, 1885: 100.
Elachista subquadrella Ragonot, 1889: 106.

9–13 mm. Male (Pl.-figs. 30, 31). Head greyish brown to ochreous brown, neck tufts, tegulae and thorax similar, posterior portion of tegulae paler. Antenna dark grey-brown, posterior side of scape light greyish. Labial palpi porrect to recurved, upperside greyish white, underside brownish. Forewing ground colour dark greyish brown, sometimes less dull along dorsum; inner white fascia from dorsum to beyond middle of wing, but never reaching to costa; a distinct, often triangular ochreous-whitish spot, at four-fifths of costa, beyond the similar tornal spot. Cilia grey-brown; cilia line distinct, blackish. Hindwing dark greyish brown, cilia greyish. Abdomen greyish brown. Hind tarsus with broad whitish rings.

Female (Pl.-fig. 32). Head yellowish white; neck tufts, tegulae and thorax brown. Scape of antenna brown, flagellum annulated cream-white/brown. Basal part of forewing paler brown than remaining part, fold and costa dark brown, at base of dorsum an ochreous white spot; costal and tornal spots distinct, white, tinged ochrous distally.

Male genitalia (Pl.-figs. 263–265). Uncus with U-shaped indentation, lobes narrow, outer margins rounded. Gnathos rounded. Tegumen narrow, a strongly sclerotized band, with many sockets, along lateral and posterior margin. Valva widest in the middle, costa straight to projecting cucullus, hump insignificant, sacculus bent before middle, without distal spine. Digitate process club-shaped. Juxta lobes apically rounded, smooth, setose. Vinculum with a short saccus, without medial ridge. Aedeagus straight, stout, tip square-ended, with two or three spoon-shaped cornuti of varying size, often extruded during preparation.

Female genitalia (Pl.-fig. 414). Apophyses of amost equal length. Antrum short, funnel-shaped, slightly pouched, ventral margin slightly curved anteriorly, dorsal wall and antrum with fine spines; colliculum broad, slightly longer than apophyses posteriores. Posterior membranous part of ductus bursae with many longitudinal folds. Signum absent.

Distribution. Widespread but generally not numerous; in all parts of Denmark and Sweden, but only in lowland areas in the most northern parts of Sweden; in Norway from Ø, Nsi and F; in Finland from almost all districts. – Otherwise from C., E. and W. Europe, including Britain, but apparently not from S. Europe.

Biology. The larva is slender, yellow-white, without pigmentation, intestine reddish; having left the mine it turns red-brown, head shining light brown, prothoracic shield made of two widely separated elongate plates. Several species of *Carex* are reported as food-plants: *C. riparia* Curt., *C. acuta* L. (Hering, 1891), *C. acutiformis* Ehrh., *C. flava* L. x *C. hostiana* D. C. (Stainton, 1879; Gudmann, 1930; Steuer, 1973). The larva starts mining in the leaf in September a few centimeters below the tip, mining downwards in a narrow, long mine. The hibernation place is not known, as the leaves decay during the winter (Gudmann, 1930). In the spring the still small larva starts a new long and slender mine, filled with a brownish, often broken, excrement line; when it reaches the stem in June it makes a short mine inside this, leaving the stem as a mature larva in late June or the beginning of July. The leaf mine is often extremely long, from a few centimeters to more than one meter (Hering, 1891). Pupation takes place on the stem or a leaf. Pupa, attached by a single girdle and the cremaster, is short, rather stout, dorsal ridge not prominent, light brown, with two dark dorso-lateral lines. The adult flies from late July to the beginning of September, but mainly in the first part of August; univoltine. The biotope is mainly wet meadows.

Note. The specific name *P. latipenella* was proposed as an unnecessary objective replacement name for *E. alpinella* Stt. by Herrich-Schäffer (1855: 308).

15. ***Elachista diederichsiella*** Hering, 1889

Plate-figs. 33, 34, 169, 266–268, 415.

Elachista diederichsiella Hering, 1889: 313.

9–11 mm. Male (Pl.-fig. 33). Frons ochreous white, clypeus and vertex grey-brown; neck tufts, tegulae and thorax dark grey-brown, posterior part of tegulae and thorax whitish. Scape of antenna proximally whitish, flagellum grey-brown. Labial palpi whitish on upperside, underside greyish brown. Forewing ground colour dark brown; basal area ochreous white, enlarged along dorsum; an ochreous white fascia from two-fifths of costa, interrupted by the dark fold, costal spot of same colour beyond a similar larger tornal spot. Cilia greyish brown; cilia line not distinct. Hindwing light grey-brown, cilia similar. Abdomen grey-brown, with ochreous anal tuft. Hind tarsus medially whitish grey, with lateral dark brown spots.

Female (Pl.-fig. 34). Head pale ochreous white. Posterior part of tegulae whitish. Basal spot on forewing prominent, inner fascia very distinct, not interrupted, widest just before fold, gradually narrowing towards dorsum, slightly curved outwards. Distal marks very prominent.

Male genitalia (Pl.-figs. 266–268). Uncus lobes very wide apart, elongate, lateral margin rounded. Gnathos small, rounded. Tegumen short and broad. Valva gradually widening distally, costa with rounded hump before the slightly projecting cucullus, sacculus without distal spine. Digitate process club-shaped, distally with almost parallel sides. Juxta lobes broadest distally, truncate, apical margin only slightly oblique, setose. Vinculum with medial ridge, saccus very short. Aedeagus straight, bulbous basally, narrowest in the middle, distal region enlarged, distal end pointed; an angular sclerotized plate and two conical cornuti, one of which may be bi- or trifid, below apex.

Female genitalia (Pl.-fig. 415). Apophyses posteriores swollen near base, apophyses of almost equal length. Posterior margin of tergite VIII deeply emarginate. Antrum short, bowl-shaped, ventral margin strong, anteriorly curved, dorsal wall and interior of antrum with minute spines; colliculum short, about half as long as apophyses posteriores. Membranous portion of ductus bursae towards colliculum with coarse teeth on sclerotized plates, often lined up in a row in one side of ductus; ductus with many longitudinal folds anterior to these. Corpus bursae posteriorly with fine internal spines. Signum elongate, coarsely dentate.

Distribution. Only from two districts in Denmark: LFM and NEZ; in Sweden from Sk. and T. Lpm.; from W. and N. Norway and from Ab and Li in Finland. – Known from many localities in E. and S. Europe, but not in the western parts or Britain.

Biology. Already outlined by Hering (1889). Larva dirty yellowish green, posteriorly more yellowish, intestine dark green, head and prothoracic shield dark brown; very similar to that of *E. albifrontella* (Hb.). It mines in a leaf of *Milium effusum* L. in April-May, mining downwards from just beyond the tip of the leaf, leaving a small green area at the tip; the first part of the mine is narrow, 2–5 mm wide, but the mine then abruptly widens forming a swollen, whitish, *Phyllonorycter*-like blotch, visible from both sides, excrement is scattered irregularly in the whole mine, which is rather short, generally not more than 2 cm, and rarely 3 cm long. Often more than one, but seldom more than four, larvae are present in a single mine. Pupa shining yellow-brown, longitudinal ridges greyish. Adults appear from early June to mid July; univoltine. The biotope is open deciduous forest, often beech forest.

Note. The type-series of *E. diederichsiella* Her., kept in Institut Zoologii, PAN, Warszawa, consists of six specimens. A male in fine condition, labelled "Z 2/6 89, Mil. eff, Damm [= Alt-Damm]", "Slide 1301♂, E. S. Nielsen" is hereby designated as lectotype.

16. ***Elachista compsa*** Traugott-Olsen, 1974
Plate-figs. 35, 36, 170, 269, 270, 416.

Elachista compsa Traugott-Olsen, 1974: 260.
Elachista holdenella auct., *nec* Stainton, 1854.

8–11 mm. Male (Pl.-fig. 35). Head white; neck tufts grey; tegulae and thorax greyish brown, coloured by dark brown-tipped scales with greyish white base. Antenna slightly ringed whitish grey/dark brown near base, distal part dark grey, finely serrate. Labial palpi porrect, curved, distal segment rather long, spear-shaped, upperside whitish grey, underside dark brown, junction between second and third segment whitish with a dark ring. Forewing long and slender, ground colour grey-brown, scales as on thorax; with an oblique, curved whitish fascia, which is narrow and often diffuse in the middle, costal spot large, almost square, white, beyond the smaller tornal spot. Cilia greyish white inside the distinct dark brown cilia line, grey beyond this. Hindwing pale greyish, cilia grey. Abdomen dark grey-brown, anal tuft ochreous. Hind tarsus slightly ringed.

Female (Pl.-fig. 36). Head silky-white. Basal third of antenna distinctly ringed. Forewing ground colour blackish brown, marks shining silky-white, fascia wider.

Male genitalia (Pl.-figs. 269, 270). Uncus lobes very small, very wide apart. Gnathos rounded. Tegumen short and broad. Sides of valva almost parallel, costa with a smooth, distal curved hump, cucullus rounded. Digitate process widest in the middle, apex blunt. Juxta lobes widest distally, apical margins oblique, setose and irregular. Vinculum with a medial sclerotized ridge, saccus short, narrow. Aedeagus long, curved, with 3–4 long, slender, pointed cornuti of almost equal length, often extruded during preparation.

Female genitalia (Pl.-fig. 416). Apophyses of almost equal length. Antrum funnel-shaped, gradually tapering into colliculum, ventral margin anteriorly curved, strongly sclerotized, forming two short rims, no spines present; colliculum narrow, long, twice as long as apophyses posteriores, often with a small, triangular, sclerotized plate at anterior end. Signum narrow, elongate, dentate.

Distribution. In Denmark from EJ, SZ and LFM; in the southern part of Sweden up to Upl.; not in Norway; in the southern districts in Finland. – Widespread in C. Europe, but only from few localities in W. Europe and not in Britain.

Biology. The larva is faintly yellowish, later more whitish; head and prothoracic shield reddish brown; plates elongate, broad, posteriorly enlarged, wide apart. It mines in the leaves of *Melica nutans* L. (Frey, 1885) and *M. uniflora* Retz. (Larsen, 1927); whether records from *Deschampsia caespitosa* (L.) PB. (Wocke, 1876; Hering, 1957) are reliable cannot be stated. The larva is present from late September to May and in June-July. The larva starts mining from the basal part of a leaf, first making a narrow mine, filled with excrement, to the tip of the leaf, having reached the tip it turns, now mining downwards, making a mine as wide as the blade; when full grown in November-December it hibernates in the mine from one-half to two-thirds down the leaf. Pupation takes place on a leaf in the late spring. Larvae of the summer brood feed in a similar way, and there the excrement line from the upward-going narrow mine can often be seen in the downward-going, wide mine. According to Steuer (1976a) the size and abundance of individuals of the first generation depend on the severity of the winter, and after a severe winter only few and small specimens appear. The adult occurs from late April and during May, and again in July and early August, bivoltine; univoltine in northern Fennoscandia. The biotope is open deciduous woods and wood edges. The adults often fly in the late afternoon and can often be seen resting on tree-trunks.

17. ***Elachista elegans*** Frey, 1859

Plate-figs. 37, 38, 171, 271, 272, 417, 418, 509.

Elachista elegans Frey, 1859: 239.

6.5–8 mm. Male (Pl.-fig. 37). Face white; scales on vertex, neck tufts, tegulae and thorax with white bases and grey tips, posterior part of tegulae greyish white. Antenna ringed white/greyish brown. Labial palpi white above, greyish on underside, but third segment white on base and apex. Forewing short and broad; ground colour dark grey, darker beyond fascia, mottled white by scales with white bases and dark grey tips; a whitish, narrow, outwards oblique fascia from two-fifths of costa to dorsum, widest at costa, narrowing towards dorsum, often interrupted in fold; costal spot narrow, triangular, white, just beyond similar tornal spot. Cilia grey, cilia line distinct, blackish, scattered scales with white bases and long blackish tips beyond. Abdomen grey. Hind tarsus ringed grey/white.

Female (Pl.-fig. 38). Tegulae and thorax darker. Forewing ground colour blackish grey, with slightly shiny, distinct white, fascia and spots.

Male genitalia (Pl.-figs. 271, 272). Uncus lobes narrow and wide apart. Gnathos transversely oval. Valva gradually widening towards cucullus, costa with a small, smooth distal hump, distal end of sacculus produced into a small spine below cucullus. Digitate process slender, sides parallel, apex blunt. Juxta lobes medially produced, apical margin oblique, sparsely setose. Vinculum without medial ridge, produced into a short, narrow saccus. Aedeagus long, cylindrical, gradually tapering, with two small distinct cornuti.

Female genitalia (Pl.-figs. 417, 418). Apophyses slender, posterior pair enlarged at base, anterior pair close, posteriores and anteriores of almost equal length. Antrum smooth, funnel-shaped, slightly pouched towards colliculum, ventral margin anteriorly curved, rims short, dorsal wall with scattered minute spines behind ventral margin; colliculum narrow, long, more than twice as long as apophyses posteriores. Ductus bursae contains 3–4 small oval plates, each with a small eccentric tooth, near the pouched inception of ductus seminalis. Corpus bursae with fine internal spines, signum elongate, medially enlarged, dentate.

Distribution. Not in Denmark, as previous records by Larsen (1927) and Traugott-Olsen (1973) are based on misidentifications; in Sweden in several districts in the southern part up to Vrm. and Upl.; not in Norway; in Finland from a few districts in the southern part. – Otherwise only from C. Europe.

Biology. According to Martini (1912c, 1916) the larva is yellowish green, dorsally dark green, with a whitish midline, head yellowish brown, prothoracic shield more yellowish. It mines in the leaves of *Calamagrostis arundinacea* (L.) Roth., ***Bromus, Dactylis*** and ***Milium effusum*** L. are also reported (Frey, 1859; Hering, 1957; Wörz, 1957). The mine (Pl.-fig. 509) is long, often more than 20 cm, on the underside of the leaf, but visible from upperside too. The larva starts mining in the basal part of the leaf, mining upwards along the middle vein; the first few centimeters of the mine are narrow, 1–1 1/2 mm, but it then gradually widens towards the edge of the leaf, leaving only a small lateral portion of mesophyll, but the width of the mine can vary; if not full grown at 5–10 cm from the tip, the larva turns here and makes a similar mine downwards until mature. The excrement is deposited in a rather characteristic line along the side of the mine, in some mines only along one side; in others, the excrement-line is interrupted, alternately situated on both sides, in a few mines the blackish excrement is deposited in a haphazard way. Pupation at base of a leaf. Pupa yellowish brown, with whitish dorsal line and yellowish lateral lines. Bivoltine; larvae present during April-May and July, adult from mid-June to late July and again from late July to mid-August in southern Scandinavia; univoltine in northern parts, adult present from late July to mid-August.

18. ***Elachista luticomella*** Zeller, 1839
Plate-figs. 39, 40, 172, 273–275, 419, 510.

Tinea guttella Hübner, 1796: 65, pl. 26, fig. 176, *nec* Fabricius, 1781.
Elachista luticomella Zeller, 1839: 212.
Elachista flavicomella Stainton, 1856: 39.

10–11 mm. Male (Pl.-fig. 39). Head pale to bright yellow, neck tufts mixed with brownish; tegulae and thorax blackish brown. Antenna grey-brown, scales slightly raised. Labial palpi almost straight, drooping, yellowish white, underside of second segment sometimes with scattered brownish scales. Forewing ground colour dark greyish brown; a yellowish white, slightly outwards bent fascia from two-fifths of costa, a small, yellowish, square spot at four-fifths of costa, beyond an indistinct tornal spot of same colour. Cilia grey-brown, cilia line darker grey, indistinct. Hindwing grey-brown, cilia similar. Abdomen dark brown to blackish, with yellowish anal tuft. Hind tarsus light greyish.

Female (Pl.-fig. 40). A little larger than male. Distal third of antenna strongly serrate. Forewing ground colour brownish to blackish, basal area lighter than distal part; fascia wide, almost straight, distal spots more prominent. Abdomen dark grey, anal tuft grey or yellowish grey.

Male genitalia (Pl.-figs. 273–275). Uncus lobes oval, less than the width of one lobe apart. Gnathos very small, round. Tegumen short, broad. Valva slightly bent, somewhat dilated towards truncate cucullus, sacculus with a very small distal spine. Digitate process club-shaped, widest beyond middle, apex blunt. Juxta lobes dilated, rounded, apical margins with short setae, without lateral processes. Vinculum short, without medial ridge, produced into a rather long, narrow saccus. Aedeagus long, bent at one-quarter from base, gradually tapering, with one big spoon-shaped cornutus.

Female genitalia (Pl.-fig. 419). Apophyses long and strong, of equal length. Antrum pouched, ventral margin deep, narrow, U-shaped, finely spined inside, dorsal wall finely spined in two elongate patches; colliculum broad, short, shorter than apophyses posteriores. Corpus bursae with minute internal spines, signum elongate, medially broadened, distally dentate.

Distribution. From almost all districts in Denmark; in many districts in S. Sweden up to Vrm. and Upl.; from Norway only a single specimen in the Zoological Museum, Bergen, which cannot be referred to district owing to an unsatisfactory locality label; from a few districts in SW. Finland. – Known from almost all parts of Europe.

Biology. Larva long and slender, lemon-coloured, head and prothoracic shield light brown (Frey, 1859); thus very similar to larva of *E. atricomella* Stt., which is paler yellowish. It mines from November to May in the leaves and stems of *Dactylis glomerata* L. (Frey, 1859), *Festuca gigantea* (L.) Vill., *F. pratensis* Huds., *Poa pratensis* L. and *Milium effusum* (Martini, 1916; Wörz, 1957), in which it makes a narrow mine in the leaves before mining down into the stem (Pl.-fig. 510); it often mines several leaves before entering the stem (Hering, 1957); the infested plants turn yellowish and wilt. The mine in the stem can easily be seen against the light. Pupa light brown, short and rounded, dorsal and lateral ridges weak. The adults fly from late June, throughout July to early August; probably univoltine. The biotope is open deciduous forests, wood edges and meadows.

Note. Specimens reported as *griseella* by Larsen (1916: 234; 1927: 146) are misidentifications of *luticomella*. *Elachista griseella* (Duponchel, 1843) is a distinct C. and S. European species.

19. ***Elachista albifrontella*** (Hübner, 1817)

Plate-figs. 41, 42, 173, 276, 277, 420, 421, 536.

Tinea albifrontella Hübner, [1817], pl. 64, fig. 432.
Tinea quadrella Fabricius *sensu* Haworth, 1828: 582.

8–9 mm. Male (Pl.-fig. 41). Head and neck tufts shining white; tegulae and thorax dark grey-brown, posterior edges not whitish. Antenna dark grey-brown, scales raised. Labial palpi slightly curved, porrect, whitish grey, second segment with a narrow grey-brown streak on underside. Forewing ground colour blackish grey, marks distinct, shining white; outwardly oblique fascia narrows towards fold, often interrupted at fold and perpendicular to dorsum from fold, widest at costa which it intercepts at two-fifths; costal spot large, beyond the smaller, triangular, tornal spot. Cilia grey-brown; cilia line distinct, black, consisting of scales with long, white bases. Hindwing grey-brown, cilia paler. Abdomen grey, with ochreous anal tuft. Hind tarsus grey, with a white ring on distal end of each segment.

Female (Pl.-fig. 42). Tegulae and thorax blackish. Antenna with slightly raised scales on distal half, distal region tinged with greyish. Forewing ground colour black, marks shining white; fascia curved outwards, wider than in male, not interrupted, distal spots larger and more distinct. Abdomen grey, anal tuft similar.

Male genitalia (Pl.-figs. 276, 277). Uncus lobes large, very close, emargination V-shaped. Gnathos rounded. Tegumen narrow. Valva narrow at base, gradually widening outwards, costa with short, prominent hump, cucullus rounded towards costa and at a right angle to sacculus. Digitate process club-shaped, base very narrow. Juxta lobes medially produced, pointed, apical margins oblique, setose. Vinculum rounded, short, with a long, narrow, distinct saccus; no medial ridge present. Aedeagus long, slender, straight, gradually tapering, with one slender, long and slightly bent cornutus.

Female genitalia (Pl.-figs. 420, 421). Apophyses long and slender, sinuous, of equal length. Antrum smooth, funnel-shaped, ventral margin anteriorly curved, dorsal wall spined before and behind ventral margin; colliculum very narrow and long, twice as long as apophyses posteriores. Signum elongate, bent, dentate.

Distribution. Common and known from almost all districts in Denmark and Sweden; in Norway up to MR; from almost all districts in Finland. – From almost all parts of Europe.

Biology. Described by Stainton (1858c). Larva pale whitish yellow, head pale brown, ventral part dark brown, prothoracic shield divided into several small sclerotized brown plates; very similar to that of *E. humilis* Z. As food-plants are reported *Holcus mollis* L., *H. lanatus* L., *Deschampsia caespitosa* (L.) PB., *Dactylis glomerata* L., *Koeleria cristata* (L.) Pers., *Milium effusum* L., *Agrostis, Agropyrum, Avena, Festuca, Phleum, Poa* and *Triticum* (Stainton, 1858c; Sønderup, 1949; Hering, 1957), and it is probably polyphagous on grasses. It starts mining downwards from the tip of the leaf, making a whitish blotch occupying whole width of the blade; the mine can be as long as the blade, or very short, when the same larva mines several leaves. Small lines of brownish green excrement are deposited inside the mine. It mines from September to May. Pupation outside the mine; pupa pale brown. Apparently univoltine, adults from early June to late July. A very common species in meadows, woods and gardens.

20. ***Elachista bifasciella*** Treitschke, 1833
Plate-figs. 43, 44, 174, 278–280, 422, 423, 511, 512.

Elachista bifasciella Treitschke, 1833: 182.
Elachista binella Zeller, 1839: 213.

7–9 mm. Male (Pl.-fig. 43). Head whitish; neck tufts, thorax and anterior region of tegulae dark grey, posterior part of tegulae whitish. Antenna thick, plain lead-grey, scales slightly raised in distal region. Labial palpi drooping, almost straight, underside lead-grey, upperside of second segment whitish grey, third segment light grey. Forewing ground colour blackish with distinct white marks; base white, enlarged at dorsum, narrow at costa; a white, broad, outward-curved fascia with irregular sides just before middle; a wider fascia of same colour beyond middle, from middle of wing to tornus often reaching base of cilia, inner margin of fascia strongly outwardly angled; sub-apical area of ground colour. Cilia blackish grey; cilia line not present. Hindwing brownish grey, cilia similar. Abdomen dark grey, anal tuft ochreous. Hind tarsus grey.

Female (Pl.-fig. 44). Very similar to male. White basal spot larger; fasciae slightly wider, distal fascia straighter.

Male genitalia (Pl.-figs. 278–280). Uncus lobes elongate, rectangular, diverging, wide apart. Gnathos elongate, teardrop-shaped. Tegumen narrowest proximally, sides strongly sclerotized, Valva narrow at base, costa almost straight, hump insignificant, cucullus projecting dorsally, ventrally at a right angle to the distally slightly projecting sacculus, sacculus bent before middle. Digitate process narrow, slender, ending in a small knob. Juxta lobes medially produced, apical margins oblique, setose. Vinculum triangular, without medial ridge; saccus short and stout. Aedeagus slightly bent in the middle, distal end indented, without cornuti.

Female genitalia (Pl.-figs. 422, 423). Apophyses slender, posteriores longer than anteriores. Posterior margin of tergite VIII emarginate. Antrum funnel-shaped, pouched or gradually tapering into colliculum, ventral margin deeply U-shaped, often more strongly sclerotized, dorsal wall with fine spines behind ventral margin; colliculum narrow, often with a longitudinal fold, less than twice as long as apophyses posteriores. Corpus bursae with or without minute spines; signum a very weak, elongate, finely dentate plate, or absent.

Distribution. In Scandinavia only reported from Denmark, where it occurs in several localities in Jutland; recently also reported from F and LFM. – Known from many parts of Europe, but not from Spain or Britain; only few records are available from E. Europe.

Biology. Recently described by Steuer (1976a). Larva rather short and thick, yellowish, transparent. It mines from April to early June in the leaves of *Deschampsia caespitosa* (L.) PB. and *Holcus mollis* L., where these plants grow in rather dark, shady places in coniferous forests. It starts mining downwards a few mm., leaving no mesophyll present in *D. caespitosa* (Pl.-figs. 511, 512) and only small "islands" in *H. mollis*. It mines at least half down the blade, and often reaches the base. The transparent larva is easily seen inside the leaf ahead of the excrement line, which often is double in the broad-leaved *H. mollis;* the larva often changes leaf. *Agrostis gigantea* Roth., *Corynephorus canescens* (L.) PB., *Deschampsia flexuosa* (L.) Trin., *Festuca ovina* L., *F. gigantea* (L.) Vill., *Milium effusum* L. and *Brachypodium* are also reported as food-plants (Lhomme, 1951; Wörz, 1957), but according to Steuer (1976a) records from at least *M. effusum* and *F. gigantea* are not reliable. Pupation takes place on the food-plant, and the pupa is easy to see. The adults fly from early June to late June; univoltine. The biotope is, as mentioned above, shady, cold, dark coniferous forest. Frey (1859) considered it an alpine species. The adult fly from noon to dusk, often in small sunny spots; during the day it is often seen resting on the food-plants.

Note. The S. and E. European *Elachista dimicatella* Rebel, 1903 (syn.: *E. niphadophanes* Meyrick, 1938) has recently been reported from Czechoslovakia and given full treatment by Steuer (1976a). Superficially it is very similar to *E. bifasciella,* but it is larger and the outer fascia is more strongly angled. The larva mines in *Sesleria caerulea* (L.) Ard. in June, and the adults appear in late June or early July.

21. ***Elachista nobilella*** Zeller, 1839

Plate-figs. 45, 46, 175, 281, 282, 424.

Elachista nobilella Zeller, 1839: 213.
Elachista subnobilella Rössler, 1866: 378.

6.5–8 mm. Male (Pl.-fig. 45). Head shining metallic silver; neck tufts, tegulae and thorax greenish bronze-brown, posterior margins shining silver. Antenna dark brown, with slightly bluish or greenish sheen, distal segments greyish white, distal third slightly serrate. Labial palpi drooping and only weakly curved, rather short, dark brown beneath, upperside light grey, especially towards base. Forewing ground colour shining bronze-brown, with shining metallic silvery marks, often with a bluish or greenish reflection; basal shining silvery area enlarged along dorsum, narrow towards costa; an outwards curved fascia before middle, often narrow or interrupted in the fold; costal spot triangular, just beyond opposite tornal spot, a third elongate spot present between apex and tornus at base of cilia. Cilia dark grey-brown; cilia line rather distinct, blackish. Hindwing light grey-brown, cilia similar. Abdomen shining lead-grey, anal tuft yellowish. Hind tarsus whitish, with broad, light grey rings.

Female (Pl.-fig. 46). Very similar to male, but fascia never interrupted and slightly wider than in male, distal spots more prominent.

Male genitalia (Pl.-figs. 281, 282). Uncus lobes short and small, wide apart. Gnathos prominent, transversely elongate. Tegumen short and broad. Base of valva narrow, costa strong, with a smooth distal hump before the rounded cucullus. Digitate process short, stout, club-shaped. Juxta lobes with pointed medial projection, narrow, apical margins very oblique, setose. Vinculum strong and long, produced into a short saccus. Base of aedeagus distinctly bulbous, otherwise straight, gradually tapering, distal end bifurcate; cornuti absent.

Female genitalia (Pl.-fig. 424). Papillae anales two large, leaf-like flattened, heavily sclerotized, finely setose plates. Apophyses short and strong, of almost equal length. Sternite and tergite VIII very narrow, strongly sclerotized. Antrum long, funnel-shaped, lateral margins slightly curved, ventral margin narrow, anteriorly curved, dorsal wall minutely spined behind ventral margin; colliculum narrow, of almost same length as antrum and longer than apophyses posteriores. Corpus bursae with faint, fine internal spines. Signum a semicircular plate with 2–3 prominent and several small teeth.

Distribution. From almost all districts in Denmark; in Sweden up to Upl.; from several districts in Norway and in the southern parts of Finland. – Widespread on the Continent, but not from the southern parts or Britain.

Biology. Steuer (1976a) has recently outlined the biology of this species. Larva slender, yellow; head brown, prothoracic shield gradually widening posteriorly, brown, distal parts dark brown. Reported from several grasses, most often *Deschampsia flexuosa* (L.) Trin., but also *Bromus, Festuca, Agrostis stolonifera* L., *Holcus lanatus* L., *Dactylis* and *Carex,* according to Sønderup (1949), Hering (1957) and Wörz (1957). It mines during April-May. The mine is short, flat, whitish and looks very much like that of *E. bifasciella* Tr., but the larva of *E. nobilella* generally pupates 2–3 weeks earlier than *E. bifasciella* Tr. Pupation takes place on the slender stem or leaf of the food-plant, and the pupa is easy to recognize; it is short, light brown, dorsal and lateral ridges not prominent. The adults fly from early June to early July; univoltine. The biotope is open, half-shaded places in deciduous forests, and never or rarely in dark coniferous woods where *E. bifasciella* occurs. It is often seen flying in the sunshine from the afternoon until dusk.

22. ***Elachista apicipunctella*** Stainton, 1849
Plate-figs. 47, 48, 176, 284–286, 425, 513, 514.

Elachista apicipunctella Stainton, 1849: 26.

10–11 mm. Male (Pl.-fig. 47). Head silky, shining, bone-white; neck tufts greyish brown; tegulae and thorax dark grey-brown, tegulae posteriorly coloured as head. Antenna dark grey-brown, distal half paler, finely serrate. Labial palpi drooping, only second segment slightly curved, bone-white, underside tinged with greyish. Forewing rather narrow, ground colour shining bronzy greyish brown, with silky white shining marks; an outwards oblique fascia, widest on costa and narrow or interrupted in the fold, perpendicular to dorsum, before middle of costa; costal spot beyond middle prominent and triangular, faintly tinged yellowish distally, almost opposite a smaller tornal spot; a third small elongate streak along base of cilia, often divided into two, between tornal spot and apex. Cilia greyish, cilia line distinct, brown. Hindwing light grey-brown, cilia similar. Abdomen grey or grey-brown, anal tuft whitish. Hind tarsus shining grey-white.

Female (Pl.-fig. 48). Face yellowish white; basal portion of antenna blackish grey and distal third light grey. Neck tufts, tegulae and thorax shining black. Ground colour of forewing shining bronzy blackish brown, base mixed with silky shining scales. Abdomen dark grey-brown.

Male genitalia (Pl.-figs. 284–286). Uncus lobes of medium size, incision U-shaped. Gnathos rather small, transversely elongate. Tegumen with parallel sides. Costa of valva strong, distal hump rounded and not distinct, sacculus projecting into a little distinct spine below the small, rounded cucullus. Digitate process club-shaped, broad distally. Juxta lobes medially produced, rounded towards the oblique, setose apical margins. Vinculum with a short, blunt saccus, without medial ridge. Aedeagus straight, base bulbous, distal end strongly pointed; without cornuti.

Female genitalia (Pl.-fig. 425). Apophyses slender, of almost equal length. Antrum elongate, funnel-shaped, gradually tapering into colliculum, ventral margin almost straight, only slightly anteriorly curved, dorsal wall strongly spined both before and behind ventral margin; colliculum rather wide, twice as long as apophyses posteriores. Signum an elongate, 8-shaped, dentate plate.

Distribution. In several districts in Denmark, but not known from W. Jutland; from many districts in Sweden, except for Lpm.-districts; also in western Norway and from all parts of Finland. – From almost all parts of Europe.

Biology. Larva yellowish, later whitish, prothoracic shield made up of two elongate plates with posterior ends laterally enlarged, L-shaped. Many genera of grasses are reported by Hering (1957) as food-plants: *Brachypodium, Dactylis, Deschampsia, Festuca, Holcus* and *Milium* and Steuer (1976a) has reported *Festuca altissima* All., *F. gigantea* (L.) Vill., *Brachypodium silvaticum* (Huds.) PB., *Milium effusum* L., *Poa nemoralis* L., *Melica nutans* L. and *Luzula pilosa* (L.) Willd. The larva mines during the autumn until December, and hibernates in the mine when mature, pupating outside the mine in the spring. Often more than one larva is present in a single mine. The larva starts mining just under the tip of the leaf, making first a narrow mine which gradually widens towards the base; it is characterized by its very irregular sides and by the often divided proximal part (Pl.-figs. 513, 514). The mine is yellowish green, not transparent, owing to small uneaten "islands" of mesophyll. The excrement is deposited in several irregular, interrupted lines. Hibernation occurs in the central part of the mine. Pupation takes place in the early spring on stones and stems. The pupa is smooth, longitudinal ridges very low, yellowish brown. The adults fly from late April to late June; univoltine. The biotope is open places in parks and deciduous forests. Adults may often be observed resting on tree-trunks.

The *pulchella*-subgroup

Mainly greyish species. Sexual dimorphism pronounced, females with forewing generally light from base to fascia, and dark beyond.

Venation (Pl.-figs. 177–186). Forewing with R_4 and R_5 on a long stalk, R_3 free or from base of stalk R_{4+5}, M_1 from stalk R_{4+5}, M_2 free. Rs simple in hindwing.

Colliculum generally widening anteriorly.

23. ***Elachista subnigrella*** Douglas, 1853

Plate-figs. 49, 50, 177, 287-289, 426, 518.

Elachista subnigrella Douglas, 1853: 210.

7.5–8.5 mm. Male (Pl.-fig. 49). Head, tegulae and thorax grey, mottled brownish by scales with light grey base and brownish grey tip, posterior margins of tegulae light grey. Antenna brownish grey. Labial palpi slightly curved, porrect or drooping, upperside whitish, underside brownish grey. Forewing ground colour dark brownish grey, mottled, scales as on head and thorax, with a very indistinct or absent, slightly oblique, pale fascia before the middle, very indistinct light spots beyond middle at costa and tornus, costal spot slightly beyond the tornal spot. Base of cilia white before the distinct blackish cilia line, cilia grey beyond. Hindwing light grey, cilia paler. Abdomen grey, anal tuft ochreous. Hind tarsus grey with white rings.

Female (Pl.-fig. 50). Basal part of antenna ringed white/grey, distal part plain grey. Colour of head, tegulae and thorax dark grey. Forewing with many black or dark grey scales and distinct white fascia and spots. Cilia line very distinct, black.

Male genitalia (Pl.-figs. 287–289). Uncus lobes very close, large and rounded. Tegumen very short, strongly sclerotized. Gnathos rounded. Valva curved, widening distally, costa strong, concave, hump insignificant, cucullus large, rounded. Digitate process only slightly widening distally, apex blunt. Juxta lobes short, truncate, apical margins almost straight, setose. Vinculum with short medial sclerotized ridge and a prominent, narrow saccus. Aedeagus slightly bent, tapering towards the indented distal end; without cornuti.

Female genitalia (Pl.-fig. 426). Apophyses slender, of almost equal length. Antrum funnel-shaped, gradually tapering into colliculum, ventral margin almost straight, only slightly anteriorly curved, dorsal wall strongly spined mainly anteriorly to ventral margin; colliculum rather wide, slightly longer than apophyses posteriores. Corpus bursae finely spined inside, signum narrow, slightly bent, dentate.

Distribution. In Denmark only known from Zealand; in Sweden from the southern and eastern districts up to Upl.; only from few districts in S. Norway; not known in Finland. – Widespread on the Continent and in Britain.

Biology. The larva is pale yellow. Head and prothoracic shield pale brown. According to Stainton (1858c) the larva mines in *Bromus erectus* Huds.; also *Avena* and *Festuca* are reported. It makes a flat, narrow mine (Pl.-fig. 518), which first is pale yellowish green, but later turns purplish. The excrement forms a nearly continuous line. The larva often mines more than one leaf. Pupation takes place where the mined leaf branches off from the stem; pupa attached by a single girdle and the cremaster. Bivoltine; the larva mines from April to mid-May and during July, and the adults fly from late May to mid-June and again in August. Adults fly slowly and low around the food-plants. The biotope is rather dry, open calcareous grounds.

Elachista reuttiana Frey, 1859

Plate-figs. 51, 52, 178, 290, 291, 427, 519.

Poeciloptilia obscurella Herrich-Schäffer, 1855: 300, 307 (nom. praeocc.).
Elachista herrichii Frey, 1859: 238.
Elachista reuttiana Frey, 1859: 251.
Elachista confluella Rössler, 1866: 380.

8–9 mm. Male (Pl.-fig. 51). Head whitish grey, vertex almost pure white; neck tufts, tegulae and thorax whitish, mottled with grey by white bases and dark tips of scales;

posterior edges of tegulae and thorax slightly lighter. Scape of antenna blackish, flagellum dark grey, slightly annulated. Labial palpi porrect and slightly curved, upperside whitish, underside greyish, base of third segment with whitish underside. Forewing ground colour dark grey to blackish, mottled by scales similarly coloured to those on thorax; a whitish, slightly outward-curved fascia to dorsum from two-fifths of costa; costal and tornal spots whitish, elongate rectangular; a third small, indistinct, whitish spot distally between the two spots. Cilia grey at apex and tornus, white along termen; cilia line distinct, black, from costal spot to tornus. Hindwing grey, with cilia of same colour. Abdomen light grey, anal tuft ochreous. Hind tarsus dark grey with indistinct rings.

Female (Pl.-fig. 52). Neck tufts, tegulae and thorax dark grey to blackish, mottled. Forewing more contrasting owing to more blackish scales beyond the fascia and more distinct white fascia and spots; costal spot confluent with the small intermedial spot. Hindwing pale grey, cilia similar.

Male genitalia (Pl.-figs. 290, 291). Uncus with deep, V-shaped indentation, lobes elongate, sides almost parallel. Gnathos small, rounded. Valva widening distally, costa strong, distally with a distinct, angulate hump; cucullus truncate. Digitate process slender, with parallel sides. Juxta lobes rounded medially, apical margins rounded, with short setae. Vinculum with a medial ridge, gradually produced into a short, blunt saccus. Aedeagus almost straight, tapering, with two spoon-shaped cornuti, one half as long as the other.

Female genitalia (Pl.-fig. 427). Apophyses rather short, of almost equal length. Antrum funnel-shaped, gradually tapering into colliculum, ventral margin U-shaped, reaching to middle of antrum, dorsal wall with minute spines before and behind ventral margin, colliculum gradually widening at both ends, twice as long as apophyses posteriores. Membranous portion of ductus bursae wide, longitudinally folded, with 4-8 triangular, sclerotized teeth towards colliculum. Corpus bursae with minute internal spines, signum elongate, curved, finely dentate.

Distribution. Not recorded from Scandinavia or Britain, but widespread in C. Europe.

Biology. Immature stages and biology were described by Hoffmann (1893), Martini (1916) and Wörz (1957). The larva is deep yellow, with a whitish dorsal line, head light brown, sclerites medially concave. The larva mines from the autumn to April-May and during July to early August in *Koeleria cristata* (L.) Pers.; *Koeleria glauca* (Schrad.) DC., *Agrostis* and *Holcus* are also reported as food-plants (Wörz, 1957). After hibernation the mine is fine, straight and narrow, situated in the lowest part of the leaf, yellowish green and not transparent, as all the mesophyll has not been eaten. The larva then mines upwards, widening the mine to the width of the blade; the mine is now swollen, yellowish to whitish, transparent and on the underside of the leaf, and the distal part of the leaf now wilts and turns dark reddish brown; the excrement is scattered irregularly inside the mine (Pl.-fig. 519). Pupation takes place on the lower concave side of a leaf; the pupa is light yellowish brown, with distinct dorsal and lateral whitish ridges. The adult fly from early May to late June and again in August, but the two broods are apparently not distinctly separated, and all stages can be collected at any time during the summer. The biotope is open dry places on calcareous ground; the larvae are mainly found on plants standing in the shadow of shrubs and trees.

Note. Parenti (1977) adopted *herrichii* as the valid name of this species owing to page precedence to *reuttiana* in Frey (1859), but it has not been possible to include this change in the present paper.

24. ***Elachista orstadii*** Palm, 1943
Plate-figs. 53, 54, 179, 292, 293, 428.

Elachista orstadii Palm, 1943: 26.

7–9 mm. Male (Pl.-fig. 53). Head, neck tufts and tegulae light brownish grey, frons and posterior margins of tegulae paler, thorax dark grey. Scape of antenna light grey, flagellum dark grey, slightly annulated light grey. Labial palpi porrect, white above, with grey-brown streak beneath. Forewing ground colour beige to brownish grey, mottled by scales with pale bases and dark tips; dark tipped scales present in the distal part of the fold, and in streaks above and distal to fold; wing marks very indistinct: a pale, blurred, almost straight fascia before middle; beyond the middle costal and tornal spots are only indicated by slightly lighter scales, many dark tipped scales before apex. Cilia whitish grey, cilia line fine, dark brown. Hindwing yellowish grey, cilia similar. Abdomen shining light grey, anal tuft similar. Hind tarsus blackish with whitish annulation.

Female (Pl.-fig. 54). Antenna distinctly annulated in basal part. Head, tegulae, thorax and inner part of forewing light beige-brown, forewing mottled blackish beyond fascia; outer margin of fascia distinct, fascia almost straight, white; distal marks distinct, white; costal spot elongate, narrow, inwardly oblique, beyond triangular tornal spot. Cilia whitish along termen, greyish from costal spot to apex and around tornus.

Male genitalia (Pl.-figs. 292, 293). Uncus with V-shaped indentation, lobes slightly tapering towards apex, medial margin slightly concave. Gnathos small. Valva with almost parallel sides, costal hump small, cucullus rounded towards costa and almost right angled to sacculus. Digitate process widening towards blunt apex. Juxta lobes truncate, distally broadening, apical margins straight, setose. Vinculum gradually produced into a short, tapering saccus, without medial ridge. Aedeagus straight, tapering, distal end bifurcate, sclerotized.

Female genitalia (Pl.-fig. 428). Apophyses slender, posteriores slightly longer than anteriores. Antrum funnel-shaped, gradually tapering into colliculum, ventral margin wide, U-shaped, slightly reinforced, dorsal wall strongly spined; colliculum narrow in the middle, slightly widening towards both ends, twice as long as apophyses posteriores. Corpus bursae with sparse, minute internal spines, signum angulate, medially enlarged, distally dentate.

Distribution. In Scandinavia only from Denmark: SJ, EJ, NEJ, NEZ, and in Sweden known from Sk. to Upl. – Otherwise only known from Scotland.

Biology. Immature stages and food-plant not known. Probably univoltine; the adults are recorded from mid-May to late June, and mainly fly on open dry grasslands, both near the sea and inland.

25. ***Elachista ingvarella*** Traugott-Olsen, 1974
Plate-figs. 55–57, 180, 294, 295, 429–441.

Elachista ingvarella Traugott-Olsen, 1974: 264.

7–9 mm. Male (Pl.-fig. 55). Head shining ash-grey, face almost white; neck tufts and thorax lead grey, posterior edge of tegulae whitish grey. Scape and antenna and pedicel coloured as head, flagellum plain grey-brown, distal half with slightly raised scales. Labial palpi slightly curved, upperside white, underside of second segment ochreous-grey and third segment dark grey. Forewing ground colour light brownish grey, mottled by scales with whitish grey bases and brownish grey tips; scattered dark tipped scales along fold and termen; pattern very indistinct, made up of an indistinct straight pale fascia before the middle and a pale costal spot beyond a similar tornal spot; distally, between these spots, a tiny lighter spot. Cilia grey, slightly lighter at apex, cilia line indistinct, consisting of few scattered dark brown-tipped scales. Hindwing grey, cilia similar. Abdomen grey, with ochreous anal tuft. Hind tarsus medially whitish, laterally with scattered dark grey scales.

Female (Pl.-figs. 56, 57). Face whitish; neck tufts, tegulae and thorax mottled grey-brown/white by white scales with dark grey-brown tips. Proximal third of flagellum distinctly annulated white/dark brown, middle third plain grey-brown and distal third slightly annulated brown/grey-brown. Labial palpi with whitish upperside and dark brown underside. Forewing ground colour whitish grey from base to fascia, mottled dark brown, dark brown beyond fascia, slightly mottled with few white based scales; pattern distinct, fascia white, of rather uniform width, margins irregular. Costal spot a distinct, elongate white streak, beyond similar triangular tornal spot. Cilia whitish grey; cilia line indistinct, blackish.

Male genitalia (Pl.-figs. 294, 295). Uncus with U-shaped indentation, lobes large, narrowing towards base. Costa of valva strong, with faint hump, cucullus rounded towards costa and almost at right angle to sacculus, which is produced distally into a small projecting spine below cucullus. Sides of digitate process almost parallel. Juxta lobes medially produced, rounded, apical margins curved, setose, with small lateral processes. Vinculum without medial ridge, produced into a long slender saccus. Aedeagus long, slightly bent, tapering towards the indented distal end, with two small spoon-shaped cornuti.

Female genitalia (Pl.-figs. 429–441). Apophyses slender, of almost equal length. Antrum smooth, funnel-shaped, gradually tapering into colliculum, ventral margin deeply U-shaped, rims long, often set with a pair of short posterior setae, ventral margin often reinforced, dorsal wall generally minutely spined behind ventral margin and rarely also before; colliculum narrow in the middle, widening anteriorly, less than twice as long as apophyses posteriores. Inception of ductus seminalis in a small pouch; ductus bursae with a varying number of small sclerotized plates with triangular teeth near inception of ductus seminalis. Corpus bursae minutely spined inside, signum elongate, medially enlarged, dentate.

Distribution. Not known in Denmark; known from several districts in the northern part of Sweden down to Hrj.; in Norway only from SFi, and in Finland only from Le.

Biology. Nothing is known about the immature stages and the food-plants. The adults are caught from early to mid-July; probably univoltine. The biotope seems to be moorland in the coniferous forest and birch forest zones.

26. ***Elachista krogeri*** Svensson, 1976
Plate-figs. 59, 60, 181, 296, 297, 442.

Elachista krogeri Svensson, 1976a: 198.

8–10 mm. Male (Pl.-fig. 59). Head and neck tufts light brownish grey; tegulae and thorax dark grey, posterior margins slightly lighter. Antenna dark grey, darkest towards base. Labial palpi drooping to porrect, slightly curved, third segment tapering proximally, upperside whitish grey, underside dark grey. Forewing ground colour generally dark grey, but varying from brownish to blackish, with blackish scales in fold; pattern very indistinct, a slightly lighter fascia before middle, often only distinct as a light patch in the fold between the dark scales; costal spot large, triangular, whitish, faintly tinged with yellowish along costa, slightly beyond the almost opposite tornal spot; often a few whitish scales towards termen between costal and tornal spots. Cilia with bases ochreous-grey, light grey beyond the blackish cilia line. Hindwing grey, cilia similar. Abdomen dark grey, anal tuft lighter. Hind tarsus dark grey, segments lighter distally.

Female (Pl.-fig. 60). Scape of antenna anteriorly ochreous-brown, posterior part of scape and flagellum dark grey. Forewing ground colour dark grey to blackish grey with distinct white marks; fascia straight, narrow at costa, wide in the fold; costal and tornal spots connected by a few whitish scales.

Male genitalia (Pl.-figs. 296, 297). Uncus with U-shaped indentation, lobes broad, Gnathos rounded, underside slightly indented. Costa of valva strong, hump smooth and long, cucullus rounded towards costa and at a right angle to sacculus, which slightly projects distally. Digitate process club-shaped, broad, blunt. Medial margins of juxta lobes straight, at a sharp angle to the setose apical margins. Vinculum with a weak medial ridge, tapering into a short, stout saccus. Aedeagus gently curved, base bulbous, distal end rounded, deeply cleft, with one spoon-shaped cornutus and beyond this a cluster of 2–4 very small teeth.

Female genitalia (Pl.-fig. 442). Apophyses slender, of almost equal length. Antrum funnel-shaped, smoothly tapering into colliculum, ventral margin U-shaped, slightly reinforced, rims strongly widening posteriorly, each with a pair of small setae, dorsal wall with fine spines behind ventral margin; colliculum narrow, slightly more than twice as long as apophyses posteriores. Membranous portion of ductus bursae towards colliculum with 1–3 small oval, sclerotized plates. Corpus bursae with very few spines, signum elongate, dentate.

Distribution. Only known from the extreme northern part of Sweden: Nb., T.Lpm., and Finland: Le.

Biology. Food-plant and immature stages unknown. The moth appears in late June and early July, flying low over the vegetation in the late dusk in moist places near rivers (Svensson, 1976a, b).

27. ***Elachista nielswolffi*** Svensson, 1976
Plate-figs. 58, 182, 298, 299.

Elachista nielswolffi Svensson, 1976a: 203.

7–9 mm. (Pl.-fig. 58). Frons light grey, clypeus dark grey; neck tufts, tegulae and thorax dark grey, posterior margins lighter. Antenna grey, slightly darker towards base, scales on distal half slightly raised. Labial palpi porrect, curved, third segment short, upperside whitish, underside dark grey, junction between the two distal segments whitish. Forewing ground colour shining grey, mottled with bluish and tinged lighter grey by blackish-tipped scales with bluish white bases, with darker scales in the fold; a slightly outward-curved whitish fascia before middle, wide in the middle, tapering towards costa and dorsum; costal and tornal spots more distinct, triangular, costal spot ochreous white, slightly beyond the more whitish tornal spot; the spots are sometimes connected by whitish scales, forming an outward-bent fascia. Cilia light bluish grey, with an indistinct cilia line. Hindwing grey, cilia similar. Abdomen grey, anal tuft paler.

Female. Unknown.

Male genitalia (Pl.-figs. 298, 299). Uncus with broad U-shaped indentation, lobes narrow, elongate. Gnathos small, rounded. Sides of valva almost parallel, costa strong with a large, smooth hump which is more prominent than the small, rounded cucullus, sacculus slightly bent in the middle, projecting into a small spine below cucullus. Digitate process club-shaped, blunt. Medial margins of juxta lobes straight, meeting the rounded, setose apical margins at a sharp angle. Vinculum with a distinct medial ridge, produced into a short, tapering saccus. Aedeagus curved near base, tapering distally to narrow, rounded apex which is deeply split into a strongly and a weakly sclerotized part; without cornuti.

Female genitalia. Unknown.

Distribution. Only known from Jmt., Sweden and TRi, Norway.

Biology. Immature stages not known. Adults collected from late June to late July. According to Svensson (1976a, b) the males fly late in the evening low down over the vegetation, often hiding in it. The species is only known from the upper subalpine region, where the only grasses found in the presently known localities are *Deschampsia flexuosa* (L.) Trin. and *Nardus stricta* L., one of which probably is the food-plant.

28. ***Elachista pomerana*** Frey, 1870

Plate-figs. 61, 62, 183, 300, 301, 443.

Elachista pomerana Frey, 1870: 282.

8–10 mm. Male (Pl.-fig. 61). Head brownish grey, shining; neck tufts, tegulae and thorax dark brown anteriorly, whitish grey along posterior margins. Scape of antenna dark grey on upperside, underside yellowish grey, flagellum uniformly dark grey. Labial palpi porrect, upperside whitish grey, slightly tinged yellowish, underside of second segment dark grey, underside of third segment as upperside, but with a few scattered grey scales. Forewing ground colour shining grey, with scattered, blackish, grey-tipped scales, especially along base of costa, in the fold and beyond fascia; before middle a white, irregular, angulate fascia outwardly oblique from costa to above fold, narrow or interrupted in fold; costal spot triangular, distinct, white, tinged beige along costa, beyond the small triangular tornal spot, which is tinged beige along dorsum. Cilia plain grey; cilia line distinct, black. Hindwing grey, cilia grey, slightly beige-tinged. Abdomen grey. Hind tarsus light grey, darkest distally.

Female (Pl.-fig. 62). Tegulae and thorax dark grey. Forewing ground colour darker and the white marks more distinct; fascia oblique from costa tofold, angulate in fold, not tapering towards costa, inconspicuous towards dorsum.

Male genitalia (Pl.-figs. 300, 301). Uncus with wide U-shaped indentation, lobes narrow, outer margin angled in the middle. Gnathos small. Valva slightly tapering proximally, costa strong with large, abrupt hump, cucullus truncate, sacculus with a small distal spine below cucullus. Digitate process rather slender, blunt. Juxta lobes truncate, apical margins almost straight, setose. Vinculum with distinct medial ridge, tapering into a short blunt saccus. Aedeagus slightly bent, tapering distally to slightly indented distal end, with two cornuti, one elongate and thorn-shaped and one small, often double- or triple-toothed.

Female genitalia (Pl.-fig. 443). Apophyses rather slender, of almost equal length. Antrum funnel-shaped, pouched towards colliculum, ventral margin U-shaped, rims and lateral margins parallel, dorsal wall strongly spined before and behind ventral margin, colliculum slightly widening anteriorly, twice as long as apophyses posteriores. Inception of ductus seminalis pouched, ductus bursae here sometimes with 1–2 triangular teeth or oval sclerotized plates, each with an eccentric tooth. Corpus bursae finely spined inside, signum elongate, medially enlarged, distally dentate.

Distribution. Known from E. Denmark, but local; known from Sk. to Nb. in Sweden, but from only few districts; in Norway from the most southern part and from Nn in the northern part; known from many districts in W. and S. Finland. – Reported from E. and C. Europe to Italy, and recently also from Scotland (Pelham-Clinton, pers. comm.).

Biology. Immature stages first reliably described by Hering (1891). Larva broad, yellowish, flecked with white, intestine dark greyish green, head and prothoracic shield light brown, laterally darker. As food-plants *Poa pratensis* L., *Phalaris arundinacea* L., *Glyceria fluitans* (L.) R. Br., *Calamagrostis* and *Avena* are reported (Wocke, 1876; Büttner, 1880; Hering, 1957). The larva starts mining from the tip of the leaf and occupies the whole width of the leaf; later it mines downwards, making a mine up to 6 cm long and 2–3 mm wide. At first the excrement is scattered irregularly in the whole mine, but is later deposited in a narrow, fine line, which first is dark green, but later turns blackish. The larva is present during April to mid-May. Pupation takes place in May in a rather dense web on a leaf, often very close to the water surface. The pupa is dark brown with lighter brown to greyish, distinct elongate ridges. The adults occur from early June to mid-July; univoltine. The biotope is damp areas around ponds and streams.

29. ***Elachista humilis*** Zeller, 1850

Plate-figs. 63, 64, 184, 302–304, 444, 520.

Elachista humilis Zeller, 1850: 201.
Elachista occultella Douglas, 1850: 7.
Poeciloptilia humiliella Herrich-Schäffer, 1855: 308.
Elachista perplexella Stainton, 1858b: 308.
Elachista airae Stainton, 1858b: 308.

9–10 mm. Male (Pl.-fig. 63). Head, neck tufts and tegulae dark grey-brown, thorax slightly darker. Antenna plain dark grey, flagellum with slightly raised scales. Labial palpi rather long, slender, drooping to porrect, curved, upperside whitish grey, underside brownish grey, basal part of third segment lighter. Forewing ground colour dark brownish grey, fascia generally weak or absent, but can be visible as a few lighter scales near costa and in fold; costal spot triangular, distinct, white, tinged beige along costa, just opposite to the smaller similar-coloured tornal spot, which can be prolonged along dorsum. Cilia dark grey, whitish grey along tornal spot, cilia line indistinct, made up by scattered dark-tipped scales. Hindwing light grey, cilia similar. Abdómen dark grey, anal tuft ochreous grey. Hind tarsus medially light grey, laterally dark grey, slightly ringed.

Female (Pl.-fig. 64). Antenna blackish grey, scales not raised. Forewing ground colour blackish brown, basal part lighter. Fascia wide, straight, white, slightly outward-oblique from costa, costal spot white, square, opposite the smaller white tornal spot. Cilia as in male.

Male genitalia (Pl.-figs. 302–304). Uncus lobes rather small, wide apart. Gnathos rounded. Valva widening distally, costa strong, with abrupt hump towards truncate cucullus, sacculus forming a small spine below cucullus. Digitate process club-shaped, base very narrow, distal part broad, blunt. Juxta lobes distally wide, apical margins broad, rounded, setose. Vinculum elongate, triangular, with a very distinct medial sclerotized ridge and tapering into a short, stout saccus. Aedeagus curved near bulbous base, otherwise narrow, distal end indented, with two small cornuti, of which one often possesses two or more small points, often extruded during preparation.

Female genitalia (Pl.-fig. 444). Apophyses rather strong, of almost equal length. Antrum smooth, funnel-shaped, ventral margin wide, U-shaped, strongly sclerotized, rims distinct, dorsal wall with few minute spines behind ventral margin and inside antrum; colliculum slightly widening anteriorly, less than twice as long as apophyses posteriores. Signum angulate, medially enlarged, distally dentate.

Distribution. Reported from almost all districts in Denmark and Sweden; from few districts in W. and N. Norway and widespread in many districts in Finland. – Well known from C. and W. parts of the Continent and Britain.

Biology. Larva dull yellowish, slightly tinged with greenish, with a lighter dorsal line, head light brown, faintly sclerotized. As food-plants *Deschampsia caespitosa* (L.) PB. (Pl.-fig. 520), *Poa pratensis* L., *Agrostis, Festuca, Holcus* and *Carex* are recorded (Stainton, 1858c; Hering, 1891; Sønderup, 1949; Wörz, 1957), but probably not all the latter records are reliable due to confusion with *E. pulchella* (Hw.). Apparently univoltine in N. Europe; larvae present during the spring, adults from late May to early August.

30. ***Elachista vonschantzi*** Svensson, 1976
Plate-figs. 65, 66, 185, 305, 306, 445.

Elachista vonschantzi Svensson, 1976a: 200.

8–10 mm. Male (Pl.-fig. 65). Head and neck tufts yellowish brown, mixed with grey scales; tegulae and thorax brownish, posterior margins lighter. Antenna dark brown, darkest near base, posterior side of scape yellowish brown, distal part of flagellum slightly serrate. Labial palpi porrect, slightly curved, distal segment slender, upperside light grey tinged with ochreous, underside dark grey, but base of distal segment lighter grey. Forewing ground colour dark grey sprinkled with blackish grey scales, especially along base of costa, in fold and beyond fascia; ochreous-brown spots at base of fold and particularly at base of dorsum; distal part of wing slightly similarly tinged; fascia whitish before middle, straight from costa to fold, where it forms a distinct white spot, often indistinct from fold to dorsum; costal spot large, triangular, ochreous white, opposite a small tornal spot with less ochreous, often connected to costal spot by a small white spot between tornus and apex. Base of cilia ochreous grey, distal part beyond distinct black cilia line grey. Hindwing grey, cilia similar. Abdomen grey, anal tuft lighter. Hind tarsus laterally blackish grey, medially whitish.

Female (Pl.-fig. 66). Forewing ground colour dark grey, with many scattered blackish scales before inner fascia, almost plain black beyond fascia, dots at base whitish; inner fascia white, wide, uninterrupted, narrowing towards costa, outwardly curved; costal and tornal spots confluent, forming a distinct whitish fascia sharply angled outwards below the middle.

Male genitalia (Pl.-figs. 305, 306). Uncus lobes small and very narrow, very wide apart. Gnathos rounded. Valva with almost parallel margins, hump of costa smooth, cucullus truncate, sacculus straight, slightly projecting below cucullus. Digitate process rather short, with parallel sides, blunt. Juxta lobes short, apical margins rounded, setose. Vinculum without medial ridge, produced into a short, narrow, tapering saccus. Aedeagus bulbous at base, straight, narrowest before middle, distal end narrow, indented; no cornuti present.

Female genitalia (Pl.-fig. 445). Apophyses rather strong, of almost equal length. Antrum short, bowl-shaped, constricted towards colliculum, ventral margin anteriorly curved, rims short, dorsal wall and innerside of antrum finely spined; colliculum broad, slightly widening anteriorly, one and a half times as long as apophyses posteriores. Corpus bursae with a patch of minute spines; signum elongate, dentate, distal teeth extremely long.

Distribution. Only known from few localities in the districts Nb. and Ång. in N. Sweden.

Biology. Immature stages unknown. As the adults were seen flying over *Calamagrostis neglecta* (Ehrh.) G.M.S. in the dusk, this is probably the food-plant (Svensson, 1976 a, b). The adults have been caught from late June to early July; probably univoltine. The localities are grass belts bordering the sea shore; all are near the Baltic.

31. ***Elachista pulchella*** (Haworth, 1828)
Pl.-figs. 67, 68, 186, 307, 308, 446.

Tinea pulchella Haworth, 1828: 582.
Elachista obscurella Stainton, 1849: 26.
Poeciloptilia incanella Herrich-Schäffer, 1855: 301, 308.
Elachista subobscurella Doubleday, 1859: 34.
Poeciloptilia montanella Wocke, 1862: 247.

Elachista ranenensis Strand, 1919: 123. **Syn.n.**
Elachista albimarginella Hering, 1924: 78. **Syn.n.**
Elachista postremella Dufrane, 1957: 11.

8–10 mm. Male (Pl.-fig. 67). Head and neck tufts pale beige, thorax and tegulae darker, mottled with brown. Antenna uniformly grey-brown. Labial palpi curved, third segment smooth, dull white, upperside with few scattered brownish scales, upperside of second segment brownish, underside whitish. Forewing ground colour grey-brown, mottled with dark brown tipped scales; marks very indistinct, fascia before middle only indicated by slightly lighter scales and fewer brown tipped scales, costal and tornal spots similar, but generally more distinct, pale beige, only with few dark-tipped scales, costal spot slightly beyond tornal. Cilia along termen ochreous grey, beyond tornus grey, cilia line distinct, fine, dark brown. Hindwing grey, tinged brownish, cilia similar. Abdomen grey, anal tuft yellowish white. Hind tarsus almost uniformly light grey.

Female (Pl.-fig. 68). Underside of the recurved labial palpi brownish, except for the basal part of the third segment, which is white. Forewing ground colour dark beige from base to fascia, very dark brown beyond fascia; fascia white, distinct, outwardly oblique from costa to fold and perpendicular to dorsum, margins of fascia irregular; costal and tornal spots very distinct, white, triangular, costal spot beyond tornal. Cilia grey, cilia line distinct, blackish. Hind tarsus blackish, with white spots on distal part of segments.

Male genitalia (Pl.-figs. 307, 308). Uncus with U-shaped indentation, lobes small, wide apart. Gnathos small. Valva long and narrow, sides almost parallel, hump of costa not distinct, cucullus truncate. Digitate process club-shaped, blunt. Juxta lobes medially produced, apical margins rounded to the short lateral processes. Vinculum without medial ridge, tapering into a short, stout saccus. Aedeagus wide at base, gradually tapering to square, not indented, distal end; with two rather large spoon-shaped cornuti.

Female genitalia (Pl.-fig. 446). Apophyses slender, posteriores slightly longer than anteriores. Antrum pouched, constricted towards colliculum, ventral margin reinforced, U-shaped, dorsal wall and inner side of ventral margin with fine spines; colliculum narrow, one and a half times as long as apophyses posteriores. Posterior dilated portion of membranous ductus bursae with longitudinal folds, with 2–5 small sclerotized plates inside, each with 1–3 small teeth. Corpus bursae with minute internal spines, signum elongate, bent, enlarged in the middle, distally dentate.

Distribution. Known from almost all parts of Denmark, and from many districts in Sweden, Norway, and Finland. – Apparently widespread in the northern part of the Continent and in Britain.

Biology. Only a few records of the biology are known. Stainton (1858b) described the larva as pale yellowish grey, with pale brown head, making a flat mine in *Holcus mollis* L. and other grasses. In Denmark reared from *Arrhenatherum pubescens* (Huds.) Sampaio. The pupa is light yellowish brown, longitudinal ridges greyish and between these two dorsal, dark brown lines bordered with yellowish. The larva mines during April-May and July. The adults fly from late May to late June and again during August-September; bivoltine. Mainly in open, rather dry places with shrubs or along edges of woods.

Note. The identity and nomenclature of this species was outlined by Janmoulle (1949b); further contributions are given by Janmoulle (1962) and Wolff (1970). The present authors have further examined the holotypes of *E. ranenensis* Strand and *E. albimarginella* Hering, which both appear conspecific with *E. pulchella* (Hw.).

Holotype of *E. ranenensis* Strand is labelled "Norwegen, Ranen, Hemnesberget, 8–14.VII. 03, E. Strand, S.", "Strand det, Elachista ranenensis Strand (n.sp.?)", "TYPE", "Zool. Mus. Berlin", "Genital praeparat A.9.11 76 ♂, E. Traugott-Olsen", "Elachista pulchella Hw., det. E. Traugott-Olsen" in the Humboldt Museum, Berlin.

Holotype of *E. albimarginella* Hering is labelled "Sortavala, 24.6.1920, Karvonen", "TYPE ♀, Elachista albimarginella, type det. Mart. Hering", "Mus. Zool. H:fors, spec. type no. 7008", "Elachista pulchella Hw., det. E. Traugott-Olsen" in Zoological Museum, Helsingfors.

The *cerusella*-subgroup

Ochreous white to pale grey species, with or without scattered yellowish or brownish scales.

Venation (Pl.-figs. 187–190). Forewing with R_4, R_5 and M_1 stalked, R_4 and R_5 almost coalescent.

Aedeagus with narrow, strongly sclerotized, indented distal portion; without cornuti.

32. ***Elachista anserinella*** Zeller, 1839

Plate-figs. 69, 70, 187, 309, 310, 447.

Elachista anserinella Zeller, 1839: 213.

10–11 mm. Male (Pl.-fig. 69). Head and neck tufts cream-white, thorax and tegulae similar, but anterior parts tinged ochreous. Scape of antenna and basal part of flagellum greyish white, distal part brownish grey, annulated. Labial palpi recurved, distal segment smooth, cream-white, upperside of second segment tinged greyish, underside grey-brown. Forewing ground colour cream-white, a narrow streak of greyish brown scales along costa; wing suffused with ochreous-tipped scales (which are almost absent in some specimens), suggesting a pale, outward-angled, fascia in the middle and a white, narrow, costal streak beyond a white triangular tornal spot. Cilia cream-white, cilia line distinct, dark brown, from costal spot to tornal spot. Hindwing dark grey, cilia brownish grey along costa, yellowish at apex and grey along dorsum. Abdomen whitish grey, anal tuft yellowish white. Hind tarsus whitish, with brown spot on outer side.

Female (Pl.-fig. 70). Flagellum distinctly ringed from scape to tip, white/blackish brown. Grey-brown scales along costa of forewing forming a fine line at base. Hindwing lighter grey.

Male genitalia (Pl.-figs. 309, 310). Uncus with V-shaped indentation, lobes very large and oval. Costa of valva strong, with a rounded distal hump before the slightly projecting, truncate cucullus. Digitate process very long and slender, sides irregular, distal part wide, blunt. Juxta lobes short, apical margins oblique, setose. Vinculum triangular, with medial ridge, tapering into narrow saccus. Aedeagus long, slightly bent, distally sclerotized, bifurcate, pointed.

Female genitalia (Pl.-fig. 447). Apophyses long and slender, of almost equal length, apophyses anteriores slightly curved. Antrum funnel-shaped, gradually tapering into colliculum, ventral margin widely U-shaped, strongly spined, dorsal wall with spines behind ventral margin; colliculum narrow, slightly longer than apophyses posteriores. Corpus bursae without spines or signum.

Distribution. Rare and local in Denmark: NEZ; rather common in S. Sweden: up to Boh. and Upl.; in Norway only from AK (recorded as *E.disertella* (HS.) by Haanshus (1933)); not in Finland. – In C. and E. Europe, Italy and France, but not in NW. Europe or Britain.

Biology. Apparently, very little is known about the immature stages. Martini (1912b) reported the larva from *Brachypodium pinnatum* (L.) PB., the distribution of which is comparable with the distribution of the moth in Scandinavia. The larva mines in the late spring, and the adults fly from late May to late June, mainly in early June; univoltine in N. and C. Europe.

33. ***Elachista rufocinerea*** (Haworth, 1828)
Plate-figs. 71, 72, 188, 311-313, 448.

Porrectaria rufocinerea Haworth, 1828: 535.
Porrectaria oleae Haworth, 1828: 535.
Porrectaria floslactis Haworth, 1828: 535.
Poeciloptilia rufocinerella Herrich-Schäffer, 1855: 312.

10–11 mm. Male (Pl.-fig. 71). Head and neck tufts whitish; thorax white with a medial ochreous brown line, anterior part of tegulae ochreous brown, posterior part white. Antenna light brown, tinged ochreous, underside of scape whitish. Labial palpi recurved, third segment and upperside of second segment white, underside of the latter slightly brownish. Forewing ground colour cream-white, base of costa distinct red ochreous brown; with blurred, ochreous, longitudinal streaks from base, separated towards termen by paler streaks; this pattern becomes more diffuse towards termen; apex with a characteristic triangular brown streak. Cilia whitish; cilia line fine, distinct, brownish. Hindwing dark grey with lighter cilia, which are ochreous-grey at apexes. Abdomen grey, anal tuft lighter. Hind tarsus brownish grey distally, medially light grey.

Female (Pl.-fig. 72). Antenna annulated brownish/white. Thorax and forewing more whitish, with less ochreous-brown; cilia whitish, cilia line very faint. Hindwing also whitish, cilia white. Abdomen and hind tarsus whitish.

Male genitalia (Pl.-figs. 311–313). Uncus with U-shaped indentation, lobes oval. Gnathos square. Costa of valva slightly curved, hump flat, cucullus rounded, sacculus curved. Digitate process with almost parallel sides, blunt. Apical margins of juxta lobes rounded, setose. Vinculum triangular, without medial ridge, distally with a short or indistinct saccus. Aedeagus with clearly defined bulbous base, middle portion rather narrow, slightly enlarged before indented tip.

Female genitalia (Pl.-fig. 448). Apophyses strong, of almost equal length. Antrum funnel-shaped, gradually tapering into colliculum, ventral margin slightly anteriorly curved, dorsal wall with minute spines before and behind ventral margin; colliculum very wide, shorter than apophyses posteriores. Corpus bursae without spines or signum.

Distribution. Rather common in few localities in SW. Denmark, but otherwise rare and local; not in Sweden, Norway or Finland. – Known from NW. Germany, Britain, Holland, Belgium, France, Italy, SE. Europe to Asia Minor; not in NE. Europe.

Biology. Stainton (1858c) described the larva as pale dull yellow, dorsally greenish grey, head brown with darker mouthparts, prothoracic shield brown, the two plates only slightly apart. As food-plant only *Holcus mollis* L. is reported, in which the larva makes broad flat, transparent mines in the leaves; the larva can mine both upwards and downwards, and even if the width of the mine may vary, it frequently occupies the whole breadth of the blade, which always remains flat. Excrement is greyish, small in quantity, irregularly scattered throughout the mine. The larva readily changes leaf. The species hibernates as a larva, and mines during the spring to late May. Pupation takes place under a slight cocoon on a leaf. The adults fly actively at sunset from late May to mid-June; univoltine. The species is found on moist bogs and on moist or drier heaths and slopes.

Elachista lastrella Chrétien, 1896
Plate-figs. 73, 74, 189, 314, 315, 449, 515.

Elachista lastrella Chrétien, 1896: 192.

8–10 mm. Male (Pl.-fig. 73). Head dark grey, tinged with yellowish, slightly shining, neck tufts similar; tegulae darker anteriorly, thorax grey. Antenna plain dark grey, finely serrate on distal three-quarters. Labial palpi drooping to porrect, second and third segment of equal length, dark grey, upperside of second segment tinged with brownish, third segment suffused with whitish grey, underside darker, distal part of third segment with an almost black spot. Wings narrow; forewing ground colour shining whitish grey, without marks, suffused darker along costa, dorsum and on base, fold area slightly tinged yellowish white. Cilia light yellowish white along termen, grey along costa to apex and around tornus; no cilia line present. Hindwing dark grey, cilia similar, lighter at apex. Abdomen dark grey, shining, anal tuft lighter. Hind tarsus whitish grey on inner side, blackish on outer side.

Female (Pl.-fig. 74). Head more yellowish. Antenna mixed with blackish, forming distinct proximal rings and more indistinct distal double rings. Labial palpi more yellowish, black dot often blurred or absent. Forewing yellowish white, base of costa greyish; cilia yellowish white. Hindwing blackish grey, cilia lighter, yellowish at apex.

Male genitalia (Pl.-figs. 314, 315). Uncus lobes rather narrow, very wide apart, outer margins bent in the middle. Gnathos small, rounded. Anterior margin of tegumen strong. Valva narrowest at base, costa strong at base, hump only slightly enlarged before dorsally rounded cucullus, which forms almost a right angle to sacculus; sacculus with a small distal spine. Digitate process club-shaped, setose. Juxta lobes distally rounded, setose; juxta with two drooping ventral lobes. Vinculum broad, truncate, without medial ridge, with a short distinct tapering saccus. Aedeagus with bulbous base, middle portion broad, distal end narrow, sclerotized, pointed and distinctly indented.

Female genitalia (Pl.-fig. 449). Apophyses strong, posteriores near base, longer than anteriores. Antrum funnel-shaped, gradually tapering into colliculum, ventral margin an almost straight line, dorsal wall strongly spined both before and behind ventral margin; colliculum narrow, slightly shorter than apophyses posteriores. Signum very narrow, elongate, dentate.

Distribution. Not from Scandinavia or Britain. – In France and S. Germany, but probably overlooked in many places.

Biology. Extensively outlined by Steuer (1976a). The larva is deep yellowish with green intestine, medial part of prothoracic shield narrow, weakly sclerotized, distal part enlarged, giving the appearance of four distinct plates. Apparently monophagous on *Bromus erectus* Huds. The larva starts mining in the autumn, making a long, often narrow mine on the underside of the lower leaves, first mining upwards, gradually widening the mine, but turning downwards having reached the tip; excrement in a fine central line. A larva changes leaf once or twice; the ultimate mine occupies the whole blade and is visible from both sides; excrement in parallel lines (Pl.-fig. 515). Pupation in the late spring; females pupate on the distal third of one of the upper leaves, males in a leaf-angle near the ground, in a fine, dense cocoon with semi-circular openings along the lower margins. Adult fly in late May in Germany, often in the dawn, not coming to light. The biotope is wet places on calcareous grounds.

34. ***Elachista cerusella*** (Hübner, 1796)
Plate-figs. 75, 76, 190, 316, 317, 450, 451, 516, 517.

Tinea cerusella Hübner, 1796:63, pl.27, fig. 183.
Elachista monosemiella Rössler, 1881: 325. **Syn.n.**
Elachista larseni Strand, [1927] 1925: 283.

10–12 mm. Male (Pl.-fig. 75). Head, tegulae and thorax white. Scape of antenna white, flagellum greyish white, distal part faintly serrate. Labial palpi recurved, porrect, white, second segment with a brownish streak on underside. Forewing ground colour white, base of costa with a narrow grey streak, sometimes tinged with yellowish in fold-area; marks brownish, a more or less prominent spot near base of dorsum; an outwardly oblique, more or less prominent, often interrupted fascia from middle of costa, widening towards dorsum, with an often distinct black dot on inner side in fold; a similar parallel fascia from three-quarters of costa, not always reaching costa and often interrupted before brown spot beyond tornus, sometimes with a black spot before middle. Basal part of cilia white, tinged with ochreous, cilia line faint, brownish, ending before tornus, distal parts of cilia grey. Hindwing lead-grey, cilia ochreous grey, lighter at apex. Abdomen light grey. Hind tarsus whitish.

Female (Pl.-fig. 76). Flagellum of antenna slightly ringed. Forewing pattern similar to male, but outer oblique fascia often broader.

The distinctness of fascial and black spots varies much; in newly emerged specimens they are often very distinct, but black and brownish scales are often lost during the first flights. Some forms have been named (Dufrane, 1957; Hering, 1963).

Male genitalia (Pl.-figs. 316, 317). Uncus with V-shaped indentation, lobes short and rounded, almost circular. Gnathos transversely elongate. Sides of valva almost parallel, costa strong, hump slightly enlarged before rounded cucullus. Digitate process narrow at base, club-shaped, setose, blunt. Juxta lobes distally enlarged, rounded, setose. Vinculum short, truncate, with a distinct short, blunt saccus, no medial ridge present. Aedeagus base little enlarged, gradually tapering to weakly sclerotized, indented distal end.

Female genitalia (Pl.-figs. 450, 451). Apophyses slender, apophyses posteriores angled inwards from strong basal portion, longer than apophyses anteriores. Antrum short, pouched, constricted towards colliculum, spined inside, ventral margin almost semi-circular, strong, with fine spines, dorsal wall with fine spines in an elongate patch; colliculum wide, shorter than apophyses posteriores. Signum a tiny, finely dentate plate.

Distribution. Common and known from all districts in Denmark; in Sweden only in the southern part up to Boh.; not known in Norway; in Finland from Al and N. – Well known from C. and NW. Europe and Britain, and according to Spuler (1910) also in USSR and Asia Minor.

Biology. Already outlined by Stainton (1858c). Larva yellowish green, head and prothoracic shield brown; on each side of segments two faint dorsal depressions. It is known to mine in several grasses: *Phragmites communis* Trin., *Phalaris arundinacea* L., *Festuca, Dactylis, Poa, Brachypodium* and *Holcus* (Stainton, 1858c; Sønderup, 1949; Lhomme, 1951), in which it makes large, conspicuous whitish green blotches (Pl.-figs. 516, 517), often near the tip of the blade. Excrement is mainly deposited in the upper part of the mine; one larva generally makes more than one mine. Sometimes two or three mines on a single leaf (Machin, 1880). Pupation on a leaf or stem, attached by cremaster and a single girdle. Pupa varying from light brown with whitish abdomen to almost blackish brown, ridges generally whitish grey. Bivoltine; the larvae feed during April-May and in July; adults fly from late May to early July and again in August-September. The biotope is meadows and moist grasslands.

Note. *E. monosemiella* Rössl. Holotype in Museum Wiesbaden labelled "Kollektion Dr. A. Rössler", "TYPE", "Monosemiella", "Holotypus, E. monosemiella Rössler, teste U. Parenti 1976", "Prep. Genitale ♂321, U. Parenti 1976" "Museum Wiesbaden gen.prep. 150" is a *cerusella* without brownish fasciae and with black dot present.

The *bedellella*-group

Ground colour white or whitish with or without blackish spots, plain ochreous, or beige to blackish brown with costal and tornal spots separated or confluent, or with only one fascia in the middle; generally with cilia paler beyond cilia line.

Venation (Pl.-figs. 191–214). Forewing with R_4 and R_5 almost or totally coalescent, M_2 free or coalescent with CuA_1, distal part of CuP sometimes present. Hindwing with Rs rarely with free branch to costa, cell more or less open.

Male genitalia. Uncus slightly to deeply indented, lobes of varying size and shape and always with short or minute setae, generally only with a few setae in a row along distal margin. Anterior margin of tegumen weakly or not reinforced. Costa of valva straight or convex in the middle and with a sub-apical emargination, always without distal hump. Digitate process varying. Juxta lobes often strongly specialized, ventral part of juxta often with a tongue-shaped process. Vinculum without medial ridge and saccus. Base of aedeagus slightly enlarged, with or without cornuti.

Female genitalia. Papillae anales anteriorly sclerotized. Venter VIII sometimes sclerotized. Antrum mainly membranous, colliculum a short ring close to ostium. Corpus bursae with 1–3 band or plate-shaped signa.

The *argentella*-subgroup

Ground colour white, sometimes restricted to broad fasciae by yellow or beige-brown suffusion.

Venation (Pl.-figs. 191-202). Forewing with R_4 and R_5 coalescent or very shortly forked, M_1 from middle of stalk R_{4+5}, M_2 free or coalescent with CuA_1, distal part of CuP sometimes present. Hindwing with Rs generally simple, rarely with free branch to costa.

35. ***Elachista argentella*** (Clerck, 1759)
Text-fig. 4; plate-figs. 77, 191, 192, 318, 319, 452.

[Phalaena] argentella Clerck, 1759, pl.11, fig.13.
Tinea cygnipennella Hübner, 1796: 67, pl.50, fig.207.
?Tinea alabastrella Schrank, 1802: 121.
Porrectaria cygnipennis Haworth, 1828: 536
Adela cygnella Treitschke, 1833: 117.
Aphelosetia semialbella Stephens, 1834: 288.

11–12 mm. Male (Pl.-fig. 77). Head, tegulae and thorax white, tegulae often slightly yellowish-tinged. Antenna white, scales on basal half appressed, very slightly raised on distal half. Labial palpi recurved, underside of second segment with raised scales, third segment smooth, pointed; palpi white, underside sometimes faintly tinged yellowish. Forewing plain white, slightly shining, often tinged with off-white along costa and dorsum. Cilia white; no cilia line present. Hindwing pale greyish white, underside white, cilia white. Abdomen dull whitish, mixed with light grey. Hind tarsus uniformly white.

Female. Similar to male, but forewing more narrow and hindwing lighter.

Male genitalia (Pl.-figs. 318, 319). Uncus longer than tegumen, distally rounded, distal margins sclerotized, with short setae, and with a short V-shaped incision. Gnathos elongate, oval. Valva narrow in the middle, cucullus rounded. Digitate process slender, sides almost parallel, distal end enlarged, set with fine setae. Juxta lobes short, truncate, lateral margin produced, apical margin with very few setae in the middle. Vinculum short, triangular. Aedeagus with basal and medial regions of equal width, square-ended distally; cornutus a dentate, sclerotized plate.

Female genitalia (Pl.-fig. 452). Papillae anales large, with long sparse setae. Apophyses posteriores longer than apophyses anteriores. Sternite VIII with U-shaped, strongly sclerotized medial area around ostium, minutely spined anteriorly. Ventral margin of antrum narrow, distinct, semi-circular, with two setae, dorsal wall without spines; colliculum tapering anteriorly, short, one-quarter length of apophyses posteriores. Anterior membranous portion of ductus bursae and middle part of corpus bursae each with a patch of fine spines; signum an irregular plate with dentate edges. Spermathecal duct with sclerotized ring.

Distribution. Common and known from all parts of Denmark; in Sweden from the most southern districts; not known from Norway and only known from a single locality in Finland: N, Hangö Peninsula. – Widely distributed in W. Palaearctic region.

Biology. The larva is large, dull greenish grey, dorsal line whitish; head pale brown, margins darker, prothoracic shield made up of eight small sclerites in two plates. It is reported from *Agrostis, Brachypodium sylvaticum* (Huds.) PB., *Bromus sterilis* L., *Calamagrostis, Dactylis glomerata* L., *Deschampsia, Elymus arenarius* L., *Festuca rubra* L., *Holcus mollis* L., *Koeleria, Phleum, Poa,* (Lhomme, 1951; Wörz, 1957). Stainton (1858c) reports it preferring broad-leaved species such as *Bromus* and *Dactylis,* in the leaves of which it mines downwards from the tip, during the spring until late May, making a generally short, whitish, transparent, not puckered mine as wide as the blade; the larva often changes leaf. Pupa light brown, ridges prominent, with two distinct dark brown dorso-lateral and two less distinct ventro-lateral elongate lines; attached by cremaster and a single girdle. The adults fly from mid-May to early July, mainly in the first part of June; univoltine. In grasslands, along wood edges and meadows. The adult often rests on the food-plants at sunset, flies low and sluggishly in the late afternoon and in the dusk, and also comes to light.

Note. Clerck (1759) figured *Phalaena argentella,* currently interpreted as here (Zeller, 1853a: 288). Linnaeus (1761: 367) also described a *Phalaena (Tinea) argentella,* without giving reference to Clerck (1759); the type of *argentella* L. (not that of Clerck as mentioned by Svensson (1966: 196)) probably is conspecific with *Blastotere arceuthina* Zeller, 1839, and a junior primary homonym of *P. argentella* Cl. (Karsholt & Nielsen, 1976: 25).

36. ***Elachista pollinariella*** Zeller, 1839

Plate-figs. 78, 79, 193, 320–323, 453–455.

Elachista pollinariella Zeller, 1839: 213.

8–10 mm. Male (Pl.-fig. 78). Head, neck tufts, tegulae and thorax white, anterior part of tegulae yellowish. Scape of antenna white, upperside often with brownish grey spot, flagellum varying from whitish grey to greyish brown; distal third slightly serrate. Labial palpi drooping to porrect, second segment with raised scales on underside and distally, third segment short, pointed; palpi white, underside of second segment brownish. Forewing ground colour white; proximal part of costa grey, greyish brown towards fold (this suffusion can be absent); suffusion of ochreous-tipped scales along veins and beyond middle, often forming an indistinct transverse band; varying number of fine dots on veins made up of black-tipped scales, often surrounded by ochreous. Cilia white, distal part greyish white; cilia line distinct, made up of black-tipped scales. Hindwing light grey, cilia similar. Abdomen whitish, anal tuft ochreous white. Hind tarsus light brown, with narrow white rings.

Female (Pl.-fig. 79) Flagellum of antenna distinctly annulated with narrow brown and broad white rings. Forewing with fewer black dots, ochreous suffusion more pronounced. Hindwing almost white.

Male genitalia (Pl.-figs. 320–323). Uncus almost as long as tegumen, deeply indented; uncus lobes close together, sides with a very few short setae distally. Gnathos oval. Anterior margin of tegumen deeply emarginate. Costa of valva strong, costa convex before middle, concave beyond, cucullus rounded towards costa, distal end of sacculus slightly projecting below cucullus. Digitate process with almost parallel sides, but narrower in the middle, distal third finely setose. Juxta lobes separated by V-shaped incision, lateral margins produced, apical margins medially prolonged, rounded, with a small, elongate group of setae. Vinculum short, arched. Aedeagus only weakly tapering, distal end produced into two blunt processes; cornutus a conglomerate of small teeth on an often weak plate; vesica sclerotized.

Female genitalia (Pl.-figs. 453-455). Papillae anales large, sclerotized, very sparsely setose. Apophyses strong, of almost equal length. Lateral margins and medial area of tergite VIII strongly sclerotized, medial area almost H-shaped. Ostium bursae simple, antrum not sclerotized, colliculum a very short sclerotized ring close to ostium. Signum an almost round plate with serrate edges, central part less sclerotized.

Distribution. In Denmark only a few localities on LFM; not in Norway; only a single specimen from Sweden, Upl., Knivstad, 23.vi.1924, W. Petersen leg., Zool.Inst., Lund coll.; from several districts in S. Finland up to Kb. – Widespread in E., C. and S. Europe, but not in NW. Europe and Britain.

Biology. Larva not described. Reported to mine in several grasses: *Festuca ovina* L., *F. longifolia* Thuill. and other *Festuca*-species, but also from *Trisetum flavescens* (L.) PB., *Brachypodium sylvaticum* (Huds.) PB. and *Poa pratensis* L. Hering (1891) reports the mined leaves to be distended, and Lhomme (1951) states the mine to be transparent, whitish and almost linear; according to Martini (1916) the larva mines both up- and downwards. Univoltine; the larva mines during the spring to late May and the moth is on the wing from late May to mid-July. Mainly on dry, exposed biotopes like dikes and sandfields.

Note. As pointed out by Traugott-Olsen (1973) all specimens previously reported from Sweden by Benander (1937) as *pollinariella* are misidentifications of *E. subocellea* (Steph.).

37. ***Elachista triatomea*** (Haworth, 1828)
Plate-figs. 80, 81, 194, 324, 325, 456.

Porrectaria triatomea Haworth, 1828: 535.
Elachista triatomella Morris, 1870: 227.

8–9 mm. Male (Pl.-fig. 80). Head and neck tufts white; tegulae and thorax white mixed with yellowish brown scales on anterior parts. Scape of antenna white, flagellum whitish brown. Labial palpi recurved, second segment rough on underside towards third segment, white, underside of second segment suffused with light brownish, distal end white. Forewing ground colour white, base of costa often darker greyish, with scattered pale brown or yellowish brown scales often in small blotches, not forming a distinct pattern; in the fold, just in the middle of the wing, a distinct black dot or streak, and between this and apex a similar rounded spot. Cilia white, distally whitish grey; cilia line weak, dark brown. Hindwing light grey, cilia similar. Abdomen whitish grey, anal tuft ochreous-grey. Hind tarsus white.

Female (Pl.-fig. 81) Tegulae and thorax white. Flagellum of antenna annulated brown/greyish white. Forewing white, almost without suffusion of darker scales, except for the two black dots or streaks and a few scattered yellowish brown scales along termen. Hindwing whitish, cilia similar. Abdomen pale yellowish white.

Male genitalia (Pl.-figs. 324, 325). Uncus lobes conical, distally rounded, with only a few small setae; medial margins rather strongly sclerotized. Gnathos long and slender. Tegumen narrow. Costa of valva strong, slightly convex in the middle; cucullus rounded, with rather short and strong setae, sacculus almost straight. Digitate process club-shaped, blunt, strong, distal half with fine setae. Juxta lobes slender at base, distally truncate, lateral margins strongly pointed, apical margins straight, laterally with 2–4 small setae. Vinculum strong, rounded and short. Aedeagus straight, slightly tapering towards distal end but enlarged beyond middle, with a single thorn-like cornutus.

Female genitalia (Pl.-fig. 456). Papillae anales short, with short setae. Apophyses posteriores enlarged near base, almost twice as long as apophyses anteriores. Tergite VIII with straight margins and a more strongly sclerotized U-shaped area. Antrum not sclerotized, except the narrow, distinct semi-circular ventral margin, dorsal wall with a few minute spines; colliculum a short sclerotized ring close to ostium. Anterior half of membranous ductus bursae with fine spines; corpus bursae with two patches of fine spines, no signum present.

Distribution. Known from almost all districts in Denmark; in Sweden up to Dlr. and Upl.; not in Norway; in Finland only from Al. – Otherwise from W. USSR to Britain and from C. Europe.

Biology. Larva slender, yellowish grey, slightly tinged olive-green, head pale brown, lateral margins darker, prothoracic shield lighter than head, with two yellowish brown streaks (Hering, 1891). *Festuca ovina* L., *F. rubra* L., and *F. tenuifolia* Sibth. are reported as food plants. The larva first starts mining from the tip of the leaf, making a 5–7 cm long mine downwards; if it changes mine, the new one is started a little below the former, the larva now mining upwards. At first the mine is yellowish grey with a violet reflection, later it turns whitish, and is easily seen. Univoltine; the larva mines from mid-April to mid-June and adults fly from late June to early August, but mainly in the first part of July. Mainly on open grasslands, hillsides and meadows.

Elachista collitella (Duponchel, 1843)

Plate-figs. 82, 83, 195, 326, 327, 457.

Oecophora collitella Duponchel *in* Godart, [1843] 1842: 327.
Poeciloptilia grossepunctella Herrich-Schäffer, 1855: 302, 312.

7–9 mm. Male (Pl.-fig. 82). Frons white, clypeus and vertex lead-grey, neck tufts whitish, scales tipped with ochreous grey, tegulae and thorax whitish, anteriorly suffused with greyish. Antenna dark grey, slightly annulated. Labial palpi porrect, straight, upperside white, underside lead-grey. Forewing ground colour white, forming a white fascia before middle and one beyond, otherwise with beige suffusion; basal half of costa lead-grey, beige towards fold, between fasciae grey on costa, beige towards dorsum, beyond outer fascia beige to white submarginal streak; on veins a varying number of small black dots each made up of 2–3 black-tipped scales. Cilia on costa grey along beige area beyond outer fascia, from apex to middle of termen white with greyish tips, and grey around tornus; cilia line distinct, blackish brown. Hindwing grey, cilia similar, whiter at apex. Abdomen grey, posterior margin of each segment suffused with bluish white. Hind tarsus light grey, distal part of segments with a narrow white ring.

Female (Pl.-fig. 83). Head and thorax almost white, suffused with ochreous. Antenna distinctly annulated whitish/brownish. Forewing generally paler.

Male genitalia (Pl.-figs. 326, 327). Uncus lobes elongate, triangular, most distal part set with a few minute setae. Gnathos elongate. Costa of valva strong at base, concave from before middle to cucullus; cucullus prominent, rounded; sacculus almost straight, but slightly incurved beyond middle. Digitate process club-shaped, distal portion only slightly enlarged, sparsely setose almost from base. Juxta lobes truncate, lateral margin with a short point, apical margins curved, with a small group of fine setae. Vinculum short, rounded. Aedeagus short and thick, tapering towards distal end, with a longitudinal fold in the distal portion; cornutus an elongate conglomerate of multiple elongate teeth.

Female genitalia (Pl.-fig. 457). Apophyses posteriores slightly longer than apophyses anteriores. Tergite VIII with a broad, U-shaped, more strongly sclerotized area. Antrum not sclerotized, ostium bursae without sclerotized margins, dorsal wall without spines; colliculum weak and short. Ductus bursae wide, middle portion with fine internal spines; corpus bursae without spines or signum. Receptaculum seminalis sclerotized.

Distribution. Not in Scandinavia or Britain, but otherwise widespread from Belgium and France through C. and S. Europe to Asia Minor.

Biology. Larva more or less light greyish green with side stripes, head brown, mouth-parts darker; prothoracic shield dark brown, made up two distally enlarged narrow plates (Martini, 1912c). It is known from *Festuca ovina* L., *Poa pratensis* L. and *Koeleria cristata* (L.) Pers., mainly preferring the first, in which it mines downwards from the tip; mine narrow. Bivoltine; larvae mining from autumn to late spring and in June-July; adults from mid-May to mid-June and in July. On dry grasslands.

Note: Herrich-Schäffer (1855: 312) based his description of *P. grossepunctella* on specimens collected by O. Hoffmann at Muggendorf. In the Hoffmann collection in the British Museum (Nat.Hist.) two micropinned syntypic specimens are present with a mutual label inscribed "Muggend. 3.7.55". From these a lectotype is hereby designated and further labelled "B.M. ♂ Genitalia Slide No. 19 762". The lectotype is conspecific with the type of *Oecophora collitella* Dup. examined by U. Parenti.

38. ***Elachista subocellea*** (Stephens, 1834)
Plate-figs. 84, 85, 196, 328, 329, 458.

Aphelosetia subocellea Stephens, 1834: 290.
Elachista (Ornix) flammeaepennella Costa, [1836]: [298], [313].
Elachista pollinariella Zeller *sensu* Stainton, 1851: 10.
Poeciloptilia disertella Herrich-Schäffer, 1855: 302, 311.
Elachista subcollutella Toll, 1936: 409. **Syn.n.**

8–10 mm. Male (Pl.-fig. 84). Head and neck tufts white, tegulae and thorax white, slightly suffused anteriorly with light ochreous. Antenna pale ochreous brown, faintly annulated with whitish. Labial palpi slender, porrect, slightly curved, white, underside of second segment mixed with ochreous. Forewing ground colour white, forming two fasciae, one before and one beyond the middle; pale yellowish suffusion; basal area greyish at costa, brownish towards fold and yellowish towards dorsum, dorsum often white, light yellow between fasciae; a light yellow band beyond outer fascia mixed with light brownish towards costa, separated from termen by a white subterminal streak, mixed with brown-tipped scales along termen; a varying number of scattered blackish brown dots over whole wing, each made up of one or two scales. Cilia white, most distal part light brownish grey beyond the distinct dark brown cilia line, grey around tornus. Hindwing brownish grey, cilia lighter. Abdomen grey, posterior margins of segments and anal tuft whitish. Hind tarsus light ochreous-brownish, with broad white band on distal end of segments.

Female (Pl.-fig. 85). Smaller than male. Antenna distinctly ringed narrow dark brown/broad light brownish. Forewing with less grey at costa base, light yellow suffusion more pronounced, accentuating the broad white inner and the more narrow outer fascia.

Male genitalia (Pl.-figs. 328, 329). Uncus indentation deep, V-shaped; uncus lobes almost triangular, lateral margin slightly concave; distal ends with a few minute setae. Gnathos elongate, tapering. Anterior margin of tegumen strong. Costa of valva strong, convex before middle, concave beyond middle to rounded cucullus. Digitate process club-shaped, with short setae from before middle. Juxta lobes enlarged distally, lateral margins not significantly produced, apical margins irregularly crenellated, finely setose. Vinculum short, rounded. Aedeagus short, broad, gradually tapering, with one prominent thorn-shaped cornutus, concave side serrate with fine teeth at base.

Female genitalia (Pl.-fig. 458). Apophyses very slender, posteriores almost twice as long as anteriores. Lateral and anterior parts of tergite VIII most strongly sclerotized. Posterior margin of sternite VII set with scattered long setae. Antrum not sclerotized; ventral margin a short, straight line, dorsal wall without spines; colliculum short, sclerotized with two longitudinal folds, close to ostium; posterior part of membranous ductus bursae with many longitudinal folds, anterior two-thirds and posterior area of corpus bursae with minute spines; no signum present. Receptaculum seminalis sclerotized.

Distribution. Not in Denmark; in Sweden only from Gtl.; not in Norway; in Finland only from Ta. – Widely distributed in W. Palaearctic region, but not from the SW. and Holland.

Biology. The larva is light dull green, head and prothoracic shield light brown, shield made up of two narrow, elongate, distally enlarged plates (Martini, 1912c).

Only *Brachypodium sylvaticum* (Huds.) PB. is reported as food-plant; the larva first mines upwards from a little below the tip, making a narrow *Stigmella*-like mine, turns at the tip, and mines downwards making a broad 5–8 cm long *Phyllonorycter*-like mine, which is greenish and difficult to see, very similar to the mine of *E. chrysodesmella* Zell. (Steuer, 1976a). Univoltine in N. Europe; larvae mining from late April to early June, adults present from late June to early August; probably bivoltine in S. Europe. Mainly on open exposed places, along wood edges, but also in partly shaded places in open forests.

Note. The description of *E. subcollutella* Toll was based on a single specimen, which it has not been possible to trace, as it is not present in the Toll collection at PAN, Kraków, and probably was not present in the collection when deposited there (Razowski, *in litt.*). From the detailed description and the figure there is no doubt that it is conspecific with *E. subocellea* (Stph.).

39. ***Elachista festucicolella*** Zeller, 1853
Plate-figs. 86, 87, 197, 330, 331, 459.

Elachista festucicolella Zeller, 1853: 415.

7–9 mm. Male (Pl.-fig. 86). Head, neck tufts, tegulae and thorax white. Scape of antenna white, flagellum light brown. Labial palpi short, strong, third segment very short; third segment and distal part of second segment white, basal portion of second segment light brownish. Forewing ground colour white, without any marks, but very faintly tinged with yellowish along costa, termen and dorsum. Cilia white, without cilia line. Hindwing whitish grey, cilia white. Abdomen white. Hind tarsus whitish.

Female (Pl.-fig. 87). Proximal three segments of flagellum white. Hindwing white.

Male genitalia (Pl.-figs. 330, 331). Uncus lobes rather short, distally tapering, distal ends with 2–3 minute setae. Gnathos small, flat, not elongate, with a long basal portion. Tegumen widening basally. Base of valva enlarged, costa and sacculus almost straight and parallel beyond base, but costa slightly convex before middle and sacculus concave near base, cucullus rounded. Digitate process slightly enlarged in the middle, finely setose from before middle. Juxta lobes very narrow at base, lateral margins produced into long tapering processes, medial regions rounded, enlarged, not setose, apical margin of lateral processes with minute setae; ventral regions of juxta with a narrow tongue-shaped process. Vinculum elongate, triangular, narrow, distally blunt. Aedeagus very long, curved, gradually tapering to chisel-shaped distal end; with a row of 7–9 long cornuti on a common plate, size of cornuti increasing towards distal end.

Female genitalia (Pl.-fig. 459). Mediobasal region of papillae anales strongly setose. Apophyses posteriores enlarged at one-fifth from base, about twice as long as apophyses anteriores. Antrum not sclerotized, margins of ostium not sclerotized, dorsal wall without spines; colliculum a weakly sclerotized ring close to ostium. Ductus bursae rather wide, medial portion with many transverse folds, anterior portion with minute internal spines. Signum with an almost right-angled bend in the middle, concave side with strong teeth. On pl.-fig. 459 an elongate spermatophore is present in the anterior part of the ductus.

Distribution. Not in Denmark, Norway and Finland; in Sweden only from Gtl. – Widely distributed in W. Palearctic from Wales, Britain (Pelham-Clinton, pers. comm.), Belgium and France through C. and S. Europe to Asia Minor.

Biology. Larva and mine apparently not described. The larvae mine in *Festuca ovina* L. (Lhomme, 1951), *F. sulcata* (Hack.) Patzke (Klimesch, 1961) and probably other *Festuca* species. Univoltine in N. Europe; adults fly from late May to early July, mainly in the middle of June. On open calcareous land.

Elachista nitidulella (Herrich-Schäffer, 1855)
Plate-figs. 88, 89, 198, 332, 333, 460.

Poeciloptilia nitidulella Herrich-Schäffer, 1855: 302, 314.

6–7 mm. Male (Pl.-fig. 88). Head, neck tufts, tegulae and thorax greyish white, thorax and tegulae tinged yellowish. Scape and pedicel of antenna greyish white, flagellum grey, with a whitish lustre. Labial palpi short, straight, pointed; upperside whitish, underside suffused with grey. Forewing ground colour white, with or without very faint yellowish sheen. Cilia white, slightly darker at tornus, without cilia line. Hindwing pale grey, cilia lighter grey. Abdomen pale greyish, anal tuft lighter. Inner side of hind tarsus white, outer side grey, with narrow white distal area on each segment.

Female (Pl.-fig. 89). Head grey, neck tufts greyish white, anterior part of tegulae and thorax lead-grey. Underside of labial palpi grey, often forming rings on upperside. Forewing white as in male, but basal edge of costa lead-grey.

Male genitalia (Pl.-figs. 332, 333). Uncus lobes narrow, elongate, widest in the middle, mediobasal region strongly sclerotized; distal end triangular with a few short setae. Gnathos short, enlarged transversely. Valva narrow and long, base enlarged, sides almost parallel from beyond base to rounded cucullus. Digitate process with parallel sides, blunt, setose from before middle. Juxta lobes tapering laterally into narrow processes, apical margins rounded, not setose; a narrow, tapering tongue-shaped process from ventral part of juxta. Base of vinculum broad, abruptly tapering. Aedeagus thick at base, tapering, curved, distal end pointed; with a row of pointed cornuti of increasing length.

Female genitalia (Pl.-fig. 460). Basal part of lateral margins of papillae anales setose. Apophyses posteriores about twice as long as apophyses anteriores. Tergite VIII with 8-shaped strongly sclerotized area. Antrum not sclerotized, ventral margin a very weak line; colliculum a weakly sclerotized ring well before ostium. Ductus bursae with many transverse folds at one-quarter from anterior end; anterior quarter of ductus and corpus bursae with minute internal spines; signum a finely dentate oval plate.

Distribution. Not in Scandinavia or Britain, but widely distributed in C. Europe from Belgium and France to Hungary and Italy.

Biology. Larva, food-plant and mine unknown; according to Lhomme (1951) the larvae mine during May, and adults are reported from mid-May to the end of July; probably univoltine.

Note. Until now all searches for syntypic specimens of *nitidulella* have been in vain, but a male from the Staudinger collection at the Humboldt Museum, Berlin labelled "nitid", "natl. Juttar", "Origin" (pink), "83", "Gen. präp. Gaed. 1411" confirms the present identity of this species.

40. ***Elachista dispilella*** Zeller, 1839

Plate-figs. 90, 91, 199, 334, 335, 461.

Elachista dispilella Zeller, 1839: 213.
Elachista distigmatella Frey, 1859: 302.

7–9 mm. Male (Pl.-fig. 90). Head, neck tufts, tegulae and thorax white, tegulae very faintly yellowish-tinged anteriorly. Scape and pedicel of antenna white, flagellum light brown, with a whitish lustre. Labial palpi straight to slightly curved, porrect, second segment rough, third segment smooth; palpi white above, underside of second segment brownish grey and of third segment white. Forewing ground colour white; basal edge of costa blackish grey, margins very faintly tinged yellowish; with two distinct black spots: an elongate or rounded spot in the fold beyond the middle of the wing, and a round spot displaced towards costa between this and apex; otherwise without yellowish or dark scales. Cilia white, without cilia line. Hindwing white, cilia similar. Abdomen and hind tarsus white.

Female (Pl.-fig. 91). Slightly smaller than male.

Male genitalia (Pl.-figs. 334, 335). Uncus lobes elongate, tapering, triangular, medial margins curved and basally sclerotized; distal ends with a few small setae. Gnathos large, oval. Valva narrow, long, base enlarged, beyond this sides almost parallel to rounded cucullus. Digitate process rather slender, blunt, distal part with few scattered setae. Juxta lobes short, lateral margins with short stout processes, apical margins rounded, without setae. Vinculum triangular. Aedeagus bent, tapering, distal portion long and pointed; with about five rather short, pointed cornuti on a common plate.

Female genitalia (Pl.-fig. 461). Basal part of lateral margins of papillae anales with a group of long setae. Apophyses posteriores enlarged near base, longer than anteriores. Tergite VIII with more strongly sclerotized 8-shaped area. Antrum not sclerotized, margins not sclerotized, two patches of minute spines to sides of ostium; colliculum a barrel-shaped sclerotized region just before ostium. Ductus bursae with transverse folds in the middle and just before corpus bursae, most anterior part and a small part of corpus with minute internal spines; signum weakly sclerotized, elongate, narrow, dentate.

Distribution. From several districts in Denmark; from many districts in S. Sweden up to Upl.; in Norway only from Ø; in Finland from Al and N. – Not in Britain, but otherwise widespread in Europe.

Biology. The larva is very slender, greenish white, head brown, mouth-parts darker, prothoracic shield grooved, thoracic legs brown (Frey, 1859). Zeller (1839) reported it from dry places with *Corynephorus canescens* (L.) PB., but subsequent authors all report *Festuca ovina* L. and a few other *Festuca* species as food-plants, in which the larva makes a long narrow whitish mine from the tip of the blade downwards towards the base. The pupa is slender, light yellowish brown. The larva mines during May-June (Sønderup, 1949) and the adults fly from mid-June to early August, mainly in the middle part of July; probably univoltine in Scandinavia, but otherwise reported as bivoltine (Hering, 1891; Lhomme, 1951). The adults fly over the food-plant after sunset to the late dusk until very difficult to observe; mainly on meadows.

41. ***Elachista triseriatella*** Stainton, 1854
Plate-figs. 92, 93, 200, 336, 337, 462.

Elachista triseriatella Stainton, 1854a: 261.

7–10 mm. Male (Pl.-fig. 92). Head, neck tufts, tegulae and thorax white, tegulae slightly suffused anteriorly with grey. Antenna serrated distally, scape and pedicel white, flagellum light brown. Labial palpi rather long and slender, almost straight, porrect, off-white. Forewing ground colour white, very faintly bluish-tinged, basal edge of costa blackish grey; with scattered black-brown scales in varying number in the disc and along termen in basal part of cilia; in the fold beyond the middle a small blackish dot and between this and apex a similar, often inconspicuous dot. Cilia white, no cilia line present. Hindwing whitish grey, faintly tinged bluish, cilia similar. Abdomen and hind tarsus whitish.

Female (Pl.-fig. 93). First five segments of flagellum white. Base of costa not dark grey, and generally with scattered dark scales on forewing, distal dot often absent.

Male genitalia (Pl.-figs. 336, 337). Uncus lobes short, stout, rather wide apart, mediobasal portion well sclerotized; tips rounded, with a few short setae. Gnathos long, slender, distally curved. Valva enlarged at base, narrow beyond base, with almost parallel sides to slightly enlarged rounded cucullus. Digitate process club-shaped, blunt, distal third with short setae. Juxta lobes short, narrow, not enlarged distally, apical margins with few short setae. Vinculum triangular, pointed. Aedeagus a coniform, strongly tapering tube to pointed, often strongly curved apex; without cornuti.

Female genitalia (Pl.-fig. 462). Basal part of lateral margins of papillae anales with a group of long setae. Apophyses posteriores enlarged near base, longer than apophyses anteriores. Tergite VIII with sclerotized 8-shaped area. Ostium bursae a transverse slit, dorsal wall posteriorly sclerotized and with minute spines, ventral margin membranous, straight; colliculum a short weakly sclerotized ring close to ostium. Ductus bursae without folds and spines; signum weakly sclerotized, with short broad triangular teeth.

Distribution. Not in Denmark, Norway and Finland; in Sweden from several districts in the southern part up to Vstm. and Upl., but not from the isles in the Baltic. – Otherwise from Britain, but further records from the Continent are unreliable owing to confusion with *E. dispunctella* (Dup.).

Biology. No reliable records of immature stages known; Hering (1891) reports larvae from *Festuca ovina* L. and *F. longifolia* Thuill. in April-May, but whether this concerns *E. dispunctella* (Dup.) or *E. triseriatella* Stt. cannot be stated. Probably univoltine; in Sweden caught from mid-May to mid-July.

Note. Separation of *E. triseriatella* Stt. and *E. dispunctella* (Dup.) was demonstrated by Svensson (1966).

Elachista dispunctella (Duponchel, 1843)
Plate-figs. 94, 95, 201, 338, 339, 463.

Oecophora dispunctella Duponchel *in* Godart, [1843] 1842: 333.

7–10 mm. Malc (Pl.-fig. 94). Head, neck tufts, tegulae and thorax white, tegulae anteriorly slightly suffused with greyish. Antenna smooth, scape and first segments of flagellum white, otherwise dark brownish grey. Labial palpi rather short, straight, porrect, off-white, underside of second segment slightly tinged yellowish. Forewing ground colour white, very faintly tinged bluish, with scattered black scales in rows along veins and along distal margins; a black streak in fold beyond middle, and a black dot between this and apex. Cilia white, without cilia-line. Hindwing pale greyish, cilia white around apex, faintly tinged ochreous along costa and dorsum. Abdomen greyish white. Hind tarsus white.

Female (Pl.-fig. 95). Head, tegulae and thorax suffused with greyish brown, neck tufts white. Scape of antenna brownish, flagellum distinctly ringed wide brown/ narrow white. Hindwing greyish, cilia white.

Male genitalia (Pl.-figs. 338, 339). Uncus lobes short, stout, truncate, distal end with few short setae. Gnathos teardrop-shaped. Base of valva enlarged, valva narrow with parallel sides beyond base, cucullus rounded, not enlarged. Digitate process very slender, distal third setose. Juxta lobes narrow, apically rounded, not setose. Vinculum strong, distal end shaped like a duck's beak. Aedeagus very long and slender, tapering from base to pointed distal end; without cornuti.

Female genitalia (Pl.-fig. 463). Apophyses strong, distally with a knob, posteriores about twice as long as incurved anteriores. Tergite VIII broad, posterior margin deeply emarginate. Medial portion of sternite VIII most strongly sclerotized, forming an almost circular margin around ostium bursae; sternite VII broad, triangular. Ventral margin of antrum strong, almost circular; colliculum a short weakly sclerotized ring close to ostium. Ductus bursae without folds and spines, corpus bursae with two twined bands of strong spines.

Distribution. Not known from Scandinavia. Known from Britain, France, Spain, Austria and Germany, but many records are unreliable owing to confusion with *E. triseriatella* Stt.

Biology. Very little is known about immature stages. Probably bivoltine, larvae mining in *Festuca* from September to April and in June-July; adults fly from late April to late June and in July- early September. On dry and open grasslands.

Elachista rudectella Stainton, 1851
Plate-figs. 96, 97, 202, 340, 341, 464.

Elachista rudectella Stainton, 1851: 26.

9–11 mm. Male (Pl.-fig. 96). Head creamy white, vertex often mixed with a few brownish scales; neck tufts, tegulae and thorax creamy white mixed with light brownish grey, anterior margins of tegulae white. Scape of antenna creamy white with a few darker scales, flagellum almost plain greyish brown. Labial palpi porrect, slightly curved, creamy white, underside of second segment greyish brown. Forewing ground colour bone-white, forming an outwards V-shaped fascia from one-third of costa and more or less blurred costal and tornal spots; costal spot beyond tornal; except for fascia and spots whole forewing suffused with beige-tipped scales also bordering small dark brown dots, the latter mainly in fold and terminal area. Cilia white, but from below apex to tornus most distal part dark grey; cilia line distinct, black. Hindwing pale greyish, cilia whitish, base of cilia tinged yellowish. Abdomen dark greyish, posterior margins of segments lighter, anal tuft dark grey. Inner side of hind tarsus creamy white, outer side grey with whitish tips to each segment.

Female (Pl. fig. 97). Slightly smaller. Flagellum of antenna distinctly annulated narrow grey/broader whitish on proximal half, distal half slightly serrate, similar in colour to male.

Male genitalia (Pl.-figs. 340, 341). Uncus lobes triangular, mediobasal region strongly sclerotized; distal medial margin concave, tips of lobes with several small setae. Gnathos oval. Anterior margin of tegumen broad, strongly sclerotized. Basal half of valva broad, beyond this both costa and sacculus are concave, cucullus rounded. Digitate process short, stout, distal part with short setae. Juxta lobes slender, distally wide apart, with short lateral setose processes; ventral portion of juxta with a short tongue-shaped process. Vinculum short, triangular. Aedeagus short, extremely broad, distal margin with 2—3 strongly sclerotized regions, one of which is triangular, tooth-shaped; one prominent spoon-shaped cornutus.

Female genitalia (Pl.-fig. 464). Papillae anales with long basal setae. Apophyses posteriores one and a half times as long as apophyses anteriores. Antrum wide, funnel-shaped, mainly membraneous, almost as long as apophyses posteriores, ventral margin distinct, sclerotized, dorsal margin with several small sclerotized plates, antrum with longitudinal folds and an elongate sclerotized area; colliculum not distinct. Medial area of ductus bursae with a patch of fine spines; corpus bursae with two or three bands of pointed teeth.

Distribution. Not in Scandinavia or Britain. – Otherwise widely distributed in C.Europe from France and central Germany to Italy and the Balkan Peninsula.

Biology. Immature stages described by Hoffmann (1893). Larva dark grey, head brownish, prothoracic shield blackish. It mines the upper half of the leaves of *Phleum phleoides* (L.) Karst, during the spring, making mines as wide as the blade, often reaching to the tip, the upper epidermis is often loose, brownish, giving the mine a *Phyllonorycter*-like appearance; larvae often change mines. The pupa is light yellowish brown, dorsal ridge low, lateral ridges prominent, each segment with a pair of tubercles with a short spine (probably produced spiracles). The adults fly from late May to early July in C.Europe, but Lhomme (1951) reports it as bivoltine. The biotop is dry, exposed calcareous hillsides.

Note. In the Stainton collection in the British Museum (Nat. Hist.) two syntypes are present. A specimen without abdomen labelled "494", "rudectella", "Syntype", "Syntype 2/2, Elachista rudectella Stt., teste K. Sattler, 1975" is hereby designated as lectotype.

The *bedellella*-subgroup

Forewing greyish brown, with two more or less distinct fasciae or one fascia and two spots beyond the middle.

Venation (Pl.-figs. 203–206). Forewing with R_4 and R_5 coalescent, M_1 from middle of R_{4+5}, M two-branched, CuA_2 opposite space R_1–R_2. Venter VIII sclerotized in female.

Elachista squamosella (Herrich-Schäffer, 1855)
Plate-figs. 98, 99, 203, 342, 343, 465.

Poeciloptilia squamosella Herrich-Schäffer, 1855: 300, 305.

8–9 mm. Male (Pl.-fig. 98). Head bone-white with a few grey-tipped scales on vertex; tegulae bone-white, anteriorly mottled with grey-tipped scales, posterior margins whitish; thorax mottled greyish. Flagellum of antenna finely serrate distally, faintly annulated pale greyish/brownish, darkest at base. Labial palpi porrect, slightly curved, third segment relatively short; palpi white above, third segment whitish on underside with a few darker scales, second segment distally white, basally dark grey. Forewing ground colour mottled greyish by scales with long whitish bases and almost black tips; base of dorsum faintly tinged with ochreous; a whitish fascia from two-fifths of costa bent before fold in obtuse angle, a light blurred costal spot at five-sixths, well beyond the more prominent tornal spot. Cilia grey along costa to apex, white from apex to middle termen beyond cilia line, grey from middle of termen, but whitish outside tornal spot; cilia line distinct, broad, black. Hindwing grey, cilia similar. Abdomen grey, anal tuft lighter. Hind tarsus dark grey, outer side with broad whitish distal area on each segment.

Female (Pl.-fig. 99). Annulations of antenna more distinct. Forewing fascia straighter, spots more prominent, white. Hindwing slightly lighter grey.

Male genitalia (Pl.-figs. 342, 343). Uncus lobes elongate, triangular, with short distal setae. Gnathos elongate, tapering. Anterior margin of tegumen strongly sclerotized. Proximal half of valva widest, costa convex, cucullus rather square, sacculus slightly concave. Digitate process club-shaped, narrow at base, tip swollen and with short setae. Juxta lobes with short, narrow lateral processes, medial margins rounded, slightly produced, apical margins curved, with a few setae. Vinculum very short, triangular. Aedeagus bent, gradually tapering to chisel-shaped distal end; without cornuti.

Female genitalia (Pl.-fig. 465). Papillae anales elongate, triangular. Apophyses posteriores enlarged near base, almost twice as long as apophyses anteriores. Antrum short, except for the wide V-shaped ventral margin, weakly sclerotized, dorsal wall finely spined posteriorly; colliculum short, narrow, very weakly sclerotized. Signa two weakly sclerotized patches with pointed teeth.

Distribution. Not in Scandinavia or Britain. Otherwise widely distributed in C. Europe from France and SW. Germany to Czechoslovakia and Dalmatia.

Biology. Only record of early stages is the incomplete one given by Martini (1916), according to which the larva mines similar to that of *E. luticomella* Zell. in the leaves and stem of *Carex montana* L., but this has not since been confirmed. The adults fly from early May to mid-June and from late July and August.

Notes. Herrich-Schäffer (1855: 305) did not cite *Oecophora squamosella* Duponchel [1843] 1842: 334 when describing *Poeciloptilia squamosella*. Later Joannis (1915) attributed *P. squamosella* HS. to Duponchel, but from the descriptions and figures the two taxa cannot be conspecific. Apparently no type material of *O. squamosella* Dup. is present (Joannis, 1915; Parenti, *in litt.*), and this species probably does not belong to the Elachistidae.

In the Staudinger collection, Humboldt Museum, Berlin a specimen labelled "squamosella H.S.", "origin," "Aussig ende Mai," "Gen. Præp. Gaed. 1414" is present; even if not a syntype, it further confirms the present identity.

Wallengren (1875: 76) reported *squamosella* from Sweden, but this was a misidentification of *E. subnigrella* Dougl.

42. ***Elachista bedellella*** (Sircom, 1848)
Plate-figs. 100, 101, 204, 344–346, 466, 467.

Microsetia bedellella Sircom, 1848: 2037.
Poeciloptilia truncatella Herrich-Schäffer, 1855: 300, 305.
Poeciloptilia nigrella Herrich-Schäffer, 1855: 305.
Elachista lugdunensis Frey, 1859: 291.

7–8 mm. Male (Pl.-fig. 100). Frons whitish, clypeus and vertex mottled with greyish brown; neck tufts, tegulae and thorax greyish brown, posterior margin of tegulae and thorax whitish. Antenna finely serrate distally, annulated beige/dark brown, darkest towards base. Labial palpi porrect, curved, upperside white, underside dark greyish, third segment with scattered white scales on underside and junction between second and third segments white. Forewing rather short and broad, ground colour mottled greyish brown with blurred pattern (ground colour of second brood darker than that of first brood); before middle an indistinct lighter fascia and beyond middle a similar sinuous fascia, generally only conspicuous as almost opposite costal and tornal spots. Cilia on costa greyish before apex, white from apex to middle of termen beyond cilia line, greyish around tornus and lighter along dorsum; cilia line distinct, broad, black. Hindwing greyish brown as forewing, cilia similar. Abdomen dark grey, posterior part of each segment and anal tuft lighter. Hind tarsus dark grey, with lighter grey distal rings.

Female (Pl.-fig. 101). Head, tegulae and thorax pale ochreous white. Scape of antenna as head, flagellum distinctly annulated brown/cream-white. Forewing lighter, ochreous brownish, with more distinct marks, inner fascia distinct, outer fascia often interrupted in the middle by scales of ground colour. Hindwing light greyish brown.

Male genitalia (Pl.-figs. 344–346). Uncus lobes almost oval, widest one-third from base, medial margin most strongly sclerotized, along distal part of outer margin a few (4–6) very small setae. Gnathos elongate, oval. Tegumen tapering distally, anterior margin strongly sclerotized. Costa of valva convex, margin irregular, cucullus prominent, distal margin straight, sacculus slightly curved. Digitate process very short, club-shaped, distal end swollen, setose. Juxta lobes with distinct tapering lateral processes; apical margins medially produced into long, bent, tapering processes with fine distal setae; ventral portion of anellus with a short tongue-shaped processes. Vinculum short. Aedeagus short, straight, almost bottle-shaped, distal end often obliquely folded; with one thorn-shaped cornutus.

Female genitalia (Pl.-figs. 466, 467). Papillae anales short, rounded. Apophyses posteriores about twice as long as apophyses anteriores. Tergite VIII broad, with a wide, U-shaped, rather strongly sclerotized area. Anterior margin of sternite VIII extended. Ostium bursae rounded, almost in the middle of sternite VIII, ventral margin of antrum strongly reinforced, antrum otherwise not produced; colliculum a short sclerotized ring just before ostium; ductus bursae narrow and membranous in a short portion from colliculum to inception of ductus seminalis, then widening to a broader, slightly sclerotized portion; ductus otherwise very narrow and only dilated in the middle. Corpus bursae with two elongate patches with minute spines, one of these often with a distinct rounded signum.

Distribution. From all parts of Denmark; from several districts in S. Sweden up to Upl.; not in Norway; from only few districts in S. Finland. – Otherwise from Britain and widely distributed on the Continent.

Biology. Larva yellowish at first, later greyish green, dorsally dark green, head narrow, pale brown, darkest towards mouth, prothoracic shield enlarged posteriorly, brownish (Stainton, 1858c; Hoffmann, 1893). It mines the leaves of *Arrhenatherum pratense* (L.) Sampaio, *Poa trivialis* L., *Festuca ovina* L. and *Phleum* (Sønderup, 1949; Lhomme, 1951), mining the distal half of the leaf, starting from the tip; the mine is often difficult to observe, even if the upper epidermis gets a whitish, puckered appearance and the underside epidermis becomes more or less tinged with purple; the young larva looks dark inside the mine, but, fully grown, looks yellowish; one larva often mines two or three leaves in succession. Pupation on leaf-base or stem; pupa small, light yellowish, almost circular, dorsal and lateral ridges only distinct on posterior abdominal segments, thorax with lateral tubercles, wing smooth. Bivoltine; larvae mine from September to late April, hibernating in the mine, and in June-July, adults fly from mid-May to mid-June and in the first part of August. On dry, exposed, often calcareous, open habitats.

Notes. *M. bedellella* Sirc. According to Sherborn (1940) the Sircom collection should be in the Bristol Museum, but all attempts to find the collection since have been unsuccessful, and it must now be considered lost. In order to stabilise the identity of *E. bedellella* (Sirc.) the male genitalia slide pictured by Pierce & Metcalfe (1935) is hereby designated as neotype, as it corresponds with the general accepted interpretation of *E. bedellella* (Sirc.). The slide is labelled "2647," "♂ Neotype," "Genitalia No. E1 8a, Bedellella, acv era JWM, F. N. Pierce, F.E.S., Liverpool", in the British Museum (Nat. Hist.).

P. truncatella HS. Herrich-Schäffer (1855: 305) attributed this species to Zeller, and further proposed the replacement name *nigrella*. A specimen labelled "truncatella", "Zeller Coll., Walsingham Collection, 1910-427", "B.M. ♂ Genitalia Slide No. 20388" in the British Museum (Nat. Hist.) is hereby designated as lectotype.

E. lugdunensis Frey. Specimens under this name from C. Europe are lighter grey with a more blurred pattern, but both male and female genitalia correspond to those of *bedellella*.

The specimens reported as *E. lugdunensis* Frey by Larsen (1916: 235; 1927: 148) are all *E. anserinella* Zell., while *E. lugdunensis* Frey listed by Benander (1946: 58) are *E. bedellella* (Sirc.) (Svensson, 1974: 166).

43. ***Elachista pullicomella*** Zeller, 1839
Plate-figs. 102, 103, 205, 347, 348, 468, 521, 522.

Elachista pullicomella Zeller, 1839: 212.
Elachista rectifasciella Stainton, 1851: 26.
Elachista bohemanni Wallengren, 1852: 218.
Poeciloptilia pullella Herrich-Schäffer, 1855: 300, 304. **Syn.n.**
Poeciloptilia furvicomella Herrich-Schäffer, 1855: 305.

8–10 mm. Male (Pl.-fig. 102). Head, neck tufts, tegulae and thorax dark grey-brown, face and posterior margins with lighter greyish suffusion. Antenna dark grey-brown, faintly annulated. Labial palpi porrect, slightly curved, upperside whitish grey, underside dark grey-brown. Forewing ground colour dark grey, scales light grey at base, tips blackish grey; before middle, a narrow, irregular, often blurred whitish fascia, beyond middle a whitish, narrow, triangular costal spot opposite a similarly coloured smaller tornal spot. Cilia brownish grey along costa, white outside cilia line from apex to beyond middle of termen, grey around tornus; cilia line distinct, dark brown. Hindwing grey-brown, cilia similar. Abdomen dark greyish, posterior margins and anal tuft lighter. Hind tarsus dark grey with distinct whitish rings on distal margin of each segment.

Female (Pl.-fig. 103). Face lighter. Antenna ringed whitish/dark grey. Forewing blackish grey-brown, slightly lighter at base; marks distinct, white. Cilia line blackish, more distinct.

Male genitalia (Pl.-figs. 347, 348). Uncus lobes rather broad, medial margins convex, tips with a few short setae. Gnathos elongate, oval. Anterior margin of tegumen rather narrow, sclerotized. Costa of valva convex, emarginate before cucullus, margin here rather crenellated; cucullus prominent, rounded; sacculus slightly concave. Digitate process extremely short and slender, not setose. Juxta lobes narrow, distally produced into narrow, blunt, lateral processes without setae, and into pointed, laterally serrate, setose apical processes; from ventral part of anellus a short, broad, strong tongue-shaped process. Vinculum short, triangular. Aedeagus slightly curved, gradually tapering to pointed distal end; without cornuti.

Female genitalia (Pl.-fig. 468). Papillae anales short, triangular. Apophyses with a small distal knob, posteriores about twice as long as anteriores. Tergite VIII broad. Ostium bursae at anterior margin of sternite VIII, rounded; antrum wide, membranous, dorsal wall finely spined, ventral margin U-shaped, reinforced; colliculum not distinct. Corpus bursae with fine internal spines, no signum present.

Distribution. From E. Denmark; from many districts in Sweden up to Ång.; in Norway from Ø; from several districts in S. Finland up to ObS. – Not in Britain, but widely distributed on the Continent from Holland, Belgium and France to W. USSR and the Balkan Peninsula.

Biology. Larva not described. As food-plants *Avena, Arrhenatherum pratense* (L.) Sampaio, *Dactylis glomerata* L., *Deschampsia flexuosa* (L.) Trin., *D. caespitosa* (L.) PB., *Festuca rubra* L., *F. ovina* L., *Poa pratensis* L., *Phleum* and *Trisetum flavescens* (L.) PB. are reported. The mine (Pl.-figs. 521, 522) is whitish, broad and long, often occupying the whole blade. Pupa light yellowish anteriorly, posterior abdominal segments ochreous; it is almost circular, dorsal and lateral ridges only slightly produced anteriorly, more strongly posteriorly, thorax with lateral tubercles. Bivoltine in southern Scandinavia; larvae mine during April-May and late June-July, adults fly from mid-May to late June and again in August.

Notes. *E. rectifasciella* Stt. A specimen without abdomen labelled "TYPE H.T.," "486," "Rectifasciella" in the Stainton collection of the British Museum (Nat. Hist.) is hereby designated as lectotype.

E. bohemanni Wllgr. Wallengren (1875: 75) himself synonymized *bohemanni* with *E. pullicomella* Zell., but under that label in the Wallengren collection, Zool Inst., Lund two species are present: *E. pullicomella* Zell. and *C. freyerella* (Hb.). A *pullicomella* labelled "243" "Zeller", "E. pullicomella Zell" is hereby designated as lectotype of *bohemanni*.

P. pullella HS. is proposed as an objective replacement name for *rectifasciella* Stt.

The identity of *E. furvicomella* HS. assumed here follows Rebel (1901).

The specimen reported as *E. abbreviatella* Stt. from Denmark by Larsen (1927) is *pullicomella;* the identity of *abbreviatella* will be treated elsewhere by the authors.

44. ***Elachista littoricola*** Le Marchand, 1938

Plate-figs. 104, 105, 206, 349, 350, 469.

Elachista littoricola Le Marchand, 1938: 95.

7–8 mm. Male (Pl.-fig. 104). Frons yellowish-white, clypeus dark grey and vertex greenish grey, neck tufts, tegulae and thorax dark grey, posterior margin of tegulae and thorax yellowish white. Scape of antenna dark grey, underside ochreous, flagellum faintly annulated grey brown/grey. Labial palpi porrect, slightly curved, upperside white, underside dark grey, distal end of second segment whitish. Forewing ground colour mottled greenish grey with a complete white fascia before the middle and one beyond, suffused with beige in the fold and along outer margins of fasciae; base of costa with a blackish dot and base of dorsum with a distinct beige spot; inner fascia curved outwards, widest in the fold, outer fascia inwardly oblique, sides irregular, broadest at costa. Cilia grey from outer fascia to apex, faintly bluish tinged, white beyond cilia line from apex to beyond middle of termen, grey around tornus and beige outside outer fascia. Hindwing dark grey, cilia lighter, base ochreous-tinged. Abdomen grey, posterior margin of each segment and anal tuft beige. Hind tarsus cream-white, outer side with grey dots.

Female (Pl.-fig. 105). Head, neck tufts, tegulae and thorax light beige, mixed with grey. Antenna distinctly annulated cream-white/brown. Forewing fasciae more prominent and distinct.

Male genitalia (Pl.-figs. 349, 350). Uncus lobes elongate, distally rounded, set with 2–4 minute setae, lateral margins almost straight, medial margins convex. Gnathos elongate, oval. Anterior margin of tegumen strongly sclerotized, medial region pointed posteriorly. Costa of valva convex before middle and only slightly emarginate before square cucullus, sacculus almost straight, very slightly concave. Digitate process reduced, almost absent. Juxta lobes truncate, with short lateral processes, apical margins laterally setose; ventral part of juxta with prominent tongue-shaped process. Vinculum strong, short, triangular. Base of aedeagus bulbous, medial region wide, distally square-ended; without cornuti.

Female genitalia (Pl.-fig. 469). Papillae anales short, triangular. Apophyses posteriores about twice as long as anterior pair. Tergite VIII broad. Ostium bursae almost circular, in middle of sternite VIII, margins not reinforced. Antrum membranous, colliculum not distinct. Ductus bursae with minute internal spines; corpus bursae with a single oval sclerotized plate, central part set with teeth of varying size.

Distribution. Denmark, less than ten specimens known from a few localities in LFM; not in Sweden, Norway or Finland. – Otherwise only reported from W. France (Lhomme, 1951).

Biology. Nothing is known about immature stages. The species probably is univoltine, as all specimens have been caught from mid-July to early August. All specimens, except one, have been caught in coastlands, the Danish specimens on dikes, flying over the vegetation in the late afternoon.

The *unifasciella*-subgroup

Forewing dark grey-brown or blackish with a single yellowish or white fascia in the middle or ground colour ochreous.

Venation (Pl.-figs. 207–214). Forewing with R_4 and R_5 coalescent, M_1 from middle of stalk R_{4+5}, M_2 free.

45. ***Elachista chrysodesmella*** Zeller, 1850
Plate-figs. 106, 107, 207, 351, 352, 470, 524.

Elachista chrysodesmella Zeller, 1850: 203.

7–8 mm. Male (Pl.-fig. 106). Head shining lead-grey, neck tufts, tegulae and thorax blackish grey. Scape of antenna blackish grey, distal two-thirds of flagellum slightly serrate, faintly annulated narrow blackish/broader dark grey. Labial palpi slender, porrect, curved; upperside beige-whitish, underside dark grey-brown. Forewing ground colour blackish brown; with a distinct shining yellow fascia in the middle, narrow at costa, gradually widening towards fold, margins of fascia slightly irregular. Cilia greyish brown, basally with bluish reflection, cilia line present, cilia lighter beyond line along termen than along costa and around tornus. Hindwing dark brownish, cilia brownish grey. Abdomen dark brown, with bluish reflection. Hind tarsus blackish with narrow white ring on distal end of each segment.

Female (Pl.-fig. 107). Distal third of flagellum of antenna distinctly ringed narrow blackish/broader whitish, otherwise as in male. Forewing almost black, fascia generally broader. Cilia line distinct.

Male genitalia (Pl.-figs. 351, 352). Uncus lobes triangular, outer margins straight; incision deep, V-shaped, tips of lobes with row of 3–5 fine setae. Gnathos elongate. Anterior margin of tegumen slightly sclerotized. Valva almost elongate oval, costa slightly concave beyond middle, cucullus rounded, sacculus slightly curved. Digitate process very slender, slightly enlarged distally, with very few setae. Juxta lobes short, apical margin rounded medially, then running almost straight to the short lateral processes, apical margin setose; a narrow tongue-shaped process from ventral part of juxta. Vinculum triangular. Aedeagus long, base slightly bulbous, medial portion and distal end slightly enlarged; without cornuti.

Female genitalia (Pl.-fig. 470). Papillae anales short, with long ventral setae. Apophyses posteriores more than twice as long as apophyses anteriores. Tergite VIII weakly sclerotized, anterior margin slightly more sclerotized. Antrum not sclerotized, margins not distinct, dorsal wall with minute spines; colliculum a short, weakly sclerotized zone close to ostium. Ductus and corpus bursae not spined, signa two weakly sclerotized small patches each with 2–3 small medial teeth.

Distribution. Not in Denmark, Norway or Finland; in Sweden only from Gtl. – Not in Britain, widely distributed on the Continent from Belgium and France to Poland, Czechoslovakia, S. Europe and Asia Minor.

Biology. Larva dull yellow, head and prothoracic shield light brownish, shield rather weakly sclerotized, not longitudinally divided, narrow in the middle, distal ends rounded and enlarged. As food-plants *Brachypodium pinnatum* (L.) PB., *Holcus, Poa trivialis* L. and *Carex montana* L. are reported, the last is probably erroneous; Steuer (1976a) reports only the first in which the larva starts mining from the middle of the blade, making a very fine narrow mine upwards, turning before the tip and then making a rather short, broad, transparent blotch with excrement in the distal end (Pl.-fig. 524). The blotch resembles a *Phyllonorycter*-mine and is very similar to the mine of *E. subocellea* (Stph.) which is generally present 3–4 weeks later. Pupation takes place on the ground. Bivoltine; larva from early spring to May and during July-early August; adults fly from mid-May to late June and in August. Mainly reported from exposed, south-facing hillsides with little shade.

46. ***Elachista megerlella*** (Hübner, 1810)
Plate-figs. 108, 109, 208, 353, 354, 471, 472.

Tinea megerlella Hübner, [1810], pl. 44, fig. 307.
Tinea albinella Linnaeus *sensu* Haworth, 1828: 581.
Elachista obliquella Stainton, 1854a: 258.

8–10 mm. Male (Pl.-fig. 108). Head grey; neck tufts, tegulae and thorax dark grey. Scape of antenna dark grey, distal two-thirds of flagellum serrate, flagellum grey-brown, faintly annulated, underside lighter. Labial palpi rather short, descending, slightly curved, upperside pale greyish, underside dark grey-brown. Forewing narrow, ground colour grey-brown, slightly lighter beyond fascia, mottled by scales with dark brown tips; medial fascia yellowish white, outwardly more yellow, narrow, inner margin irregular, outer margin blurred. Cilia grey along costa to apex, pale greyish beyond cilia line from apex to middle of termen, grey around tornus; cilia line rather distinct, dark brown. Hindwing greyish brown, cilia similar. Abdomen light greyish brown. Hind tarsus darker greyish brown, each segment with a narrow white ring distally.

Wörz (1957) reports males with interrupted or reduced fasciae, but it is questionable whether he correctly identified the present species.

Female (Pl.-fig. 109). Head pale ochreous grey, frons paler, mottled with brownish on vertex. Scape of antenna ochreous white with a prominent dark brown spot on upperside of base; flagellum distinctly annulated narrow dark brown/broader bone-white. Forewing broader, tinged with ochreous-brownish.

Male genitalia (Pl.-figs. 353, 354). Uncus lobes broad, tips rounded and with several small setae. Gnathos narrow, elongate. Anterior margin of tegumen reinforced. Valva widening to abrupt sub-apical emargination before dorsally rounded cucullus which is almost at right angles to the sacculus. Digitate process strong, narrow in the middle, distal end enlarged, with short setae. Juxta lobes prominent, apical margins produced medially into elongate tapering setose processes, lateral processes obtuse. Vinculum short, rounded. Aedeagus long and slender, curved, gradually tapering, blunt, without cornuti.

Female genitalia (Pl.-figs. 471, 472). Papillae anales short. Apophyses posteriores one and a half times as long as apophyses anteriores. Antrum triangular, membraneous, ventral margin slightly anteriorly curved, set with spines, dorsal wall with minute spines; colliculum short, sclerotized. Signa two small sclerotized plates, each with 2–6 pointed teeth.

Distribution. In Denmark only from LFM; in Sweden from Öl. and Gtl.; in Norway from HOy and STi; not in Finland. – Otherwise in Britain and C. and S. Europe.

Biology. The larva is dull greenish grey, head pale brown, mouth darker, prothoracic shield divided, irregular, dark brown (Stainton, 1858c). As food-plants *Brachypodium sylvaticum* (Huds.) PB., *Bromus erectus* Huds., *Calamagrostis, Carex acuta* L., *C. elata* All., *C. spicata* Huds., *C. sylvatica* Huds., *C. ornithopoda* Willd., *Deschampsia caespitosa* (L.) PB., *Koeleria cristata* (L.) Pers., *Milium, Poa chaixii* Vill. and *Sesleria caerulea* (L.) Ard. are reported, but as mentioned by Steuer (1976a: 167) records of at least *Deschampsia* and *Sesleria* probably relate to *E. revinctella* Zell. The eggs are laid on the upper surface of the leaf of the food-plant, frequently 2–3 eggs on the same leaf. The larva starts mining upwards, forming a slender, almost linear brown mine, in which excrement is deposited in the lower part. It soon leaves the original mine, starting a new mine on an adjoining leaf; this mine is often as broad as the blade, the upperside puckered; the larva generally mines upwards (Hering, 1957), more rarely downwards (Stainton, 1858c). Pupation on leaf or stem, attached by a single girdle and cremaster. Bivoltine; larvae from autumn to mid-May and in July, adults from late May to mid-July and in August, but generation distribution is Scandinavia is not quite clear. Mainly in shady places and along wood edges.

47. ***Elachista cingillella*** (Herrich-Schäffer, 1855)
Plate-figs. 110, 111, 209, 355, 473, 474.

Poeciloptilia cingillella Herrich-Schäffer, 1855: 299, 303.
Elachista densicornuella Hodgkinson, 1879: 56.

8.5–9.5 mm. Male (Pl.-fig. 110). Head pale grey, frons paler; neck tufts, tegulae and thorax dark grey-brown. Antenna faintly serrate distally, dark grey-brown, slightly annulated. Labial palpi slender, porrect, cream-white above, dark grey-brown beneath. Forewing ground colour dark grey with medial fascia; fascia white, faintly yellowish-tinged, almost straight, narrow and of almost equal width from costa to dorsum, bound by darker suffusion. Cilia distinct black basally, cilia line dark grey with scattered dark scales, beyond cilia line whitish from apex to mid-termen, otherwise grey. Hindwing dark grey, cilia greyish. Abdomen grey, anal tuft lighter. Hind tarsus blackish, each segment with a distal whitish ring.

Female (Pl.-fig. 111). Forewing fascia slightly broader.

Male genitalia (Pl.-fig. 355). Uncus lobes broad, rounded, with a few short distal setae. Gnathos narrow, small. Anterior margin of tegumen not reinforced. Valva widest in the middle, costa strongly convex in the middle, emarginate before the truncate cucullus, sacculus almost straight. Digitate process dorsally enlarged beyond middle, ventral margin straight, tapering to short end, enlarged portion setose. Juxta lobes narrow, elongate, distal end enlarged, with very short lateral processes and short, setose, produced medial regions. Vinculum short, rounded. Aedeagus long, curved, gradually tapering, distal end chisel-shaped.

Female genitalia (Pl.-figs. 473, 474). Papillae anales short. Apophyses posteriores about twice as long as apophyses anteriores. Tergite VIII weakly sclerotized around ostium bursae, antrum not developed, except for ventral margin and rims, dorsal wall and rims finely spined, ventral margin not spined; colliculum very short, weakly sclerotized. Corpus bursae with scattered fine spines, signa two elongate, irregular sclerotized plates, each with a varying number (2–7) of triangular, pointed teeth.

Distribution. Not in Denmark; from three districts in southern Sweden; in Norway from TEy; not in Finland. – In Britain and from C., S. and SE. Europe.

Biology. Only record of the life history is that given by Hering (1891: 208) and subsequently quoted. The larva mines in the base of the leaves of *Milium effusum* L., making a mine similar to that of *E. apicipunctella* Stt., from September to April and in July – early August; the moth is on the wing from mid-May to early July and in August – early September; bivoltine.

48. ***Elachista unifasciella*** (Haworth, 1828)

Text-fig. 43; Plate-figs. 112, 113, 210, 356, 357, 475, 525.

Tinea unifasciella Haworth, 1828: 584.

9–10 mm. Male (Pl.-fig. 112). Head blackish brown, frons lighter brown; neck tufts, tegulae and thorax blackish brown. Antenna plain blackish brown, distal half of flagellum serrate. Labial palpi porrect, slightly curved, cream-white above, base of third segment dark brown, underside dark brown. Forewing rather broad, ground colour blackish brown; middle fascia distinct, pale yellowish, of equal width from costa to fold, slightly enlarged from fold to dorsum. Cilia blackish brown, but whitish around apex beyond the indistinct cilia line. Hindwing lighter blackish brown than forewing, cilia similar. Abdomen almost black. Hind tarsus black, rings pale yellowish white.

Female (Pl.-fig. 113). Distal part of antenna faintly annulated. Forewing almost black; fascia broader, only very slightly widening towards dorsum. Cilia line more distinct.

Male genitalia (Pl.-figs. 356, 357). Uncus lobes short, broad, tips rounded, set with several setae, proximal setae longest. Anterior margin of tegumen strongly sclerotized. Valva widest before middle, costa convex but distally slightly emarginate; cucullus dorsally produced, rounded ventrally towards sacculus. Digitate process broad, club-shaped, setose. Juxta lobes medially produced and rounded, apical margins curved, setose, lateral process short. Vinculum triangular. Aedeagus with basal region swollen, gradually tapering to chisel-shaped distal end; with one prominent cornutus on an elongate plate.

Female genitalia (Pl.-fig. 475). Papillae anales short. Apophyses with distal knob, posteriores longer than anteriores. Sternite VIII with a postero-medial triangular patch of short setae. Antrum semi-circular, ventral margin narrow, strongly sclerotized, without spines, lateral rims with posterior spines; colliculum broad, longer than apophyses posteriores, separated from antrum by a short membranous region; curved and strongly sclerotized plate, with crenellations on concave surface, in posterior region of membranous ductus bursae; anterior to curved plate, ductus bursae with a transversely folded area and a minutely spined zone. Corpus bursae with three dentate signa, two triangular and one elongate.

Distribution. Mainly from eastern Denmark; in Sweden only Sk.; not in Norway or Finland. – In Britain and C. Europe.

Biology. The larva is rather thick, light yellowish, gut green; head light brown, prothoracic shield lighter brown, divided, plates almost triangular, narrow anteriorly, broad and rounded posteriorly, outer margin strongly concave. Larva mining mainly in *Dactylis glomerata* L., but also in *Brachypodium sylvaticum* (Huds.) PB.; reports of *Holcus mollis* L. unreliable. Much confusion is present in the literature concerning the biology of this species, which has, until recently, been confused with *E. gangabella* Zell. Critical reviews of the literature are given by Bradley (1963) and Steuer (1973); the latter author, moreover, gives correct information on the biology. During the autumn the larva makes a long, somewhat blistered, slightly transparent mine; in spring it mines to late May, and at this time is found in the basal leaves which lie on the ground; the mine is swollen, whitish green, opaque, and the hollowed tips of the mined leaves are puckered and shrunk, filled with excrement (Pl.-fig. 525). Pupation on or in the ground; pupa attached by a single girdle. Univoltine; adults fly during June-July. The biotope is shady woods and north-facing edges of woods.

49. ***Elachista gangabella*** Zeller, 1850

Plate-figs. 114, 115, 211, 358, 359, 476, 526.

Elachista gangabella Zeller, 1850: 202.
Elachista taeniatella Stainton, 1857: 109.
Elachista ursinella Chrétien, 1896: 192.

9–10 mm. Male (Pl.-fig. 114). Head dark grey-brown, frons greyish; neck tufts, tegulae and thorax plain dark grey-brown. Antenna uniformly dark grey-brown, distal fourth of flagellum serrate. Labial palpi porrect, slightly curved; palpi light greyish above, underside grey-brown. Forewing ground colour almost plain grey-brown; middle fascia pale yellowish, very narrow between costa and fold, often not reaching costa and more or less interrupted, widening triangularly from fold towards dorsum. Cilia similar to ground colour, not lighter at apex and along termen; cilia line distinct, dark brown. Hindwing lighter grey-brown, cilia similar. Abdomen dark grey-brown. Hind tarsus grey, without white rings.

Female (Pl.-fig. 115). Antenna dark brown. Forewing broader, ground colour dark brown; fascia deep yellow, distinct from costa to dorsum, inner margin almost straight, outer margin outwardly oblique. Cilia almost blackish grey, cilia line black.

Male genitalia (Pl.-figs. 358, 359). Uncus lobes short, medial margins convex, outer margins straight, tips rounded with several minute setae. Gnathos broad, oval to triangular. Anterior margin of tegumen not reinforced. Valva widest in the middle, margins almost parallel, costa slightly convex in the middle, cucullus rounded, sacculus slightly curved. Digitate process narrow and slender, margins parallel, not enlarged distally, with scattered setae almost from base. Juxta lobes short, truncate, apical margin straight, setose, with short lateral processes. Vinculum short, triangular. Aedeagus straight, gradually tapering, without cornuti.

Female genitalia (Pl.-fig. 476). Papillae anales short. Apophyses rather short, of almost equal length. Tergite VIII with a more strongly sclerotized U-shaped area. Ostium bursae near anterior margin of venter VIII. Antrum membranous, narrow, with many longitudinal folds, dorsal wall with fine spines; colliculum narrow, short, weakly sclerotized, two patches set with fine spines lateral to antrum. Ductus and corpus bursae without minute internal spines; signa two curved bands of narrow, pointed teeth.

Distribution. From several districts in E. Denmark; in Sweden from Sk. and Öl.; not in Norway or Finland. – In Britain and widespread on the Continent to Asia Minor.

Biology. Recently described by Steuer (1973). Larva is olive-green when feeding, greyish during hibernation. Integument strongly sculptured, with dark pigment; head and prothoracic shield dark brown, shield divided, plates triangular, broadest and rounded posteriorly, outer margin slightly concave; meso- and metathorax each with a pair of dorso-lateral, anteriorly curved, elongate areas without sculpturing and pigmentation. The food-plant is mainly *Dactylis glomerata* L., more rarely *Brachypodium sylvaticum* (Huds.) PB. and *Melica nutans* L. The larva starts mining near the tip, mining downwards; more rarely it starts mining upwards, but then it almost immediately turns round. The mine (Pl.-fig. 526) is long and slender, transparent and easy to observe; it has a characteristic web-tube mixed with excrement through the middle, into which the larva retracts when disturbed, and the margins of the mine are strongly irregular, as the larva eats small blotches away from the sides of the mine. It feeds during September-November, hibernating fully grown in the mine and quitting it in the early spring; pupation on a stem or leaf, attached by a girdle. Univoltine; adults fly from late May to mid-July. The biotope is shady forest.

50. ***Elachista subalbidella*** Schläger, 1847
Plate-figs. 116, 117, 212, 360, 361, 477, 527.

Elachista subalbidella Schläger, 1847: 241.
Elachista ochreella Stainton, 1849: 27.
Elachista heinemannii Frey, 1866: 137.
Elachista immolatella Zeller, 1868: 625.
Elachista subochrella Morris, 1870: 229.
Elachista laetella Rebel, 1930: 15. **Syn.n.**

10–13 mm. Male (Pl.-fig. 116). Head, neck tufts, tegulae and thorax light ochreous, tegulae and thorax often slightly darker. Antenna slightly annulated light brownish/light ochreous. Labial palpi porrect, slender, light ochreous, underside of second segment mixed with brownish. Forewing ground colour varying from pale yellowish to brownish-ochreous; basal edge of costa dark brown, dark suffusion on wing sometimes accentuates a lighter fascia in the middle of the wing. Cilia as ground colour or paler, no cilia line present. Hindwing grey-brown, with lighter cilia, cilia at apex yellow. Abdomen dark grey-brown. Hind tarsus yellow to light ochreous.

Female (Pl.-fig. 117). Flagellum of antenna distinctly annulated dark brown/pale ochreous.

Male genitalia (Pl.-figs. 360, 361). Uncus lobes short, broad, rounded, with short to minute setae. Gnathos short, oval. Anterior margin of tegumen broad, slightly sclerotized. Valva widest at base, costa convex, emarginate before the distinct, dorsally rounded cucullus, ventral angle almost a right-angle, distal half of sacculus concave, basal half almost straight. Digitate process prominent, base broad, middle portion widest, tapering to the long curved distal end, distal half setose. Juxta lobes elongate, narrow, distal regions produced into short, setose medial processes and short lateral processes. Aedeagus curved, thickened beyond middle, with deep slit from distal end on one side, without cornuti.

Female genitalia (Pl.-fig. 477). Papillae anales triangular. Apophyses posteriores longer than apophyses anteriores. Tergite VIII with a V-shaped, strongly sclerotized area. Antrum sclerotized, bowl-shaped, ventral margin distinct on inner side with fine spines, dorsal wall finely spined; colliculum sclerotized, short, with longitudinal folds. Signa two small patches set with long, pointed teeth.

Distribution. From almost all districts in Denmark, Sweden and Finland; scattered records in Norway up to NTi. – Widely distributed in the W. Palaearctic area from Britain to W. USSR.

Biology. The larva is dark olive-green, dark grey during hibernation; head and prothoracic shield almost black, divided, shield with parallel sides anteriorly, posteriorly strongly enlarged, rounded, distal margin concave, with a circular patch without pigment and sculpturing; meso- and metathoracic segments each with a pair of similar dorso-lateral, elongate, anteriorly curved patches. The food-plants are *Brachypodium sylvaticum* (Huds.) PB., *B. pinnatum* (L.) PB., *Poa* and *Molinea caerulea* (L.) Moench. The mine (Pl.-fig. 527) is long, narrow, flat and transparent. The larva starts mining just below the tip, mining downwards or upwards for a short period and then turning and mining to the base of the blade, gradually widening the mine; in the middle a faint web-tube mixed with excrement; sides of mine regular, smooth. The larva feeds during the autumn and hibernates in the mine which is difficult to see as the blade is brownish and wilted (Steuer, 1973). Pupation on the ground, attached by a girdle. Univoltine, adults fly from late May to late July, mainly in June. Mainly in open places on moors and in meadows.

Note. The lectotype of *E. laetella* Rbl. designated by Gaedike (1975) is a large specimen of *E. subalbidella* Schl.

51. ***Elachista revinctella*** Zeller, 1850
Plate-figs. 118, 119, 213, 362, 363, 478, 528–530.

Elachista revinctella Zeller, 1850: 202.
Elachista adscitella Stainton, 1851: 10.
Elachista abruptella Stainton, 1854a: 258.
Elachista alpella Krone, 1909: 129.
Elachista cinctella sensu auct., *nec* (Clerck, 1759).

9–11 mm. Male (Pl.-fig. 118). Head and neck tufts pale yellowish white, tegulae and thorax grey-brown. Antenna grey, faintly annulated, distal third of flagellum serrate. Labial palpi porrect to recurved, slender, pale yellowish white, underside of second segment brownish except for distal part. Forewing ground colour grey-brown, basal half darkest, lighter beyond fascia, mottled with dark brown; medial fascia pale yellowish white, especially tinged with yellow along outer margin, slightly outwards-curved, narrow or interrupted in the middle, widest at dorsum. Cilia grey along costa yellowish white outside cilia line from apex to beyond middle of termen, greyish around tornus; cilia line rather distinct, blackish. Hindwing grey, cilia grey-brown, lighter at base along dorsum. Abdomen dark grey-brown. Hind tarsus dark grey, with narrow white rings.

Female (Pl.-fig. 119). Basal part of flagellum distinctly annulated dark grey/whitish. Forewing fascia broader, narrowest in the middle, wider at dorsum.

Male genitalia (Pl.-figs. 362, 363). Uncus lobes short, rounded, with short distal setae. Gnathos short. Anterior margin of tergum not reinforced. Valva widest in the middle, costa convex in the middle, cucullus truncate, sacculus slightly concave beyond the middle. Digitate process prominent, with a strong, almost triangular, enlargement beyond middle, distally tapering to pointed tip, distal half setose. Juxta lobes narrow, gradually widening to bilobed distal end; medial lobes short, apical margins setose. Vinculum rounded, short. Aedeagus curved, widest in the middle, with deep slit from distal end on one side; without cornuti.

Female genitalia (Pl.-fig. 478). Papillae anales elongate, triangular. Apophyses posteriores almost twice as long as apophyses anteriores. Antrum wide, bowl-shaped, ventral margin narrow, distinct, with fine spines on inner side, rims broad, spined, dorsal wall finely spined; colliculum very short. Signa two very weakly sclerotized plates with strongly pointed, rather widely spaced teeth.

Distribution. From almost all districts in Denmark and Sweden; scattered in Norway, up to HOi; in southern Finland up to Sb. – Widely distributed in the W. Palaearctic region.

Biology. The larva is dull yellowish green, more or less tinged grey; head pale brown, prothoracic plate brown, divided, strongly enlarged posteriorly, prothorax with two circular patches without pigment and sculpturing, meso- and metathorax with similar elongate patches. It is reported from *Deschampsia caespitosa* (L.) PB., *Brachypodium sylvaticum* (Huds.) PB., *Sesleria caerulea* (L.) Ard., *Poa, Festuca altissima* All., *F. gigantea* (L.) Vill. and *Milium effusum* L. When mining begins in the autumn, one or often two larvae are present in the mine; on broad-leaved grasses two or three mines can be present on a single blade; the larvae mine both up- and downwards, excrement is deposited in the lower part of the mine (Pl.-fig. 529). Hibernation takes place inside the mine; in spring the mine is made in a new leaf in late April (Steuer, 1976a), the larvae now mine mainly downwards, making a rather broad, irregular, whitish, not puckered mine, in which the excrement is deposited in small heaps terminally to the larvae (Pl.-figs. 528–530); one larva can make no more than one mine in spring (Stainton, 1858c). Pupation on the ground. Bivoltine in southern Scandinavia; larva from September to May and in July; adults from late May to early July and during August. The biotope is open, shady, broad-leaved forest. Adults come to light.

52. ***Elachista bisulcella*** (Duponchel, 1843)
Plate-figs. 120, 121, 214, 364, 365, 479.

Lita bisulcella Duponchel *in* Godart, [1843] 1842: 331.
Elachista zonariella Tengström, [1848] 1847: 150.

8–10 mm. Male (Pl.-fig 120). Head pale yellowish, vertex suffused with grey-brown; neck tufts lighter greyish brown, tegulae and thorax blackish brown. Scape of antenna and base of flagellum blackish grey, flagellum slightly annulated narrow blackish/broader light grey, distal third serrate. Labial palpi porrect to recurved, upperside yellowish or ochreous, underside blackish, except for distal lighter part of second segment. Forewing ground colour dark grey-brown to blackish; medial fascia distinct, white with warm yellow suffusion along outer margin, narrow at costa, gradually widening to dorsum, inner margin almost straight, outer margin outwardly oblique. Cilia dark grey, but they are yellowish white from apex to middle of termen beyond cilia line; cilia line weak, black. Hindwing blackish, cilia dark grey. Abdomen blackish grey. Hind tarsus black with narrow whitish rings.

Female (Pl.-fig. 121). Larger than male. Flagellum annulated from base to tip black/light grey. Forewing broader, black, with a violet reflection, fascia broader, narrowest at costa, gradually widening.

Male genitalia (Pl.-figs. 364, 365). Uncus lobes short, broad, sides almost parallel, distally rounded, with several short setae. Gnathos short, almost as wide as long. Anterior margin of tegumen deeply emarginate, weakly sclerotized. Valva widest in the middle, costa convex, emarginate before truncate cucullus, sacculus distally concave, almost straight in basal half. Digitate process prominent, rather slender, enlarged beyond middle, with scattered setae almost from base. Juxta lobes elongate, narrow, gradually widening distally, truncate, apical margin rounded, medially setose, lateral processes short. Vinculum triangular. Aedeagus curved, slender, middle region slightly enlarged, with a deep slit from distal end on one side, without cornuti.

Female genitalia (Pl.-fig. 479). Papillae anales elongate, narrow. Apophyses posteriores longer than apophyses anteriores. Antrum wide, bowl-shaped, ventral margin narrow, distinct, without spines, rims broad, spined, dorsal wall with spines; colliculum short. Signa two rather small, weakly sclerotized patches set with short, almost triangular teeth.

Distribution. From a few districts in E. Denmark; scattered in S. Sweden up to Upl.; not in Norway; from all districts in S. Finland up to Oa and Tb. – From W. USSR through C. Europe to Britain.

Review of plate-figures:

Nos. 1–152: Colour plates. Specimens approx. 5.5x.
- 153–228: Wing venation.
- 229–395: Male genitalia.
- 396–497: Female genitalia.
- 498–533: Larval mines.
- 534–536: Pupae.

Abbreviations applied to slides used for figures:

BÅB: B. Å. Bengtsson, Löttorp, Sweden.
BMNH: British Museum (Natural History), London, England.
EJ: E. Jäckh, Bidingen, Germany.
ETO: E. Traugott-Olsen, Marbella, Spain.
GP: G. Pallesen, Århus, Denmark.
IS: I. Svensson, Österslöv, Sweden.
KL: K. Larsen, Copenhagen, Denmark.
MvR: M. von Schantz, Helsinki, Finland.
NLW: N. L. Wolff, Copenhagen, Denmark.
OK: O. Karsholt, Skibinge, Denmark.
OP: O. Pellmyr, Örebro, Sweden.

Fig. 1. *Mendesia farinella* (Thnbg.), ♂, Denmark, NEZ, Ørholm Fælled, 26.vi.1873.
Fig. 2. *Mendesia farinella* (Thnbg.), ♀, Sweden, Vg., Kinnekulle, 3.vi.1968.

Forewing white, upperside and underside of hindwing grey-brown. Differs from *E. argentella* (Cl.) by the strongly ciliate antenna.

Fig. 3. *Perittia herrichiella* (HS.), ♂, Denmark, NEZ, København, l. ix.1922.
Fig. 4. *Perittia herrichiella* (HS.), ♀, Denmark, NEZ, København, l. ix.1922.

Forewing grey-brown with a single spot on dorsum beyond middle. Differs from *Heliozela* spp. by presence of pecten and reduction of maxillar palpi.

Fig. 5. *Perittia obscurepunctella* (Stt.), ♂, Denmark, NEZ, Hareskov, l. vi.1921.
Fig. 6. *Perittia obscurepunctella* (Stt.), ♀, Denmark, NEZ, Hareskov, l. vi.1921.

With a more or less pronounced, dark brown streak in fold, inwards and outwards bound by lighter spots.

Fig. 7. *Stephensia brunnichella* (L.), ♂, Denmark, LFM, Eriksvold, l. 23.vii.1921.
Fig. 8. *Stephensia brunnichella* (L.), ♀, Denmark, LFM, Fyrrevænget, l. 3.viii.1933.

Marks on forewing golden metallic shining, medial fascia reaching to costa. Antenna with white portion beyond middle succeeded by dark distal end.

Fig. 9. *Elachista regificella* Sirc., ♂, Denmark, EJ, Løvenholm Skov, 5.vii.1968.
Fig. 10. *Elachista regificella* Sirc., ♀, Denmark, NEZ, Tisvilde, 2.vi.1969.

Marks on forewing golden metallic shinning, medial fascia not reaching to costa as in *S. brunnichella* (L.). Only most distal part of antenna white.

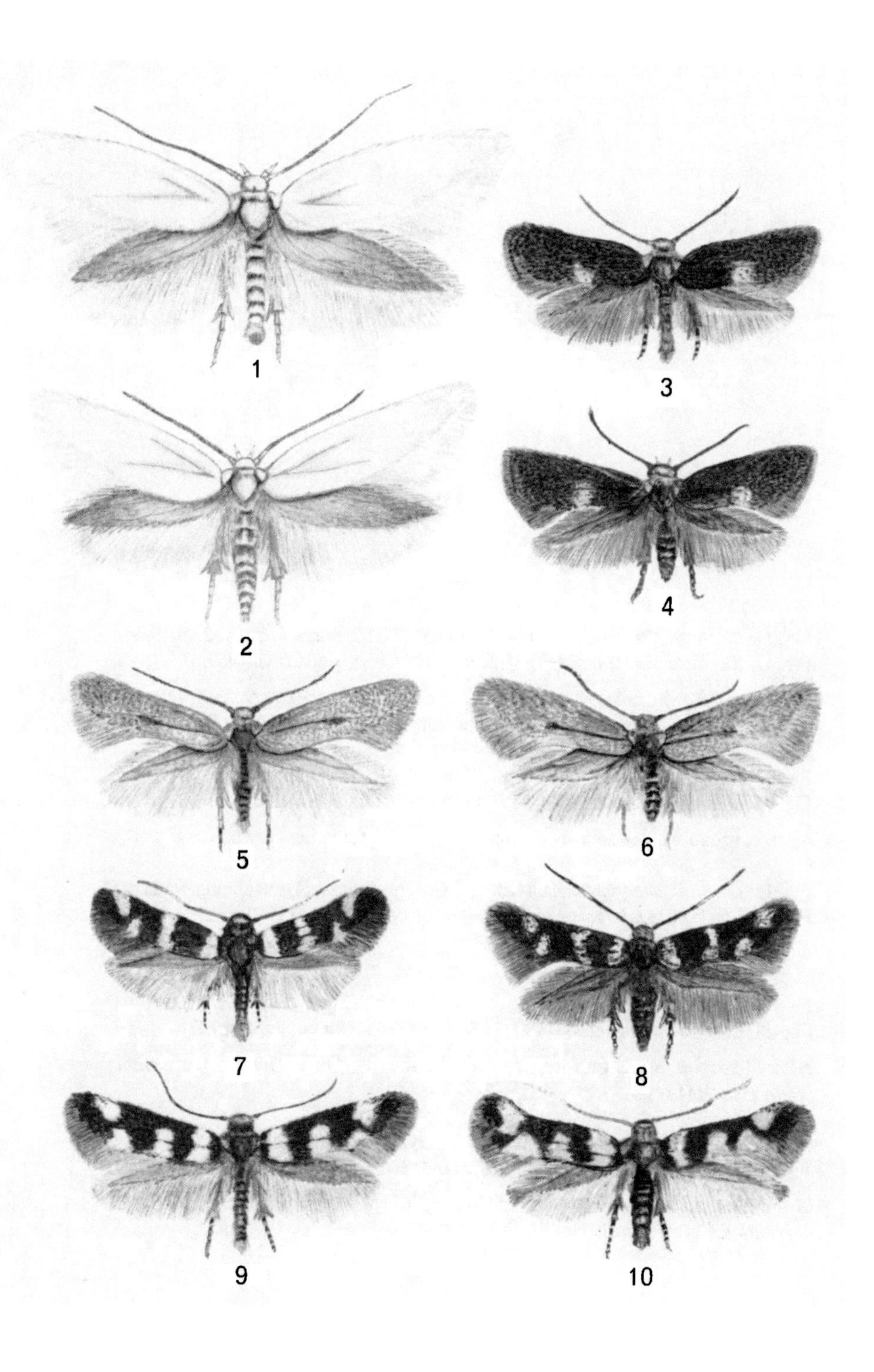
1
3
2
4
5
6
7
8
9
10

Fig. 11. *Elachista gleichenella* (F.), ♂, Denmark, NEZ, Boserup Skov, 2.vii.1969.
Fig. 12. *Elachista gleichenella* (F.), ♀, Denmark, LFM, Bøtø, 5.vii.1969.

Marks on forewing silvery metallic shining, medial fascia to costa. Antenna without white coloration. Larger and more bright shining than *E. nobilella* Zell.

Fig. 13. *Elachista pigerella* (HS.), ♂, E. Germany, Bad Blankenburg, el. 27.vii.1968.
Fig. 14. *Elachista pigerella* (HS.), ♀, E. Germany, Bad Blankenburg, el. 23.vii.1968.

A large, broad-winged species. Margin of scales on forewing very pale.

Fig. 15. *Elachista quadripunctella* (Hb.), ♂, Czechoslovakia, Uretschning, 9.vii.1972.
Fig. 16. *Elachista quadripunctella* (Hb.), ♀, Austria, Linz, 9.vii.1972.

A large species. Spots on forewing silky white.

Fig. 17. *Elachista tetragonella* (HS.), ♂, Sweden, Vg., Kinnekulle, 16.vi.1968.
Fig. 18. *Elachista tetragonella* (HS.), ♀, Sweden, Vg., Kinnekulle, 5.vi.1971.

A small species. Spots on forewing metallic shining. Differs by this from *Biselachista ornithopodella* (Frey).

Fig. 19. *Elachista biatomella* (Stt.), ♂, Denmark, NEZ, Asserbo, 16.viii.1960.
Fig. 20. *Elachista biatomella* (Stt.), ♀, England, Swanage Coast, 23.v.1887.

Ground colour of forewing varying from light grey to brownish grey. Fold paler, the two dark spots always distinct.

11
13
12
14
15
17
16
18
19
20

Fig. 21. *Elachista martinii* Hofm., ♂, E. Germany, Bad Blankenburg, ex. l. *Carex humilis.*
Fig. 22. *Elachista martinii* Hofm., ♀, E. Germany, Bad Blankenburg, ex. l. *Carex humilis.*

A very small species. Strongly mottled.

Fig. 23. *Elachista poae* Stt., ♂, Denmark, NEZ, København, l. 22.v.1922.
Fig. 24. *Elachista poae* Stt., ♀, Denmark, NEZ, København, l. 22.v.1922.

A stout species, recognizable by zigzag fascia before middle, which is bound outwards by dark suffusion.

Fig. 25. *Elachista atricomella* Stt., ♂, Denmark, LFM, Roden Skov, 13.vi.1967.
Fig. 26. *Elachista atricomella* Stt., ♀, Denmark, LFM, Roden Skov, 20.vi.1965.

A very large species. Fascia most prominent at costa and dorsum, often interrupted in the middle.

21
23
22
24
25
26

Fig. 27. *Elachista kilmunella* Stt., ♂, Sweden, Dlr., Trånstrand, 19–20. vi.1968.
Fig. 28. *Elachista kilmunella* Stt., ♀, Sweden, Hall., Knäred, 12.vi.1966.

Forewing faintly tinged with bluish. Recognizable on the white cilia from apex to middle of termed beyond the distinct cilia line.

Fig. 29. *Elachista parasella* Tr.–O., ♂, Norway, TRi, Paras, 28.vi.1972.

A rather large species. Ground colour of forewing a cold bluish grey, costal and tornal spots confluent into a zigzag fascia.

Fig. 30. *Elachista alpinella* Stt., ♂, Denmark, LFM, Krenkerup, 26.viii.1964.
Fig. 31. *Elachista alpinella* Stt., ♂, Sweden, Nb., Luleå, 29.vii.1961.
Fig. 32. *Elachista alpinella* Stt., ♀, Denmark, B, Gudhjem, 1. viii.1923.

Fascia never completely reaching to costa. Costal spot beyond tornal spot, often overlapping; spots outwardly suffused with beige.

Fig. 33. *Elachista diederichsiella* Her., ♂, Denmark, NEZ, Ganløse Ore, l. 25.v.1924.
Fig. 34. *Elachista diederichsiella* Her., ♀, Denmark, NEZ, Ganløse Ore, l. 25.v.1924.

Frons, posterior margin of thorax and tegulae, and base of forewing, ochreous-white.

27
28
29
30
31
32
33
34

Fig. 35. *Elachista compsa* Tr.–O., ♂, Denmark, SZ, Jungshoved, 13.v.1971.
Fig. 36. *Elachista compsa* Tr.–O., ♀, Denmark, SZ, Jungshoved, 13.v.1971.

Head white. Forewing very slender; costal spot prominent and often squarish.

Fig. 37. *Elachista elegans* Frey, ♂, Sweden, Gtl., Rute, 18.vii.1969.
Fig. 38. *Elachista elegans* Frey, ♀, Sweden, Gtl., Norrlanda, 13.vii.1969.

Frons white. Forewing short and broad.

Fig. 39. *Elachista luticomella* Zell., ♂, Denmark, NEZ, Boserup Skov, 29.vi.1969.
Fig. 40. *Elachista luticomella* Zell., ♀, Denmark, NEZ, Karlstrup Strand, 14.vii.1969.

Head yellow.

Fig. 41. *Elachista albifrontella* (Hb.), ♂, Denmark, NEZ, Solrød Strand, 25.vi.1966.
Fig. 43. *Elachista albifrontella* (Hb.), ♀, Denmark, SZ, Slagelse Skov, 23.vi.1963.

Head and neck tufts bright white.

Fig. 43. *Elachista bifasciella* Tr., ♂, Denmark, NEJ, Rold Skov, 17.vi.1968.
Fig. 44. *Elachista bifasciella* Tr., ♀, Denmark, WJ, Grimstrup Krat, 14.vi.1968.

Head white. Forewing with silky white base and two silky white fasciae.

35
37
36
38
39
40
41
42
43
44

Fig. 45. *Elachista nobilella* Zell., ♂, Denmark, EJ, Hvidding Krat, 19.vi.1970.
Fig. 46. *Elachista nobilella* Zell., ♀, Denmark, WJ, Grimstrup Krat, 16.vi.1968.

A small species. Head metallic shining. Marks on forewing silvery shining; a small spot of same colour medially beyond costal and tornal spots.

Fig. 47. *Elachista apicipunctella* Stt., ♂, Denmark, SZ, Jungshoved, 14.vi.1970.
Fig. 48. *Elachista apicipunctella* Stt., ♀, Denmark, SZ, Jungshoved, 14.vi.1970.

Head silky white. Forewing long and slender, with silky shining fascia and spots, medially beyond these a third tiny spot.

Fig. 49. *Elachista subnigrella* Dougl., ♂, Denmark, NEZ, Karlstrup Strand, 14.vii.1969.
Fig. 50. *Elachista subnigrella* Dougl., ♀, Denmark, NEZ, Karlstrup Strand, 14.vii.1969.
A small species. Scales forming the cilia line with bright white bases and almost black tips.

Fig. 51. *Elachista reuttiana* Frey, ♂, E. Germany, Bad Blankenburg, ex. l. *Koeleria cristata.*
Fig. 52. *Elachista reuttiana* Frey, ♀, E. Germany, Bad Blankenburg, ex. l. *Koeleria cristata.*

A very mottled species. Fascia and spots white, often with a third white spot distally.

Fig. 53. *Elachista orstadii* Palm, ♂, Denmark, NEZ, Hjortekær, 1.vi.1949.
Fig. 54. *Elachista orstadii* Palm, ♀, Sweden, Vg., Kinnekulle, 6–8.vi.1966.

Male with greyish head; forewing often with three dark streaks. Female with pale greyish head; costal spot inwardly oblique, beyond tornal spot.

45
46
47
48
49
50
51
52
53
54

Fig. 55. *Elachista ingvarella* Tr.–O., ♂, Sweden, Jmt., Lundsåker, 15.vii.1962.
Fig. 56. *Elachista ingvarella* Tr.–O., ♀, Sweden, Hrj., Hamrafjället, 22.vi.1975.
Fig. 57. *Elachista ingvarella* Tr.–O., ♀, Sweden, Nb., Nedertornå, 9.vii.1955.

Male with faint fascia and spots. Female with many dark tipped scales in and along fold.

Fig. 58. *Elachista nielswolffi* Svens., ♂, Sweden, Jmt., Storulvån, 20.vii.1975.

A small species. Forewing tinged with bluish grey.

Fig. 59. *Elachista krogeri* Svens., ♂, Sweden, Nb., Kaunisjoki, 29.vi.1975.
Fig. 60. *Elachista krogeri* Svens., ♀, Sweden, Nb., Kaunisjoki, 29.vi.1975.

Fascia in male only distinct in fold, blurred towards costa and dorsum. Female with two almost complete fasciae.

Fig. 61. *Elachista pomerana* Frey, ♂, Denmark, LFM, Ulfshale, 13.viii.1971.
Fig. 62. *Elachista pomerana* Frey, ♀, Frey coll., in British Museum (Nat. Hist.).

Head grey. Fascia oblique from costa and sharply bent above fold. Costal spot only slightly beyond tornal spot.

Fig. 63. *Elachista humilis* Zell., ♂, Denmark, LFM, Roden Skov, 13.vi.1969.
Fig. 64. *Elachista humilis* Zell., ♀, Frey coll., in British Museum (Nat. Hist.).

A dark species. Fascia generally not distinct; costal and tornal spots opposite.

55
56
57
58
59
60
61
62
63
64

Fig. 65. *Elachista vonschantzi* Svens., ♂, Sweden, Nb., Båtskärsnäs, 5.vii.1975.
Fig. 66. *Elachista vonschantzi* Svens., ♀, Sweden, Nb., Båtskärsnäs, 6.vii.1975.

Two distinct, small, light spots at base of fold and dorsum. Blackish suffusion beyond distinct fascia. Female with more distinct marks, costal and tornal spots confluent, forming a fascia.

Fig. 67. *Elachista pulchella* (Hw.), ♂, Denmark, WJ, Grimstrup Krat, 6.vi.1970.
Fig. 68. *Elachista pulchella* (Hw.), ♀, Denmark, LFM, Bøtø, 30.v.1971.

Head in both sexes pale beige. Forewing of male almost without marks, of female with distinct white fascia, costal and tornal spots.

Fig. 69. *Elachista anserinella* Zell., ♂, Sweden, Sk., Kisby, 5.vi.1963.
Fig. 70. *Elachista anserinella* Zell., ♀, Sweden, Sk., Kisby, 22.v.1961.

Forewing irregularly suffused with ochreous tipped scales, without dark apical streak and plical spots.

Fig. 71. *Elachista rufocinerea* (Hw.), ♂, Denmark, SJ, Frøslev Mose, 2–4.vi.1963.
Fig. 72. *Elachista rufocinerea* (Hw.), ♀, W. Germany, N.-Weser, Beversted: Heyh, 5.v.1948.

Forewing with distinct, reddish ochreous, apical streak.

Fig. 73. *Elachista lastrella* Chrét., ♂, E. Germany, Bad Blankenburg, ex l. 3.v.1974.
Fig. 74. *Elachista lastrella* Chrét., ♀, E. Germany, Bad Blankenburg, ex l. 3.v.1974.

Forewing very narrow, plain grey in male, plain yellowish in female.

65
66
67
68
69
70
71
72
73
74

Fig. 75. *Elachista cerusella* (Hb.), ♂, Denmark, WJ, Grimstrup Krat, 14.vi.1968.
Fig. 76. *Elachista cerusella* (Hb.), ♀, Denmark, WJ, Grimstrup Krat, 14.vi.1968.

Forewing with distinct, black plical spot beyond middle.

Fig. 77. *Elachista argentella* (Cl.), ♂, Denmark, NEZ, Roskilde, 13.vi.1965.

Forewing white, upperside and underside of hindwing greyish white. Similar to *Mendesia farinella* (Thnbg.), but differs by antenna having appressed scales and cilia.

Fig. 78. *Elachista pollinariella* Zell., ♂, Denmark, LFM, Bøtø, 6.vii.1969.
Fig. 79. *Elachista pollinariella* Zell., ♀, Denmark, LFM, Bøtø, 17.vi.1962.

Some of the scales on forewing with either ochreous or black-brown tips.

Fig. 80. *Elachista triatomea* (Hw.), ♂, Denmark, NEZ, Roskilde, 20.vi.1965.
Fig. 81. *Elachista triatomea* (Hw.), ♀, Denmark, NEZ, Boserup Skov, 2.vii.1969.

Cilia line present, though not always distinct. Without scattered, black-brown tipped scales on forewing, but with two distinct black spots. Differs from *E. dispilella* Zell. by presence of cilia line.

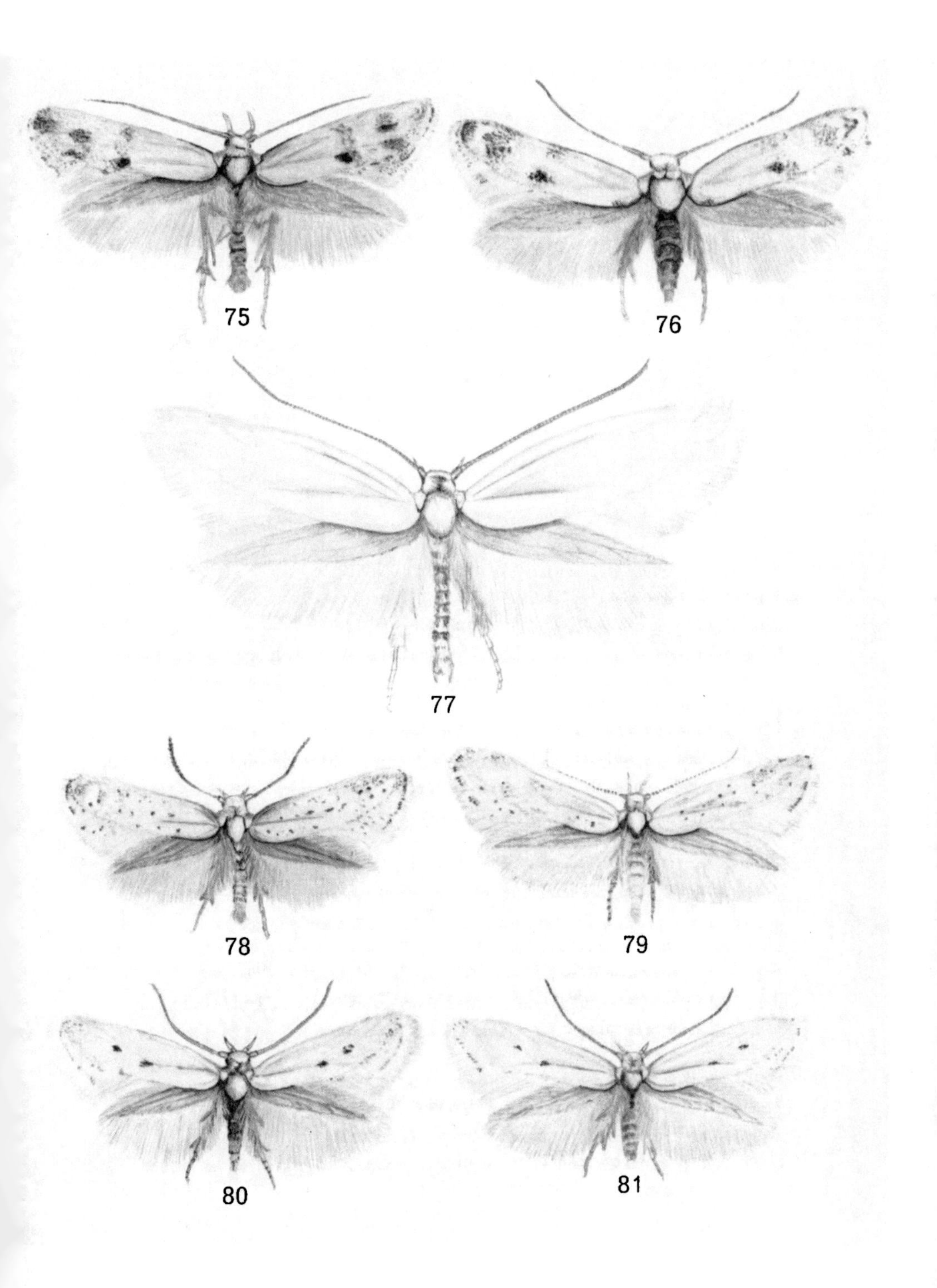
75
76
77
78
79
80
81

Fig. 82. *Elachista collitella* (Dup.), ♂, Austria, Linz, 23.v.1929.
Fig. 83. *Elachista collitella* (Dup.), ♀, Austria, Linz, 23.v.1929.

A small species. Base at costa dark grey. Many black-brown tipped scales scattered as small dots on forewing. Basal, medial and apical areas suffused with beige tipped scales.

Fig. 84. *Elachista subocellea* (Stph.), ♂, Sweden, Gtl., Öja, 23.vii.1933.
Fig. 85. *Elachista subocellea* (Stph.), ♀, Sweden, Gtl., Öja, 1.viii.1933.

Base of costa greyish. Some black-brown tipped scales scattered over forewing. Basal, medial and apical areas suffused with yellow tipped scales.

Fig. 86. *Elachista festucicolella* Zell., ♂, Sweden, Öl., Rismo, 10.vi.1967.
Fig. 87. *Elachista festucicolella* Zell., ♀, Sweden, Öl., Högserum, 24.vi.1973.

Antenna dark greyish. Forewing white, without marks and cilia line.

Fig. 88. *Elachista nitidulella* (HS.), ♂, Austria, Herzograd, 19–24.iv.1968.
Fig. 89. *Elachista nitidulella* (HS.), ♀, Austria, Herzograd, 19–24.iv.1968.

Antenna dark grey. Forewing white, without marks and cilia line. Cilia white, slightly darker tinged around tornus.

Fig. 90. *Elachista dispilella* Zell., ♂, Denmark, WJ, Fanø, 17.vii.1949.
Fig. 91. *Elachista dispilella* Zell., ♀, Sweden, Boh., Foss, 3–6.vii.1962.

Forewing white, with two prominent dark spots; otherwise without marks. Differs from *E. triatomea* (Hw.) by absence of cilia line.

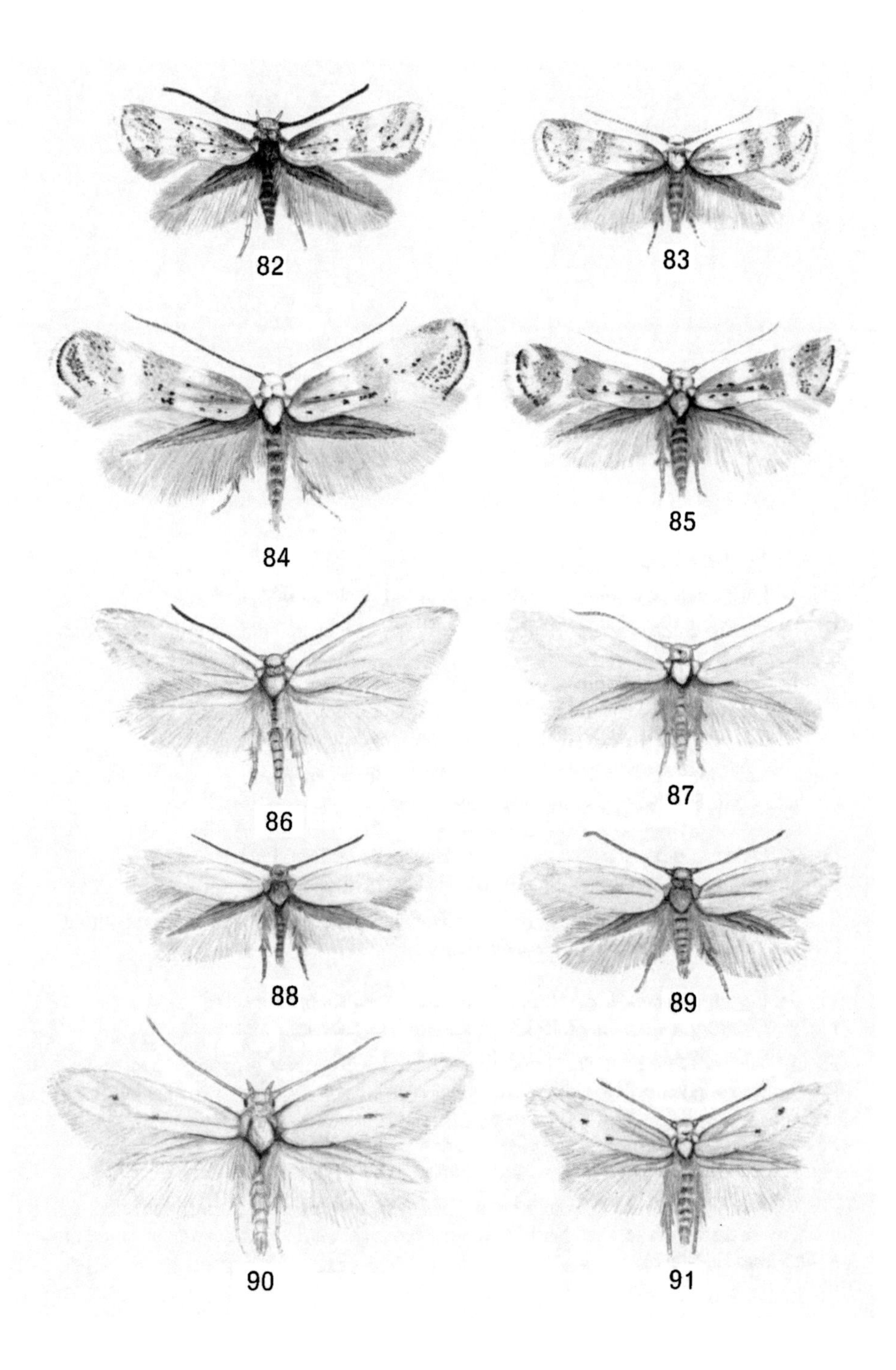
82
83
84
85
86
87
88
89
90
91

Fig. 92. *Elachista triseriatella* Stt., ♂, Sweden, Sml., Grönskärna, 2.vii.1970.
Fig. 93. *Elachista triseriatella* Stt., ♀, Sweden, Sdm., Enskede, 15.v.1960.

Very similar to *E. dispunctella* (Dup.); black dots more prominent and hindwing cold bluish grey.

Fig. 94. *Elachista dispunctella* (Dup.), ♂, Austria, Hundsheimer Berg, 30.iv.1975.
Fig. 95. *Elachista dispunctella* (Dup.), ♀, Austria, Hundsheimer Berg, 30.iv.1975.

Small black dots on forewing fine and often reduced in number compared to *E. triseriatella* Stt. Hindwing pale greyish beige.

Fig. 96. *Elachista rudectella* Stt., ♂, Austria, Hackelsberg, 15.vi.1973.
Fig. 97. *Elachista rudectella* Stt., ♀, Austria, Hackelsberg, 15.v.1966.

Scales with dark brown tip scattered over forewing, especially in apical area; without black plical spot as found in *E. cerusella* (Hb.).

Fig. 98. *Elachista squamosella* (HS.), ♂, Austria, Hackelsberg, 24.iv.1963.
Fig. 99. *Elachista squamosella* (HS.), ♀, Austria, Hackelsberg, 24.iv.1963.

Head, neck tufts and tegulae bone white mixed with a few greyish tipped scales. Forewing strongly mottled. Cilia white along termen beyond dark brown cilia line.

Fig. 100. *Elachista bedellella* (Sirc.), ♂, Denmark, LFM, Høvblege, 12.vi.1970.
Fig. 101. *Elachista bedellella* (Sirc.), ♀, Denmark, NEZ, Lynæs, 27.vii.1924.

Forewing short and broad, ground colour beige, slightly mottled. Pattern indistinct in male, more distinct in female. Without the ochreous spot at base of dorsum as found in *E. littoricola* Le March.

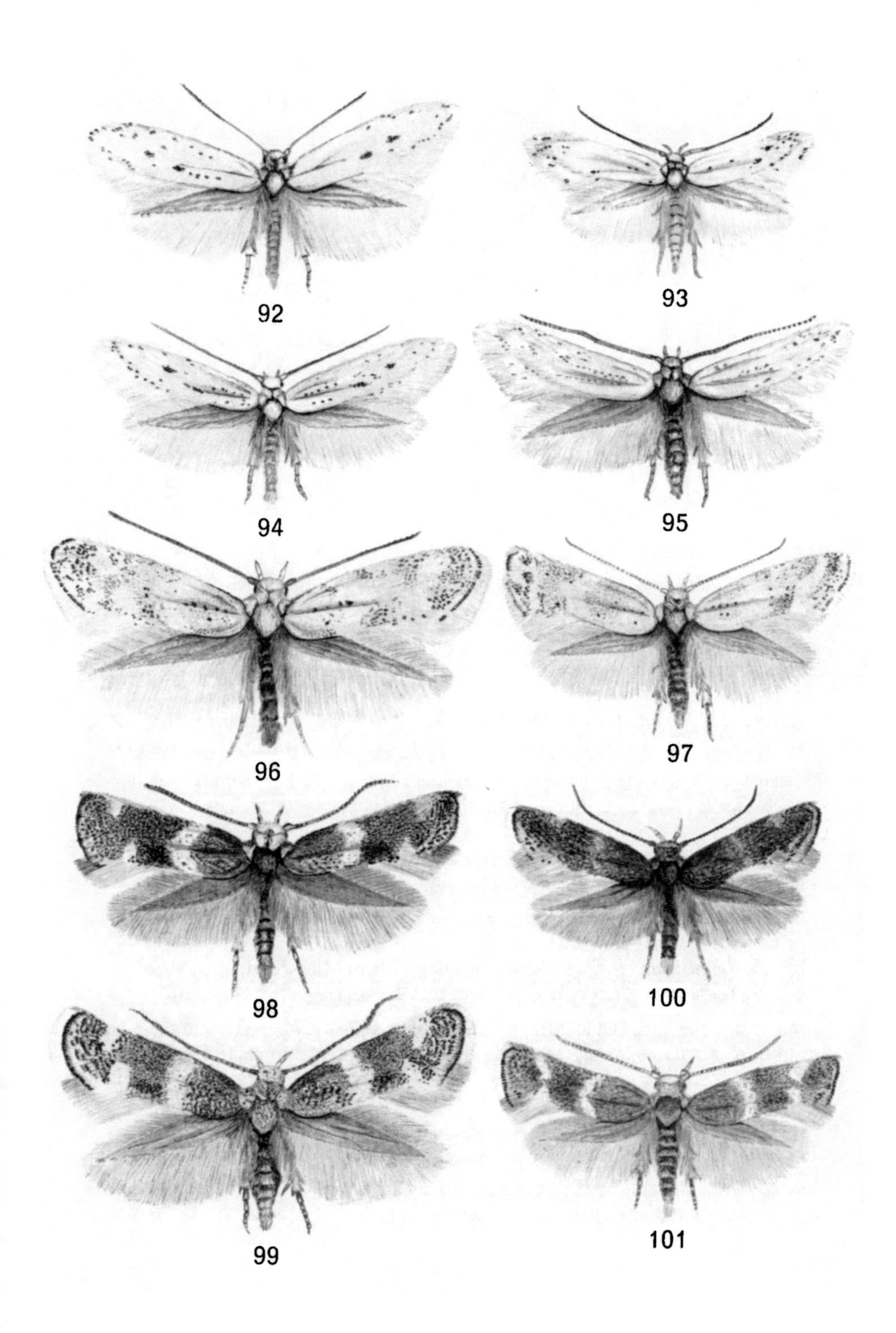
92
93
94
95
96
97
98
100
99
101

Fig. 102. *Elachista pullicomella* Zell., ♂, Denmark, NEZ, Bognæs, 5.vi.1962.
Fig. 103. *Elachista pullicomella* Zell., ♀, Denmark, NEZ, Bognæs, 1.vi.1968.

Ground colour of forewing blackish brown, with two more or less distinct, whitish fasciae, most distinct in female. Cilia white beyond the blackish cilia line.

Fig. 104. *Elachista littoricola* Le March., ♂, Denmark, LFM, Brundragene, 10.vii.1976.
Fig. 105. *Elachista littoricola* Le March., ♀, Denmark, LFM, Brundragene, 10.vii.1976.

Fasciae wide and white; cilia white beyond dark brown cilia line. Differs from female of *E. bedellella* (Sirc.) by presence of an ochreous spot at base of dorsum.

Fig. 106. *Elachista chrysodesmella* Zell., ♂, Sweden, Gtl., Norrlanda, 13.vii.1969.
Fig. 107. *Elachista chrysodesmella* Zell., ♀, Italy.

A small species. Fascia with a yellow luster. Cilia pale yellow beyond blackish cilia line.

Fig. 108. *Elachista megerlella* (Hb.), ♂, Denmark, LFM, Møns Klint, 4.vii.1960.
Fig. 109. *Elachista megerlella* (Hb.), ♀, Sweden, Öl., Vickleby, ex. l. 11.vii.1942.

Frons pale yellowish. Outer margin of fascia blurred, fascia not narrow in the middle as in *E. revinctella* Zell. Cilia pale yellowish grey along termen in apical third beyond grey brown cilia line.

Fig. 110. *Elachista cingillella* (HS.), ♂, Italy, Interneppo, 10–27.vi.1961.
Fig. 111. *Elachista cingillella* (HS.), ♀, Sweden, Vg., Kinnekulle, 6–8.vi.1966.

Fascia almost straight, bound by slightly darker suffusion. Cilia yellowish white from apex to middle of termen beyond blackish cilia line.

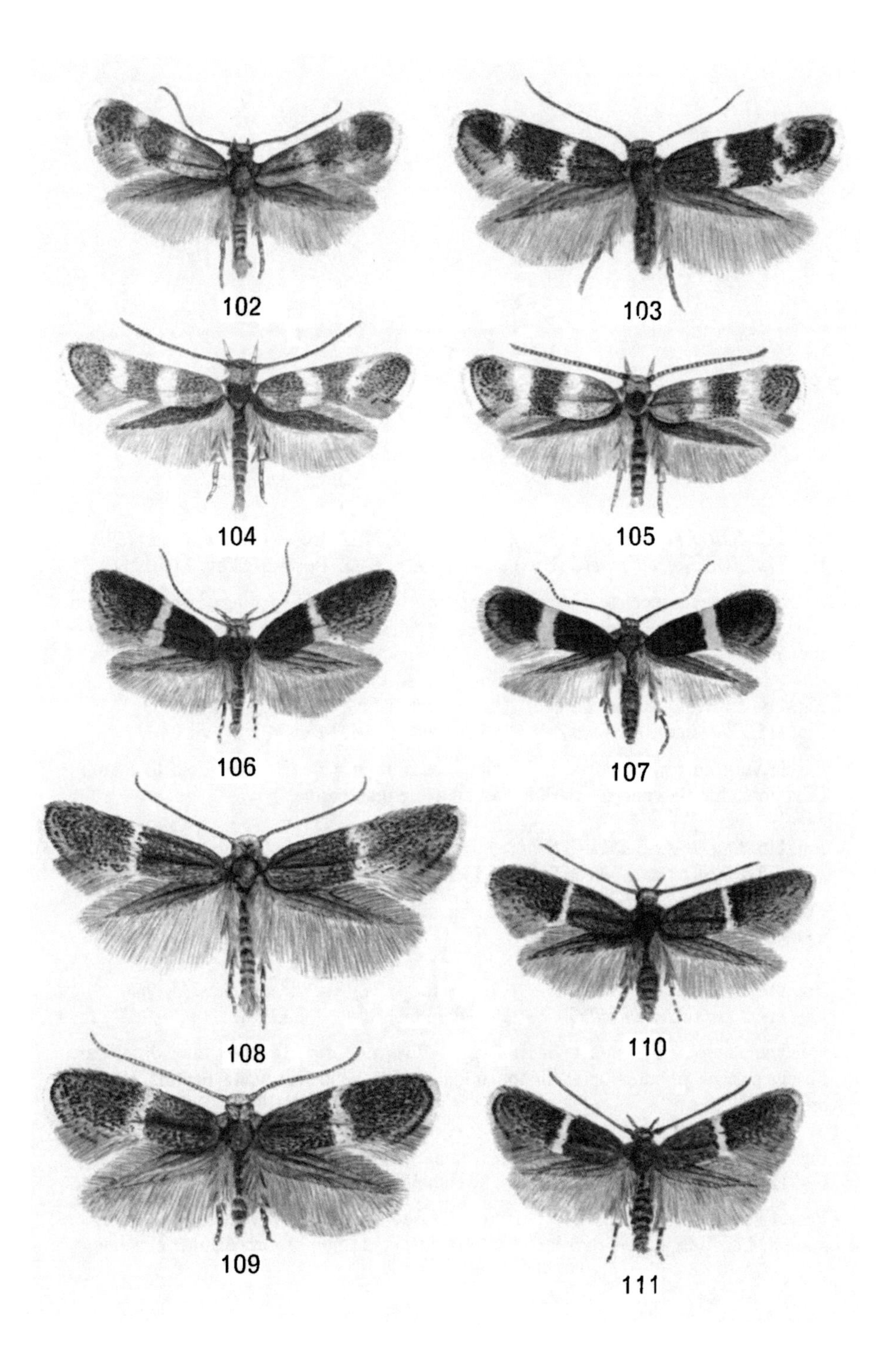
102
103
104
105
106
107
108
110
109
111

Fig. 112. *Elachista unifasciella* (Hw.), ♂, Denmark, NEZ, Boserup Skov, 27.vi.1969.
Fig. 113. *Elachista unifasciella* (Hw.), ♀, Denmark, NEZ, Boserup Skov, 2.vii.1969.

A broad-winged species. Fascia distinct from costa and widening towards dorsum. Cilia white around apex beyond indistinct cilia line; this is only visible in fresh specimens.

Fig. 114. *Elachista gangabella* Zell., ♂, Denmark, LFM, Møns Klint, 4.vii.1963.
Fig. 115. *Elachista gangabella* Zell., ♀, Denmark, LFM, Møns Klint, 4.vii.1963.

Fascia very narrow at costa, if reaching costa, then widening from fold to dorsum. Cilia not lighter beyond the dark brown cilia line than before.

Fig. 116. *Elachista subalbidella* Schl., ♂, Denmark, LFM, Virket Lyng, 6.vi.1965.
Fig. 117. *Elachista subalbidella* Schl., ♀, Denmark, EJ, Hvidding Krat, 21.vi.1969.

Forewing generally plain coloured, occasionally suffused with darker scales indicating a paler fascia.

Fig. 118. *Elachista revinctella* Zell., ♂, Denmark, NEZ, Roskilde, ex. l. 27.vi.1968.
Fig. 119. *Elachista revinctella* Zell., ♀, Sweden, Öl., Böda, 4.viii.1961.

Head and neck tufts pale ochreous-white. Base of forewing paler than distal part. Fascia narrow or interrupted in the middle. Cilia yellowish white beyond blackish brown cilia line.

Fig. 120. *Elachista bisulcella* (Dup.), ♂, Denmark, LFM, Kosteskov, 23.viii.1962.
Fig. 121. *Elachista bisulcella* (Dup.), ♀, Sweden, Sk., Löderup, 15–16.viii.1964.

Fascia widening from costa to dorsum, suffused with a warm yellowish along outer margin. Cilia pale yellowish from apex to middle of termen beyond distinct cilia line.

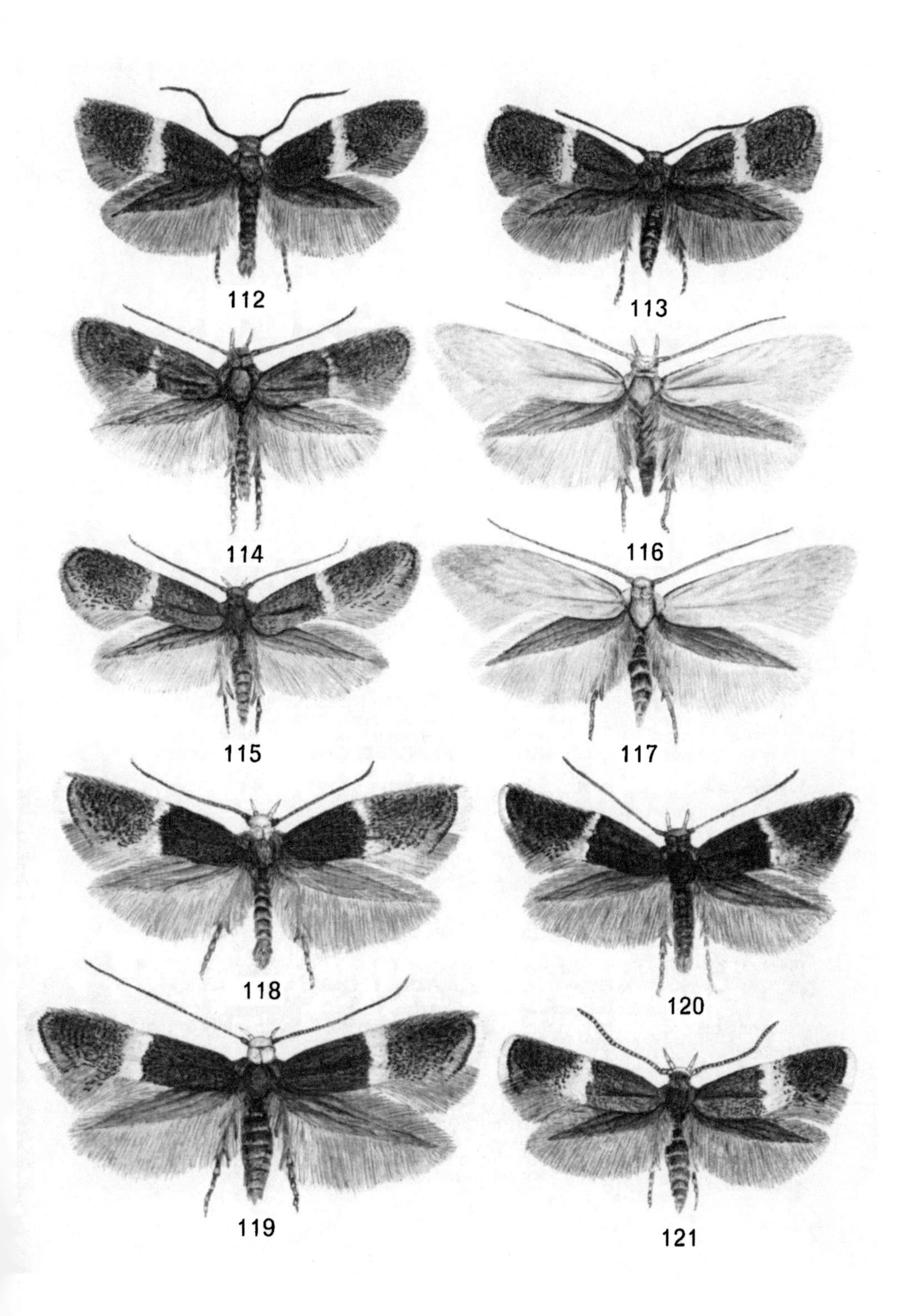
112
113
114
116
115
117
118
120
119
121

Fig. 122. *Biselachista trapeziella* (Stt.), ♂, Sweden, Vg., Billingen, 5–6.vii.1968.
Fig. 123. *Biselachista trapeziella* (Stt.), ♀, Sweden, Boh., Foss, 12.vii.1967.

Forewing almost plain dark brown with silky shining spots. Cilia line blurred.

Fig. 124. *Biselachista imatrella* (Schantz), ♂, Norway.
Wing pattern very indistinct.

Fig. 125. *Biselachista cinereopunctella* (Hw.), ♂, Italy, Interneppo, 10–27.v.1968.
Fig. 126. *Biselachista cinereopunctella* (Hw.), ♀, coll. British Museum (Nat. Hist.).

Base of forewing strongly mottled. Cilia line distinct, dark brown.

Fig. 127. *Biselachista kebneella* sp. n., ♂, Sweden, T. Lpm., Nikkaluokta, 2.vii.1974.
Fig. 128. *Biselachista kebneella* sp. n., ♀, Sweden, T. Lpm., Kåkitjårro, 9.vii.1974.

Forewing between fold and dorsum almost plain greyish. Cilia line only to middle of termen.

Fig. 129. *Biselachista ornithopodella* (Frey), ♂, Frey coll., in British Museum (Nat. Hist.).
Fig. 130. *Biselachista ornithopodella* (Frey), ♀, Frey coll., in British Museum (Nat. Hist.).

A small species. Forewing dark black-brown, spots silky shining, not metallic as in *E. tetragonella* (HS).

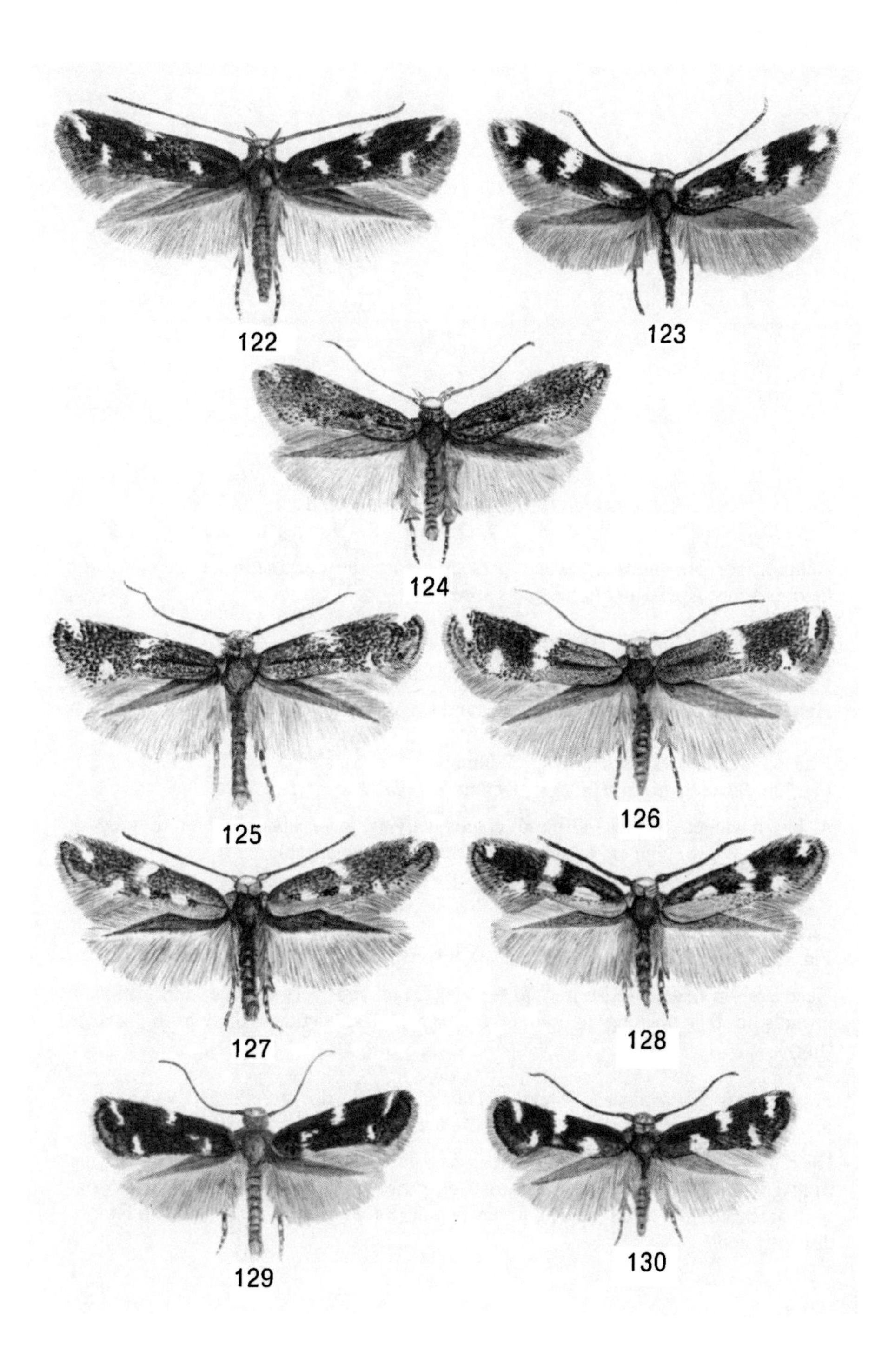
122
123
124
125
126
127
128
129
130

Fig. 131. *Biselachista serricornis* (Stt.), ♂, Denmark, LFM, Bøtø, 15.vii.1969.
Fig. 132. *Biselachista serricornis* (Stt.), ♀, Denmark, SJ, Holmegårds Mose, 19.vi.1971.

Ochreous or brownish red scales present before the dark suffusion of costa and before tornus. Apical and tornal spots absent.

Fig. 133. *Biselachista freyi* (Stgr.), ♂, Sweden, Gtl., Tingstäde, 4.vii.1969.
Fig. 134. *Biselachista freyi* (Stgr.), ♀, Sweden, Sm., Marbäck, 19.vii.1966.

Head white. Marks on forewing distinct and white.

Fig. 135. *Biselachista scirpi* (Stt.), ♂, Denmark, NEZ, Amager, l. v.1921.
Fig. 136. *Biselachista scirpi* (Stt.), ♀, Denmark, NEZ, Amager, l. v.1921.

A broad-winged species. Ground colour varying from almost white to grey or ochreous. White submarginal streak almost always recognizable.

Fig. 137. *Biselachista eleochariella* (Stt.), ♂, Denmark, NEZ, Karlstrup Strand, 22.vii.1969.
Fig. 138. *Biselachista eleochariella* (Stt.), ♀, Sweden, Sk., Hjärsås, 25–28.vii.1967.

Head grey in male, whitish grey in female. Costal spot very oblique, dark suffusion inwards to this pointing to middle of termen, less angulate than in *B. utonella* (Frey).

Fig. 139. *Biselachista utonella* (Frey), ♂, Denmark, NEZ, Karlstrup Strand, 9.vii.1969.
Fig. 140. *Biselachista utonella* (Frey), ♀, Sweden, Sk., Österslöv, 20.vi.1957.

Head greyish to cream-white, at least frons pale. Ground colour of forewing varying from grey to orange yellow or almost white; brown suffusion inside whitish costal spot arranged moreright angled to costa than in *B. eleochariella* (Stt.), and more distinctly bent.

131
132
133
135
134
136
137
138
139
140

Fig. 141. *Biselachista albidella* (Nyl.), ♂, Denmark, LFM, Horreby Lyng, 15.vii.1969.
Fig. 142. *Biselachista albidella* (Nyl.), ♀, Denmark, LFM, Horreby Lyng, 15.vii.1969.

Head white. Margin of costa basally blackish, black plical streak prominent.

Fig. 143. *Biselachista abiskoella* (Bengts.), ♂, Sweden, T. Lpm., Abisko Östra, 9.vii.1976.
Fig. 144. *Biselachista abiskoella* (Bengts.), ♀, Sweden, T. Lpm., 2 km V. Abisko, 9.vii.1976.

Male with two fasciae. Female with costal and tornal spots of same size, opposite.

Fig. 145. *Cosmiotes freyerella* (Hb.), ♂, Denmark, LFM, Vindeholme Skov, 7.vi.1969.
Fig. 146. *Cosmiotes freyerella* (Hb.), ♀, Denmark, NWZ, Bromølle, 27.v.1962.

Head, or at least frons, whitish. Neck tufts greyish.

Fig. 147. *Cosmiotes exactella* (HS.), ♂, Denmark, EJ, Løvenholm Skov, 5.vii.1968.
Fig. 148. *Cosmiotes exactella* (HS.), ♀, Denmark, NEZ, Tisvilde, 2.vi.1969.

Head and neck tufts plain grey coloured as forewing.

Fig. 149. *Cosmiotes stabilella* (Stt.), ♂, Denmark, NEZ, Kongelunden, 4.vii.1954.
Fig. 150. *Cosmiotes stabilella* (Stt.), ♀, B, Melsted, 23.vii.1928.

Head and neck tufts white.

Fig. 151. *Cosmiotes consortella* (Stt.), ♂, Denmark, LFM, Høvblege, 16.viii.1969.
Fig. 152. *Cosmiotes consortella* (Stt.), ♀, Denmark, LFM, Høvblege, 30.vii.1961.

Frons pale beige, neck tufts mottled with darker tipped scales, paler than forewing.

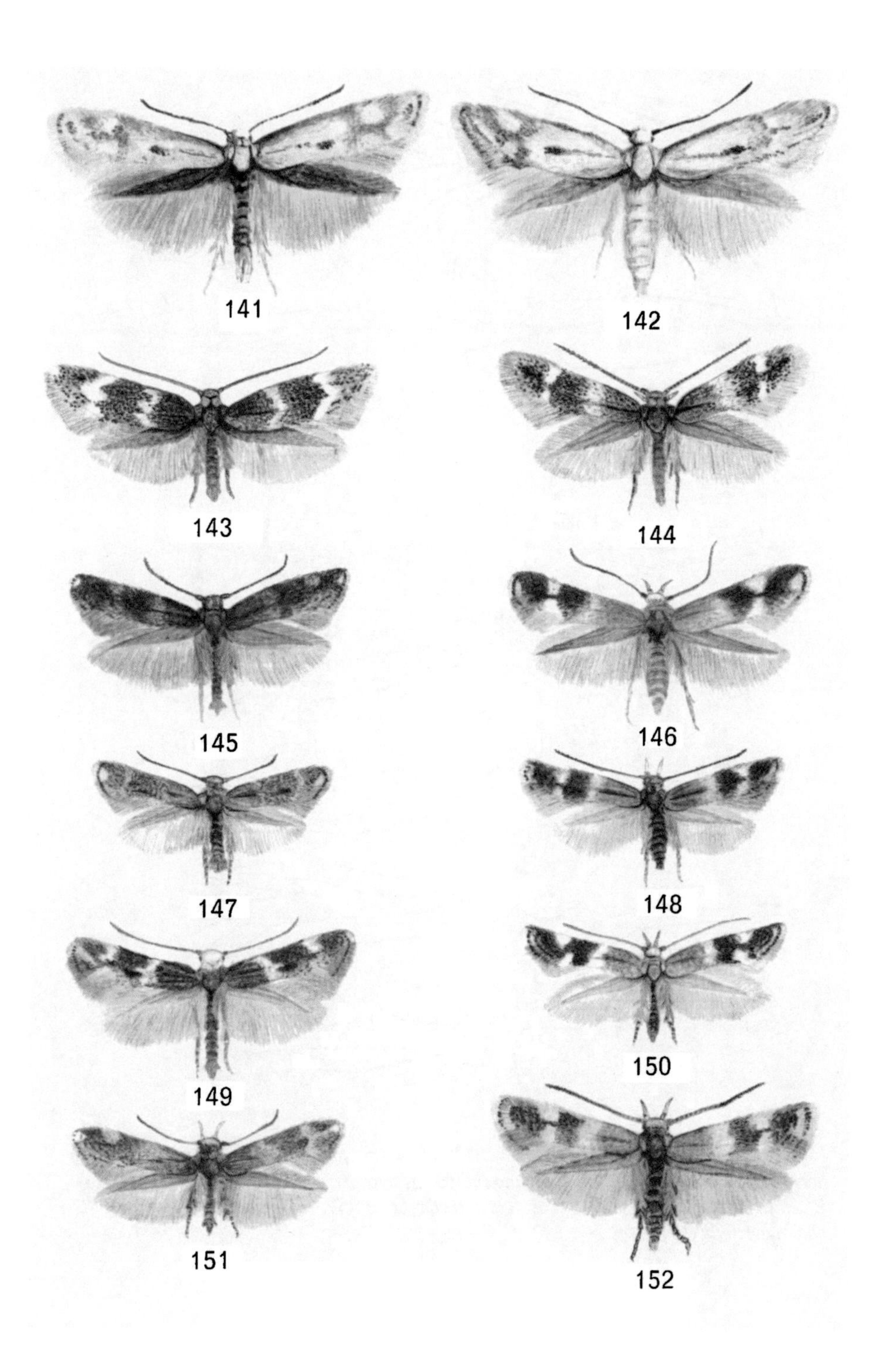
141
142
143
144
145
146
147
148
149
150
151
152

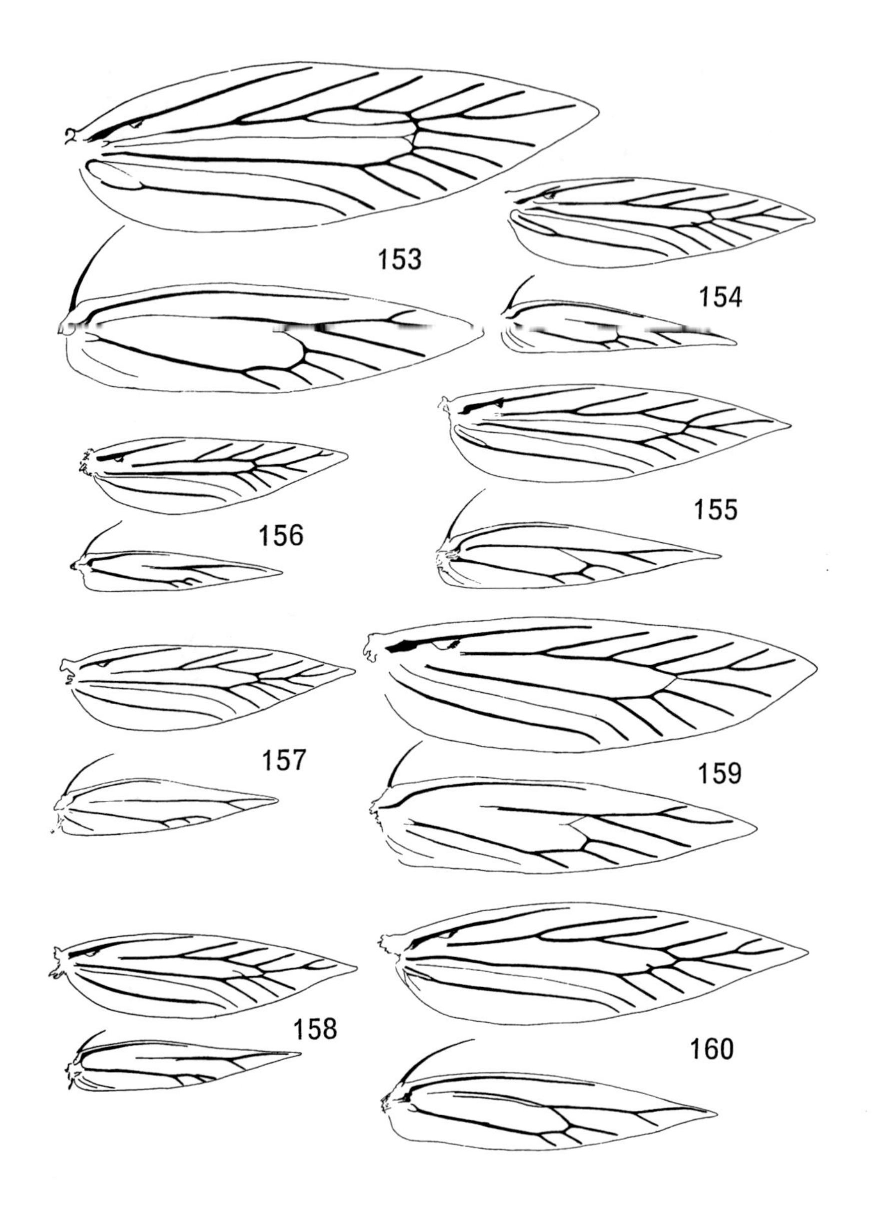

Figs. 153–160. Venation of males. – 153: *Mendesia farinella* (Thnbg.); 154: *Perittia herrichiella* (HS.); 155: *P. obscurepunctella* (Stt.); 156: *Stephensia brunnichella* (L.); 157: *Elachista regificella* Sirc.; 158: *E. gleichenella* (F.); 159: *E. pigerella* (HS.); 160: *E. quadripunctella* (Hb.).

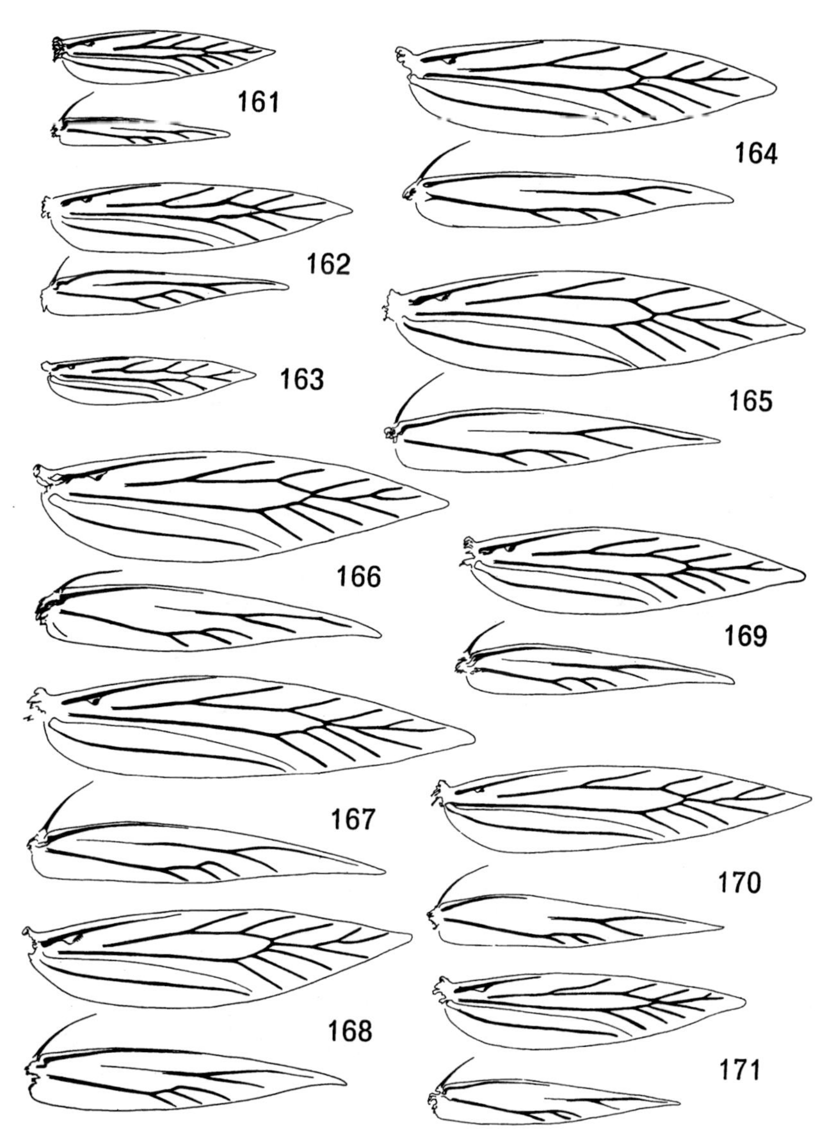

Figs 161–171. Venation of male *Elachista*. – 161: *E. tetragonella* (HS.); 162: *E. biatomella* (Stt.); 163: *E. martinii* Hofm.; 164: *E. poae* Stt.; 165: *E. atricomella* Stt.; 166: *E. kilmunella* Stt.; 167: *E. parasella* Tr.-O.; 168: *E. alpinella* Stt.; 169: *E. diederichsiella* Her.; 170: *E. compsa* Tr.-O.; 171: *E. elegans* Frey.

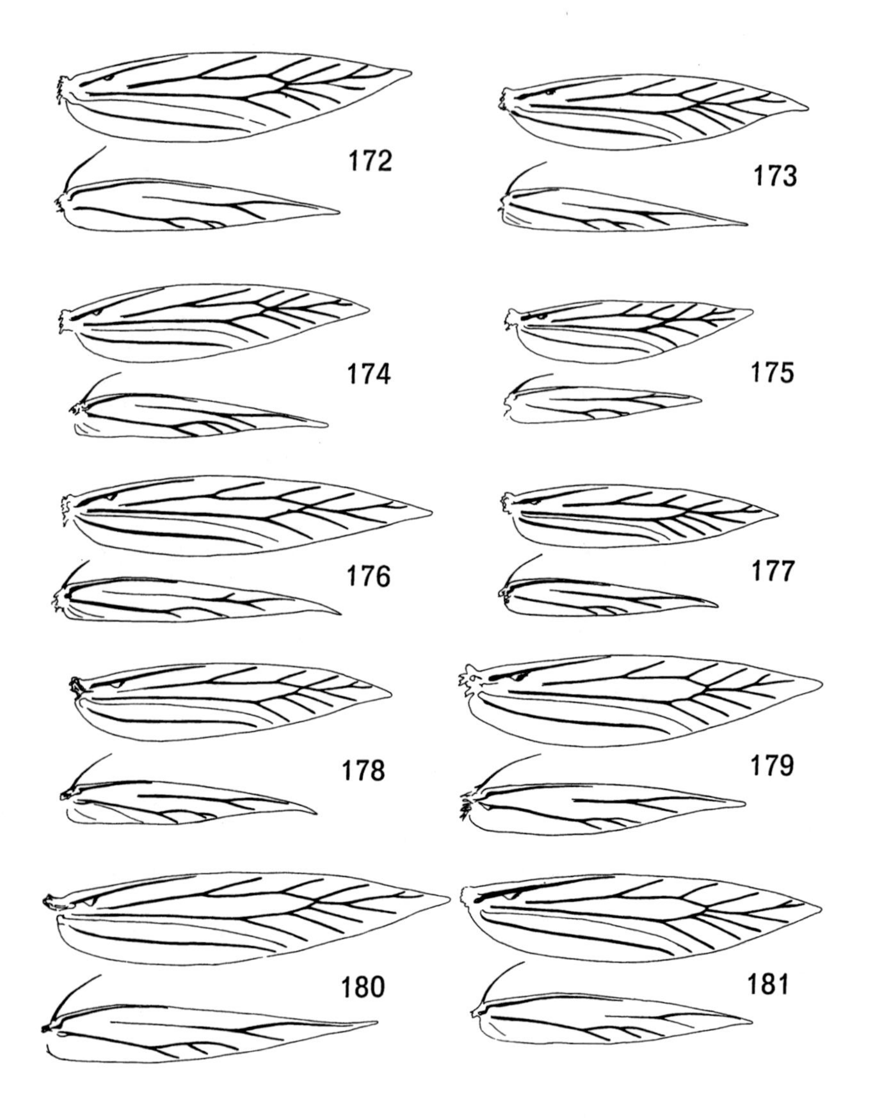

Figs. 172–181. Venation of male *Elachista*. – 172: *E. luticomella* Zell.; 173: *E. albifrontella* (Hb.); 174: *E. bifasciella* Tr.; 175: *E. nobilella* Zell.; 176: *E. apicipunctella* Stt.; 177: *E. subnigrella* Dougl.; 178: *E. reuttiana* Frey; 179: *E. orstadii* Palm; 180: *E. ingvarella* Tr.–O.; 181: *E. krogeri* Svens.

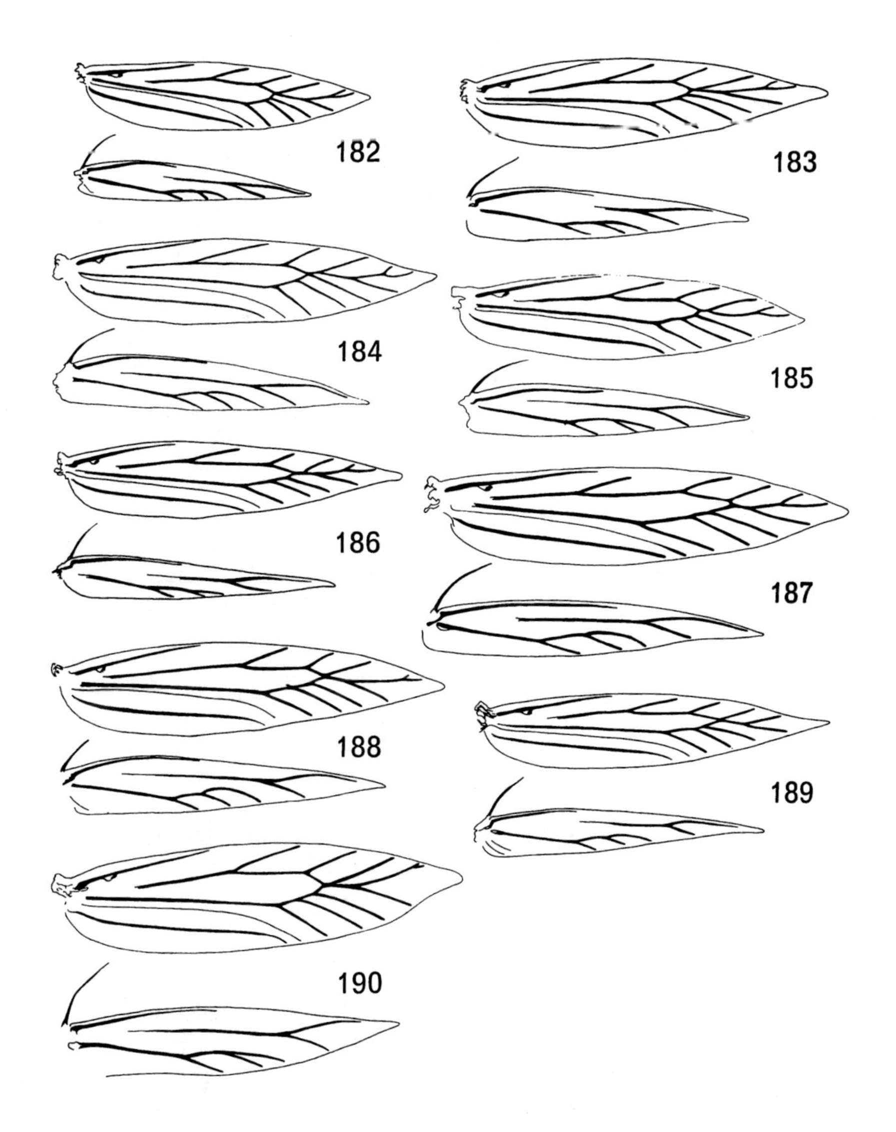

Figs. 182–190. Venation of *Elachista*. – 182: *E. nielswolffi* Svens.; 183: *E. pomerana* Frey; 184: *E. humilis* Zell.; 185: *E. vonschantzi* Svens.; 186: *E. pulchella* (Hw.); 187: *E. anserinella* Zell.; 188: *E. rufocinerea* (Hw.); 189: *E. lastrella* Chrét.; 190: *E. cerusella* (Hb.).

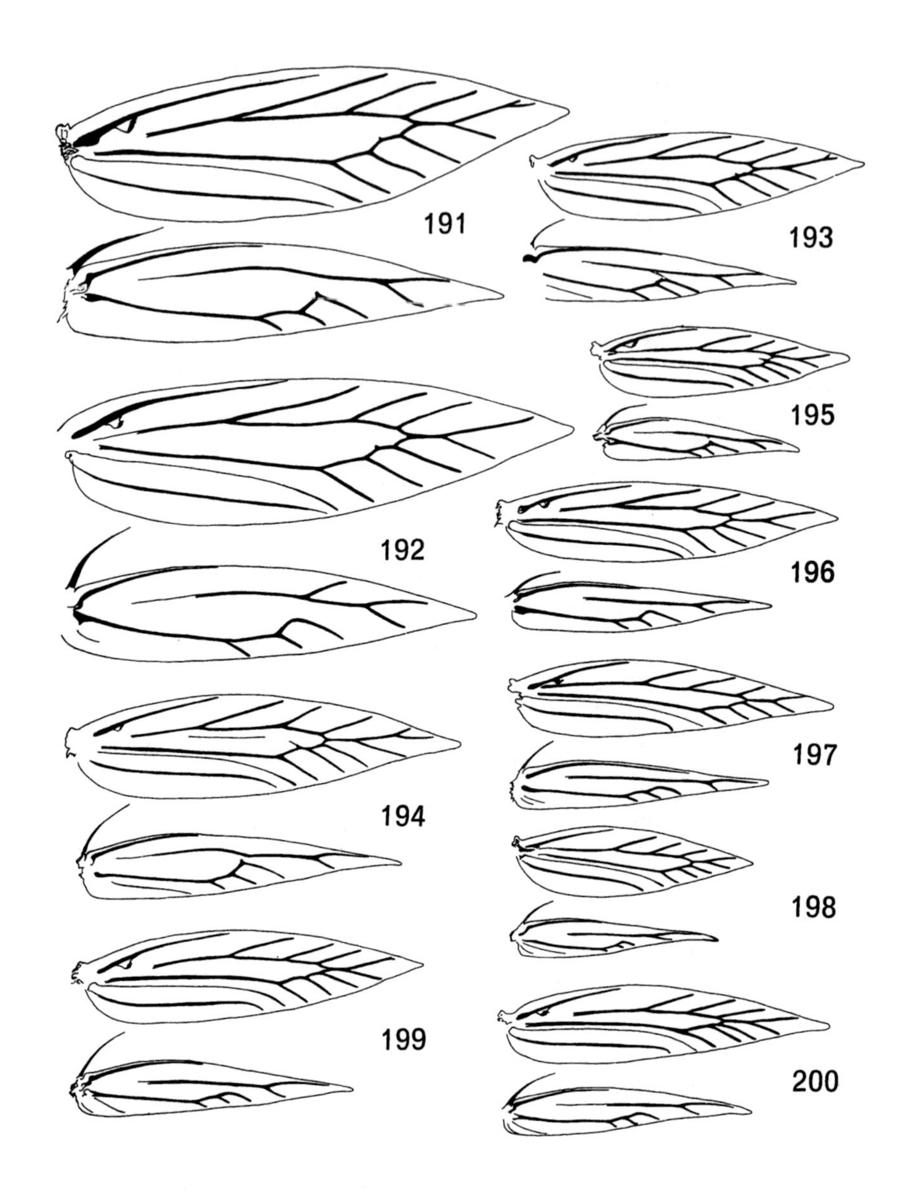

Figs. 191–200. Venation of *Elachista*. – 191, 192: *E. argentella* (Cl.); 193: *E. pollinariella* Zell.; 194: *E. triatomea* (Hw.); 195: *E. collitella* (Dup.); 196: *E. subocellea* (Stph.); 197: *E. festucicolella* Zell.; 198: *E. nitidulella* (HS.); 199: *E. dispilella* Zell.; 200: *E. triseriatella* Stt.

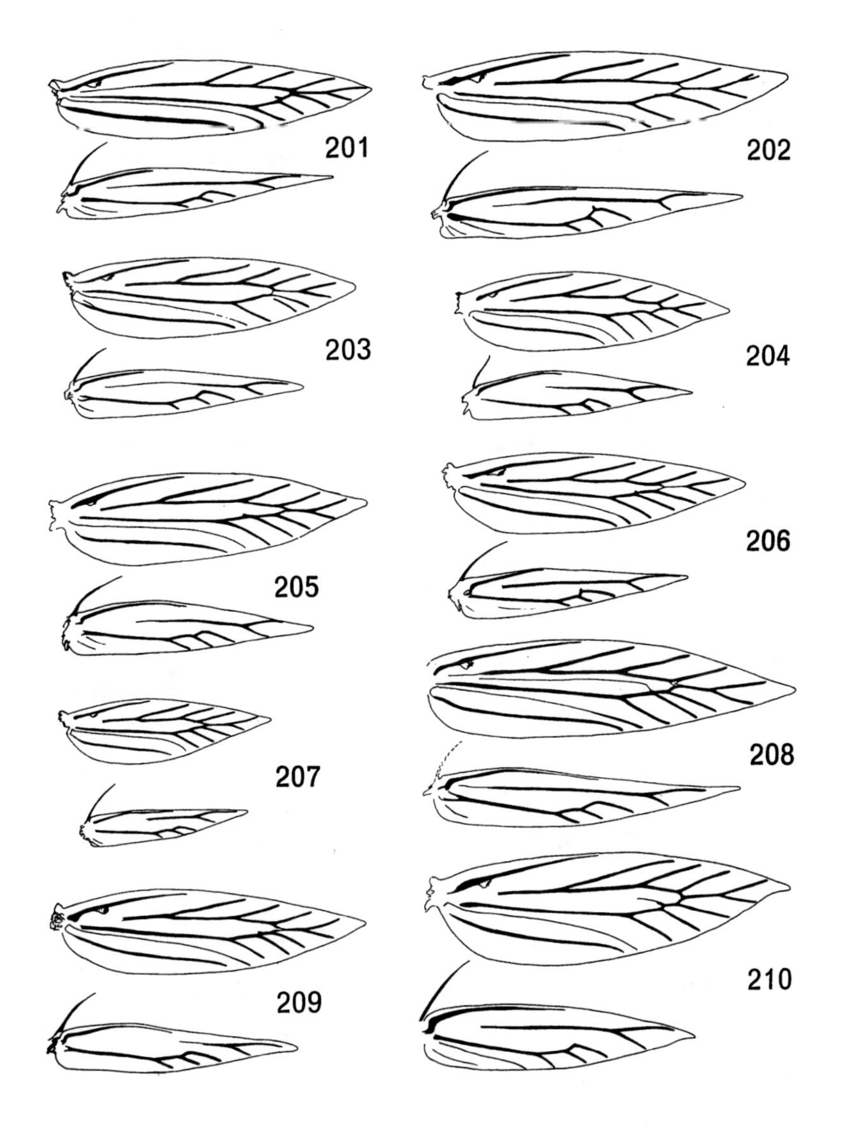

Figs. 201–210. Venation of *Elachista*. – 201: *E. dispunctella* (Dup.); 202: *E. rudectella* Stt.; 203: *E. squamosella* (HS.); 204: *E. bedellella* (Sirc.); 205: *E. pullicomella* Zell.; 206: *E. littoricola* Le March.; 207: *E. chrysodesmella* Zell.; 208: *E. megerlella* (Hb.); 209: *E. cingillella* (HS.); 210: *E. unifasciella* (Hw.).

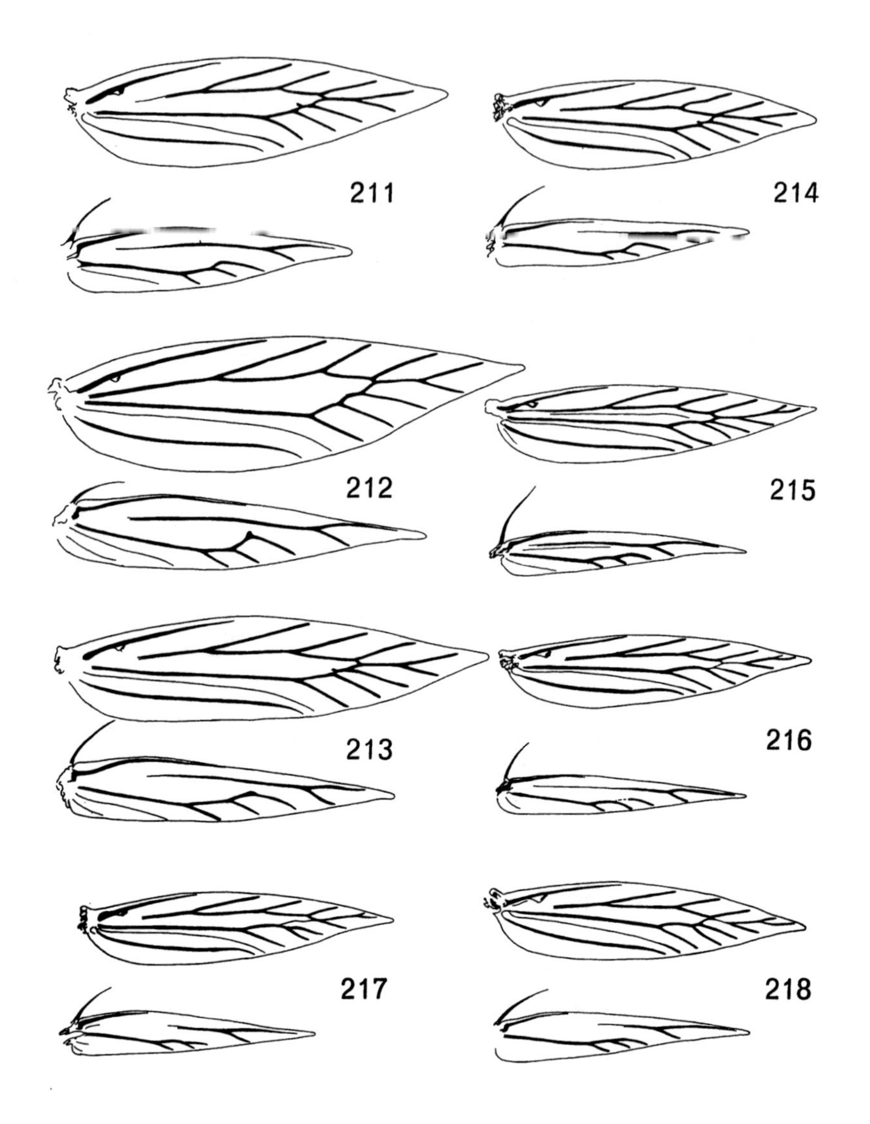

Figs. 211–218. Venation of males. – 211: *Elachista gangabella* Zell.; 212: *E. subalbidella* Schl.; 213: *E. revinctella* Zell.; 214: *E. bisulcella* (Dup.); 215: *Biselachista trapeziella* (Stt.); 216: *B. imatrella* (Schantz); 217: *B. cinereopunctella* (Hw.); 218: *B. kebneella* sp. n.

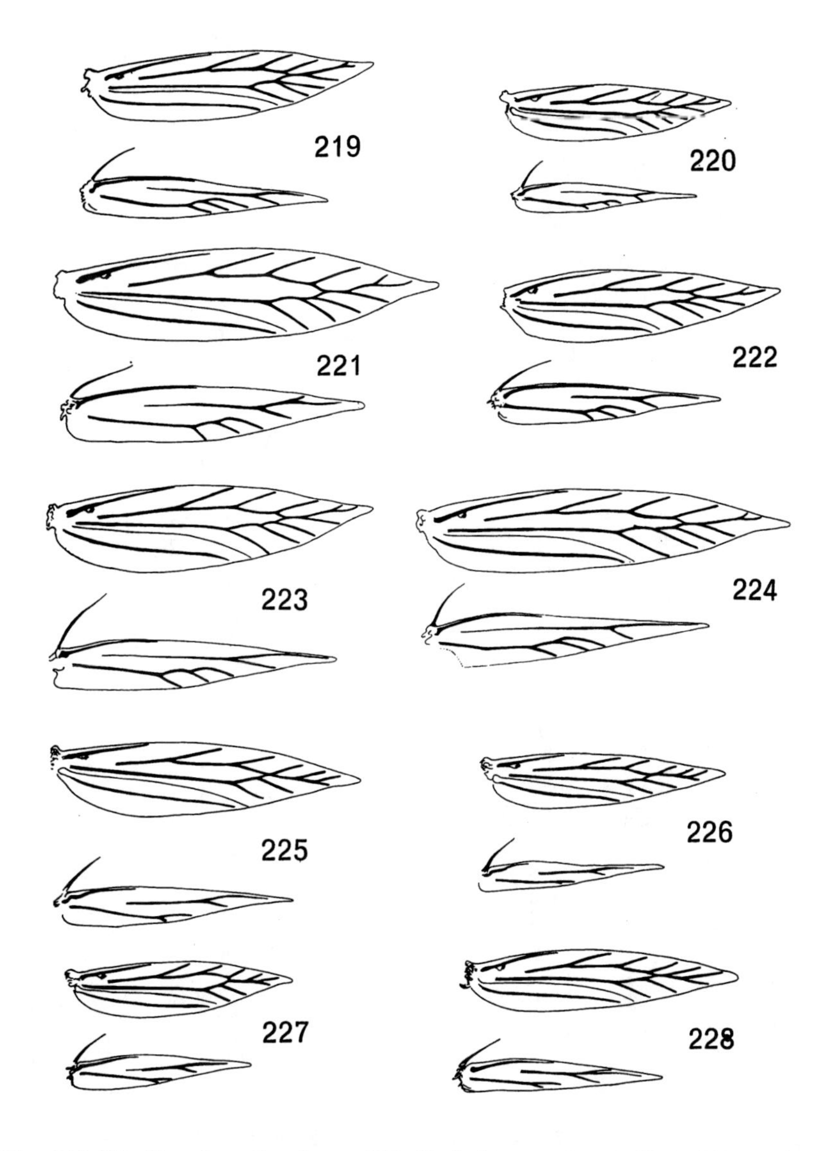

Figs. 219–228. Venation of males. – 219: *Biselachista serricornis* (Stt.); 220: *B. freyi* (Stgr.); 221: *B. scirpi* (Stt.); 222: *B. eleochariella* (Stt.); 223: *B. utonella* (Frey); 224: *B. albidella* (Nyl.); 225: *Cosmiotes freyerella* (Hb.); 226: *C. exactella* (HS.); 227: *C. stabilella* (Stt.); 228: *C. consortella* (Stt.).

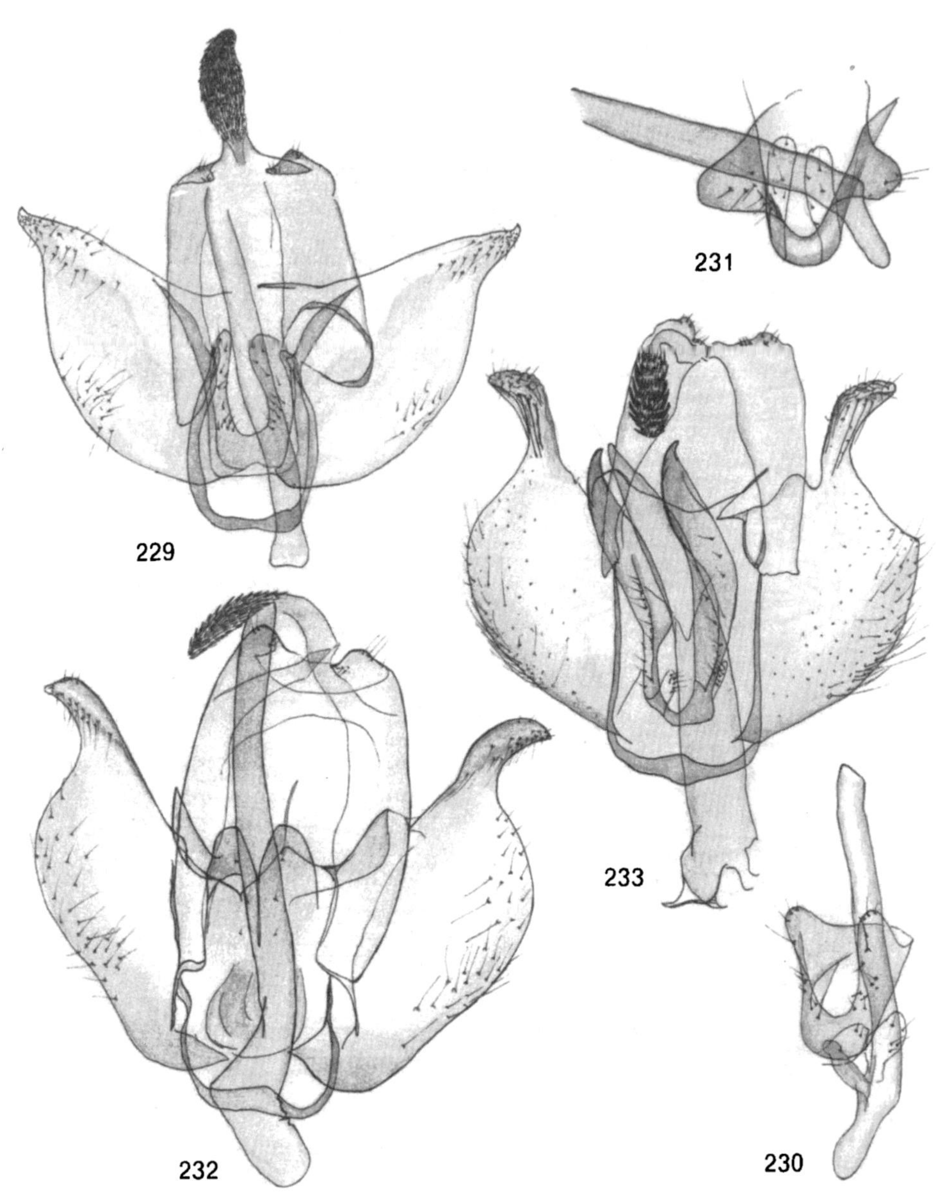

Figs. 229–231. *Mendesia farinella* (Thnbg.). – 229: male genitalia (ETO 1636); 230: digitate process, juxta, aedeagus (ETO B. 7.2.77); 231: digitate process, juxta, aedeagus (ETO A. 7.2.77).
Fig. 232. *Perittia herrichiella* (HS.), male genitalia (ETO 1489).
Fig. 233. *Perittia obscurepunctella* (Stt.), male genetalia (ETO 1634).

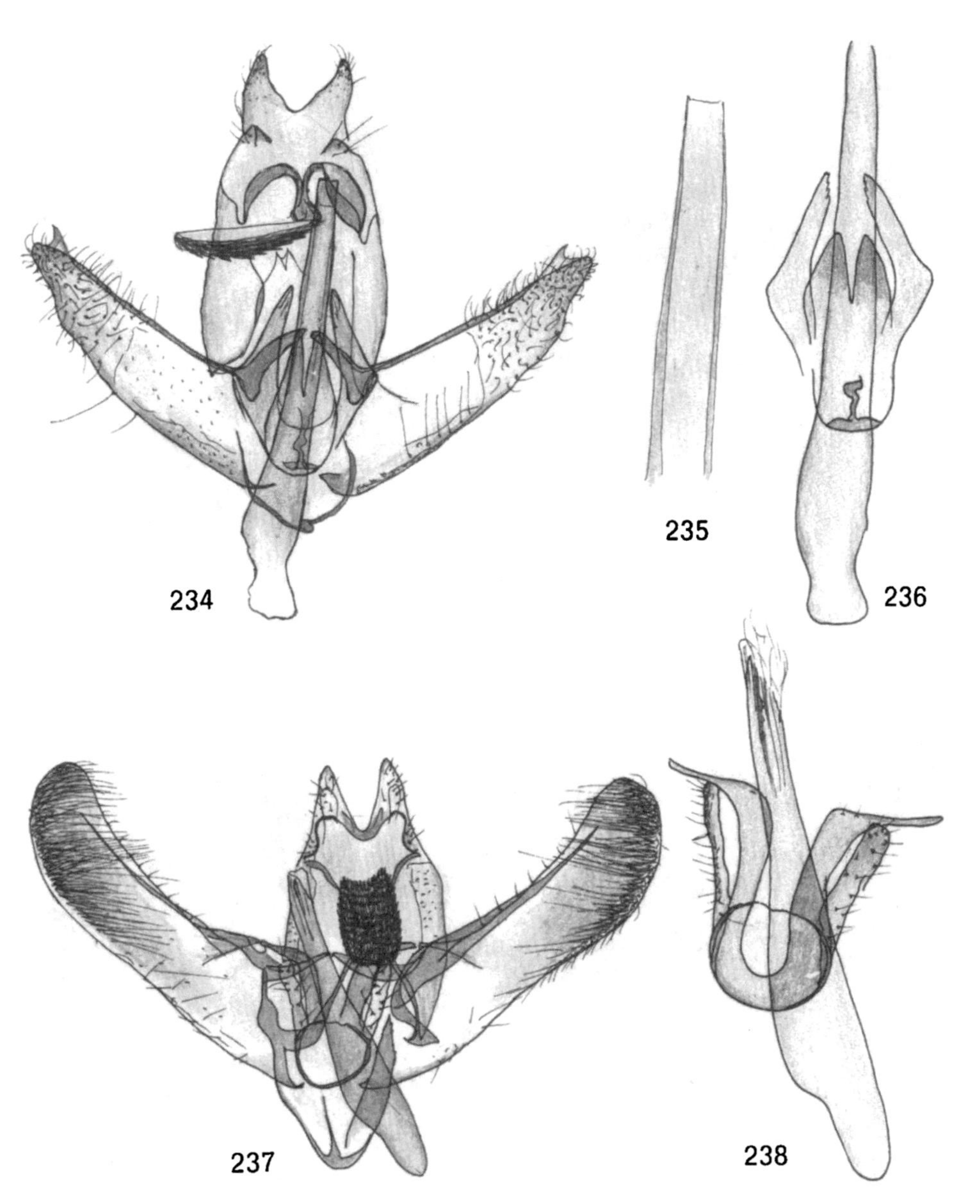

Figs. 234–236. *Stephensia brunnichella* (L.). – 234: male genitalia (ETO 1902); 235: distal end of aedeagus (ETO 1902); 236: digitate process, juxta (ETO 1902).
Figs. 237, 238. *Elachista regificella* Sirc. – 237: male genetalia (ETO 1129); 238: digitate process, juxta, aedeagus (ETO 1129).

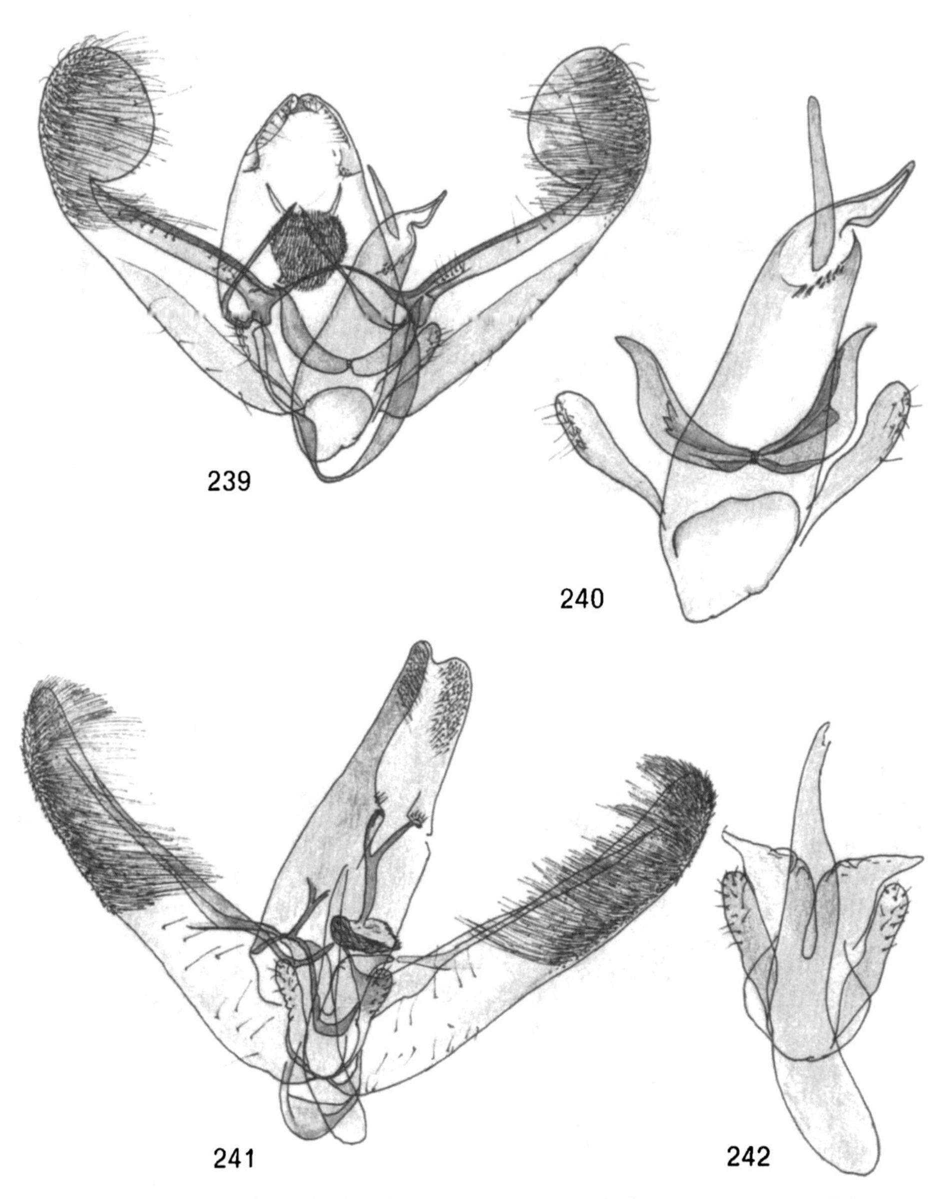

Figs. 239, 240. ***Elachista gleichenella*** (F.). – 239: male genitalia (ETO 1364); 240: digitate process, juxta, aedeagus (ETO 1364).
Figs. 241, 242. ***Elachista pigerella*** (HS.). – 241: male genitalia (ETO 2050); 242: digitate process, juxta, aedeagus (ETO 2050).

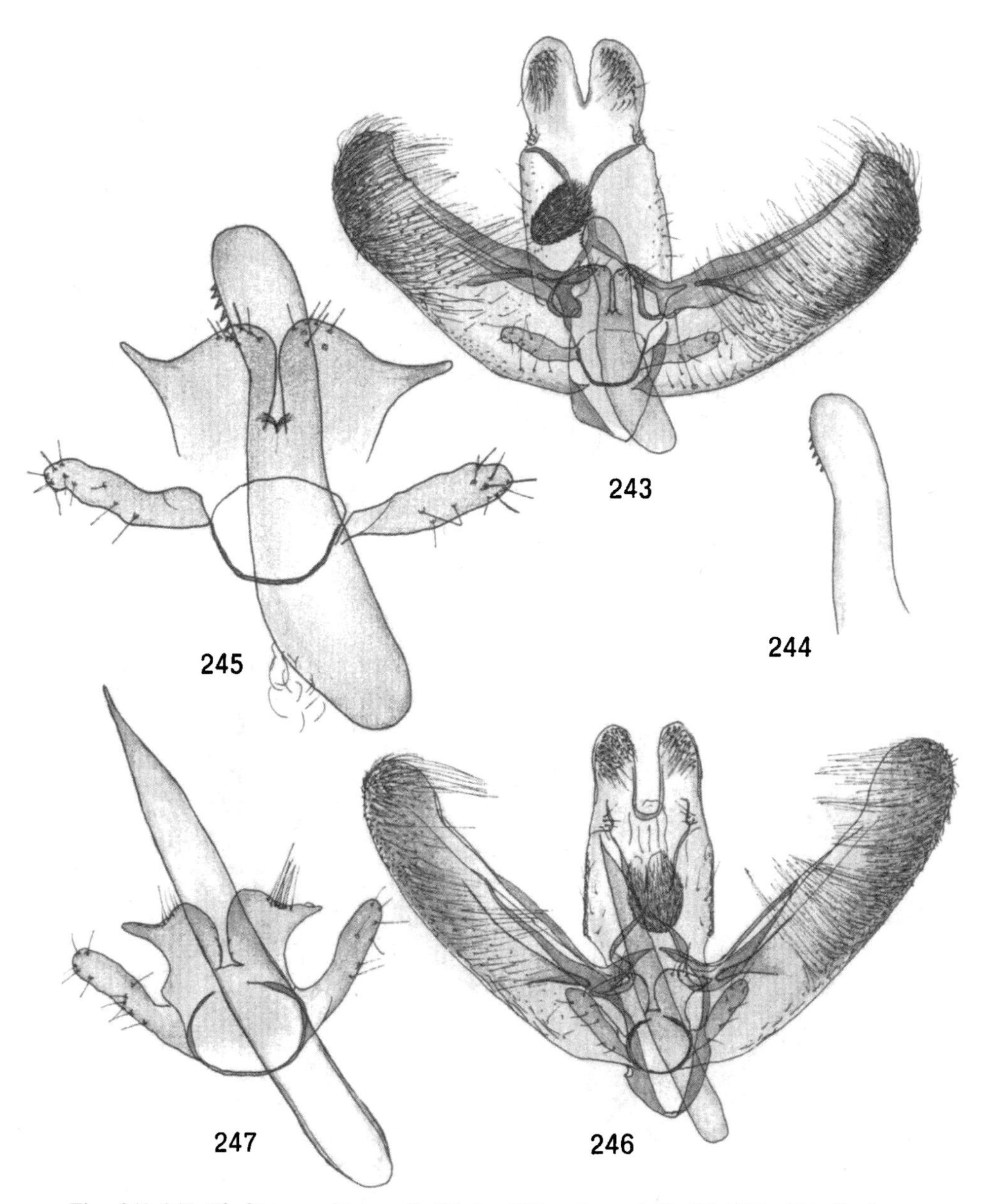

Figs. 243–245. *Elachista quadripunctella* (Hb.). – 243: male genitalia (IS 4135); 244: distal end of aedeagus (IS 4135); 245: digitate process, juxta, aedeagus (IS 4135).
Figs. 246, 247. *Elachista tetragonella* (HS.). – 246: male genitalia (ETO 1868); 247: digitate process, juxta, aedeagus (ETO 1868).

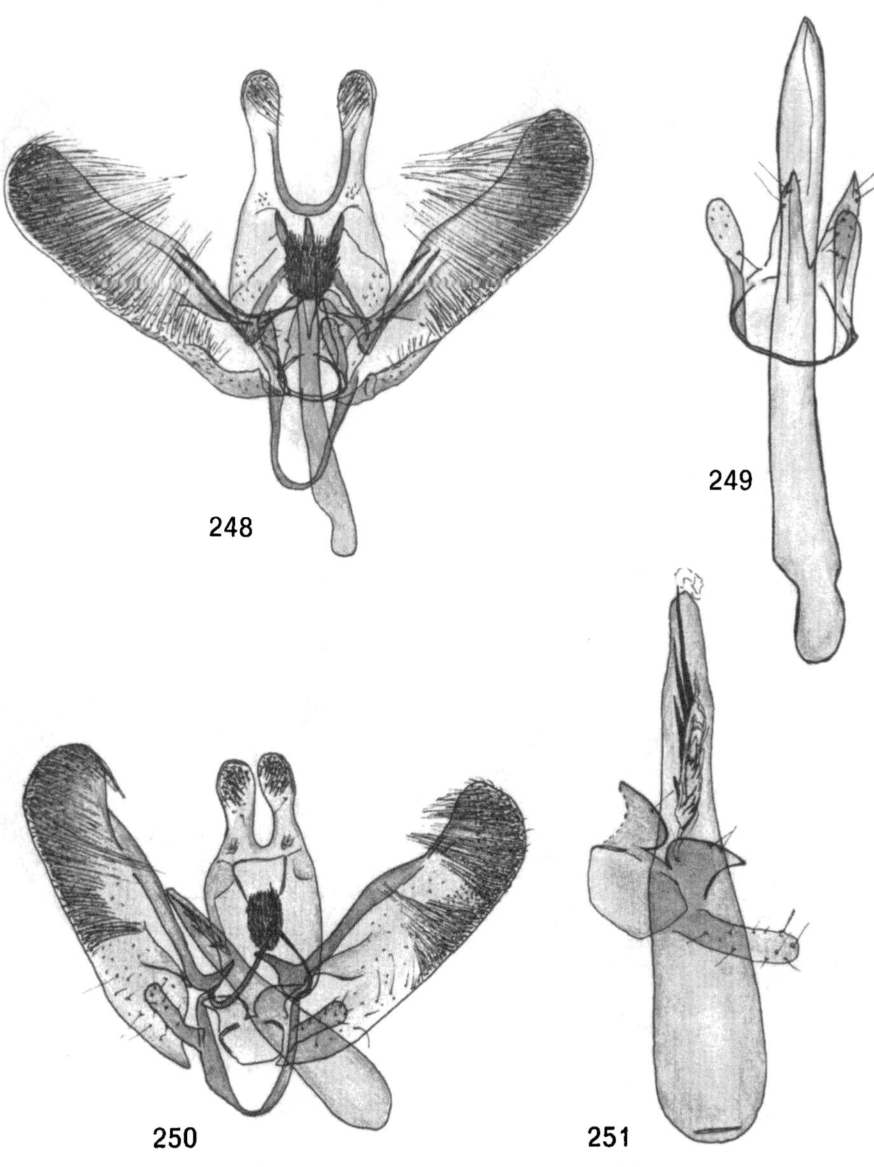

Figs. 248, 249. *Elachista biatomella* (Stt.). – 248: male genitalia (ETO 1449); 249: digitate process, juxta, aedeagus (ETO 1449).
Figs. 250, 251. *Elachista martinii* Hofm. – 250: male genitalia (ETO 2053); 251: digitate process, juxta, aedeagus (ETO 2053).

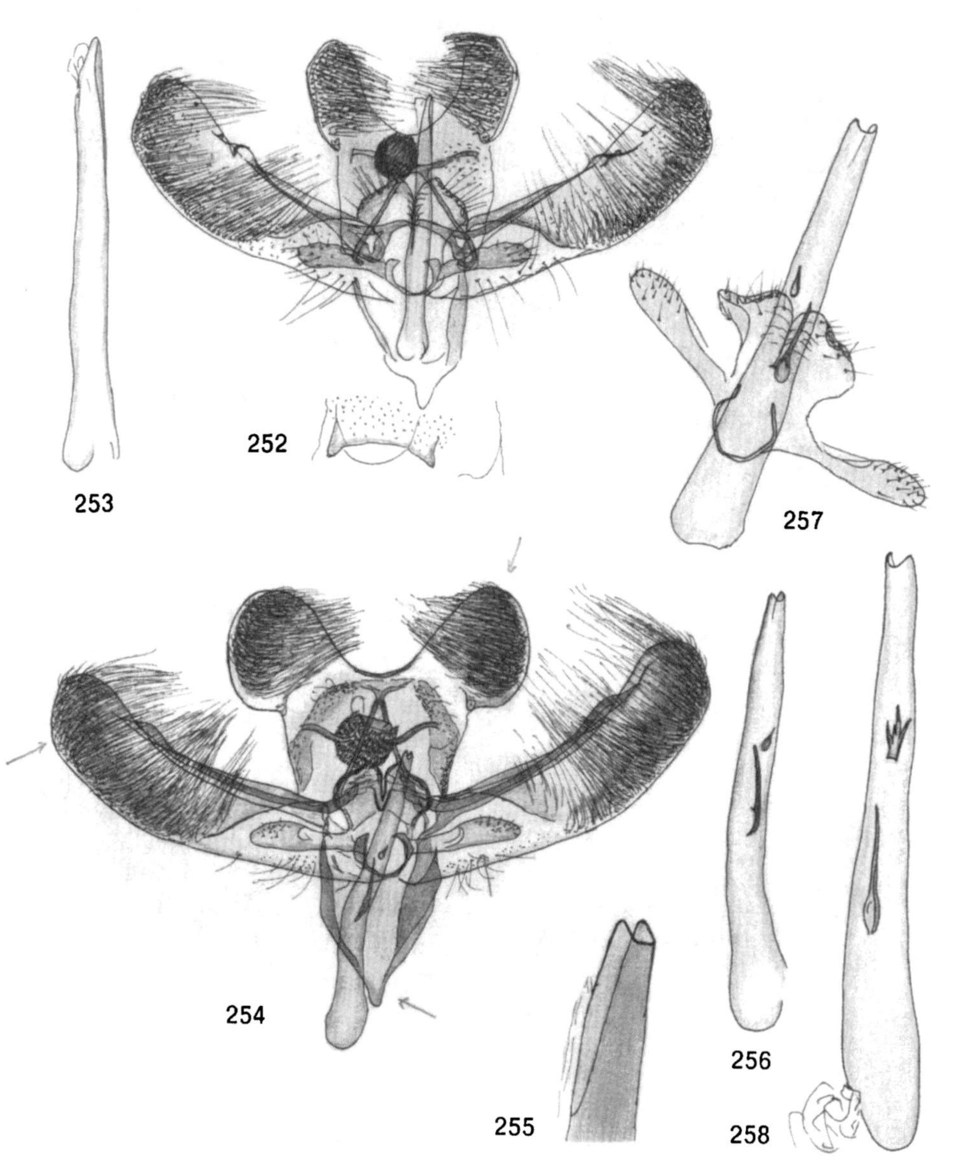

Figs. 252, 253. *Elachista poae* Stt. - 252: male genitalia (ETO 1880); 253: aedeagus (ETO 1880).
Figs. 254–258. *Elachista atricomella* Stt. - 254: male genitalia (ETO 800); 255: distal end of aedeagus (ETO 800); 256: aedeagus (ETO 800); 257: digitate process, juxta, aedeagus (ETO 1492); 258: aedeagus (BMNH 10783).

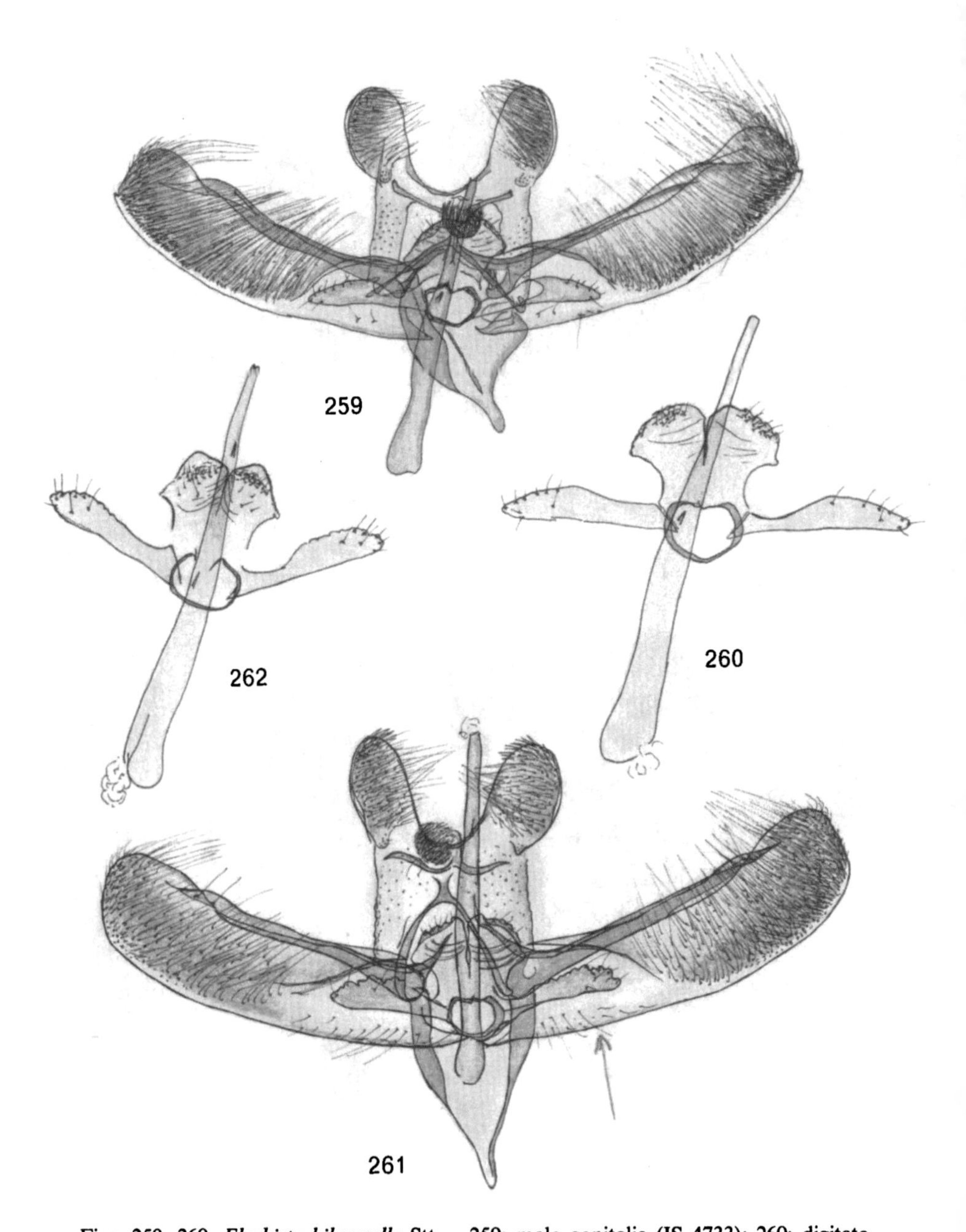

Figs. 259, 260. *Elachista kilmunella* Stt. - 259: male genitalia (IS 4733); 260: digitate process, juxta, aedeagus (IS 4733).
Figs. 261, 262. *Elachista parasella* Tr.-O. - 261: male genitalia (MvS 472); 262: digitate process, juxta, aedeagus (MvS 472).

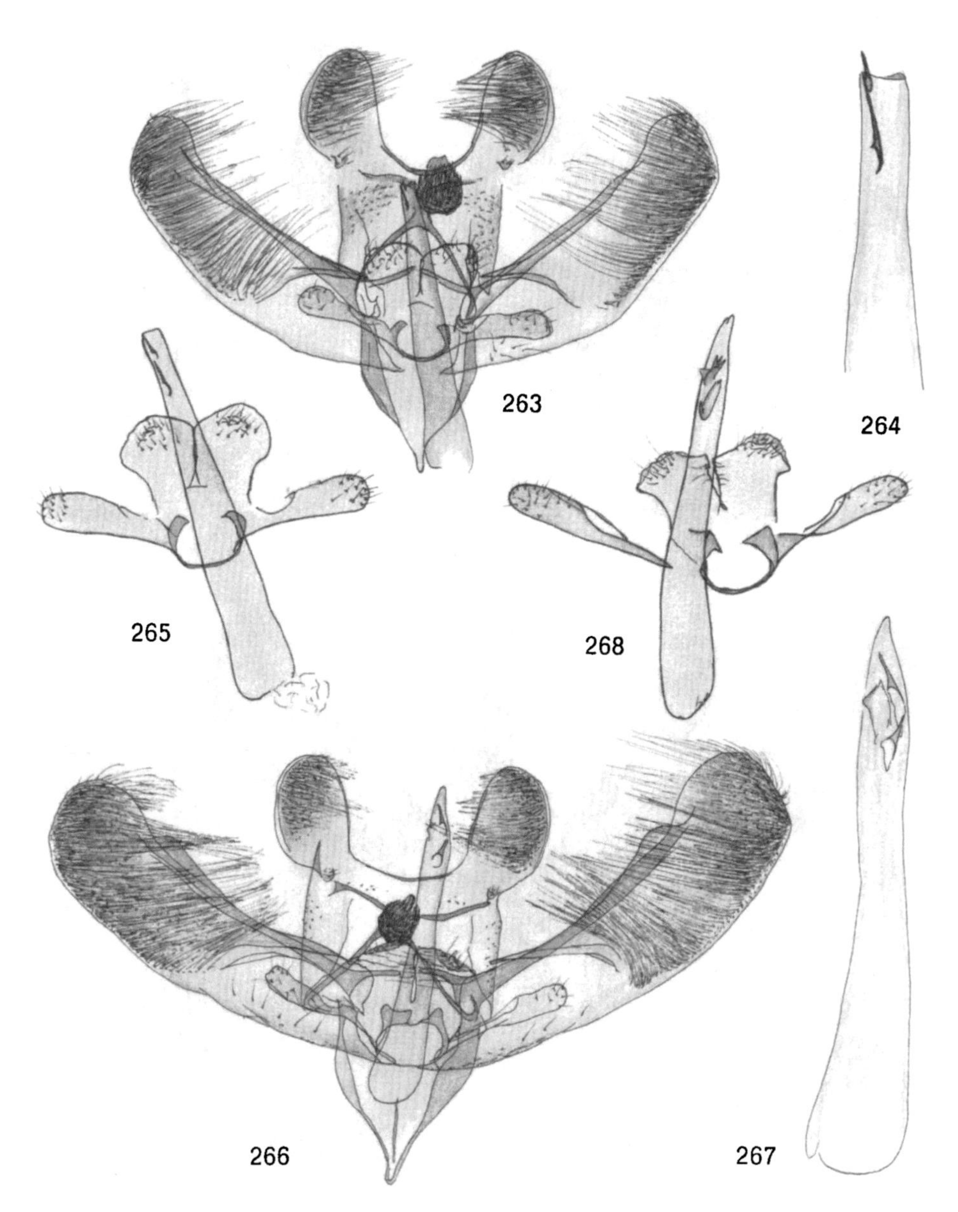

Figs. 263–265. *Elachista alpinella* Stt. – 263: male genitalia (ETO 1776) 264: distal end of aedeagus (ETO 1776); 265: digitate process, juxta, aedeagus (ETO 1776).
Figs. 266–268. *Elachista diederichsiella* Her. – 266: male genitalia (ETO 1842); 267: aedeagus (ETO 1842); 268: digitate process, juxta, aedeagus (OK 2369).

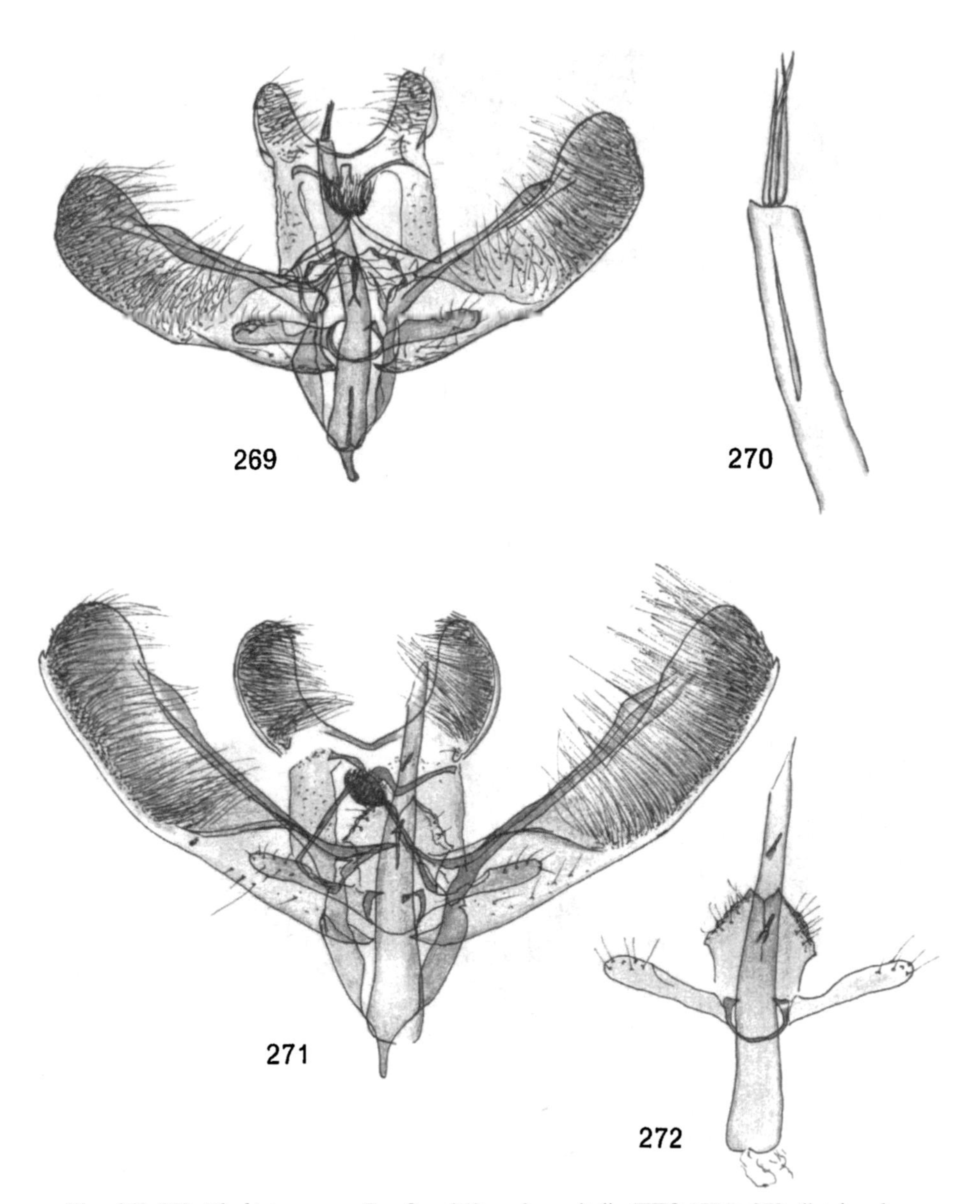

Figs. 269, 270. *Elachista compsa* Tr.–O. – 269: male genitalia (ETO 1784); 270: distal end of aedeagus (ETO 1784).
Figs. 271, 272. *Elachista elegans* Frey. – 271: male genitalia (IS 4781); 272: digitate process, juxta, aedeagus (IS 4781).

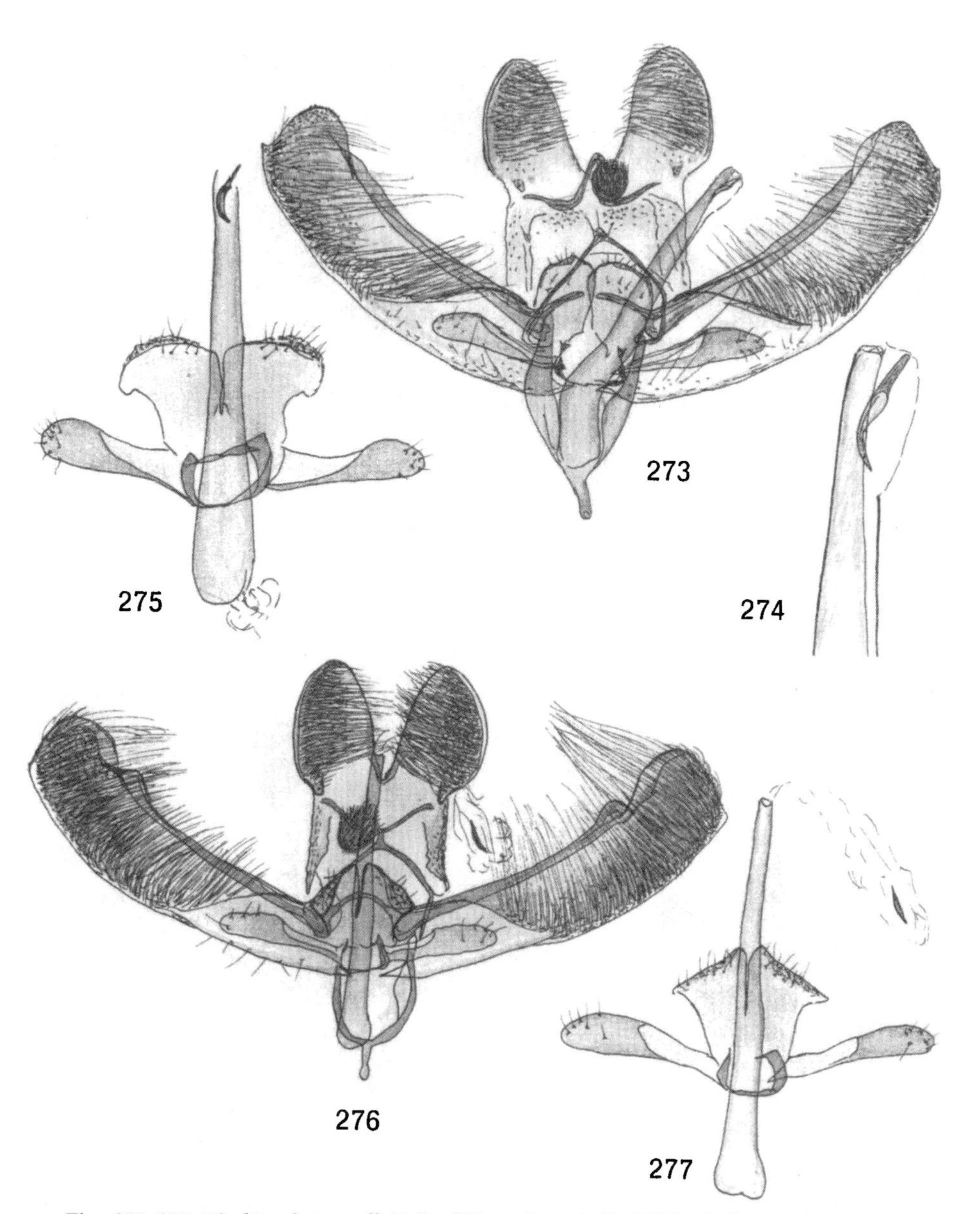

Figs. 273–275. *Elachista luticomella* Zell. – 273: male genitalia (ETO 1356); 274: distal end of aedeagus (ETO 1356); 275: digitate process, juxta, aedeagus (ETO 1333).
Figs. 276, 277. *Elachista albifrontella* (Hb.). – 276: male genitalia (ETO 590); 277: digitate process, juxta, aedeagus (ETO 590).

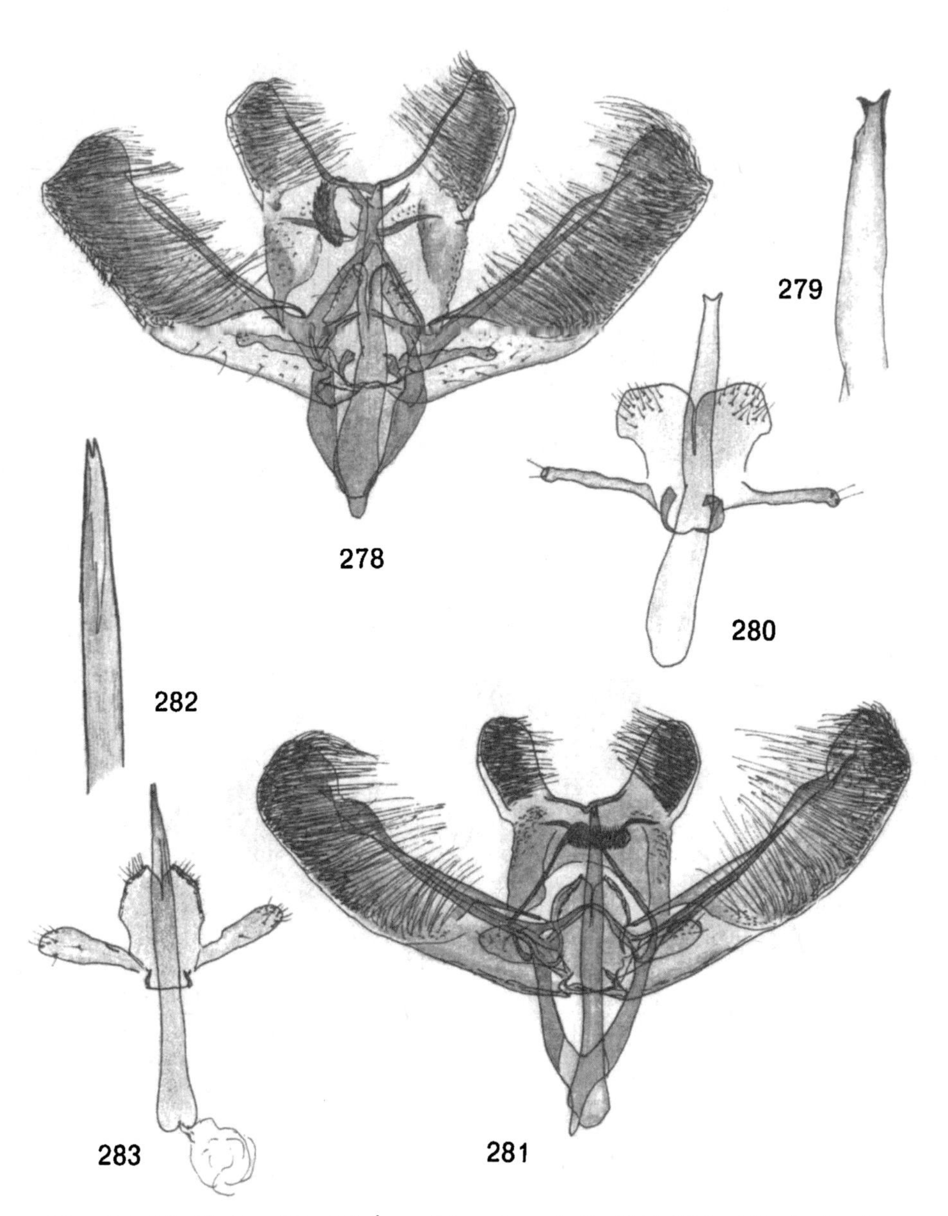

Figs. 278–280. *Elachista bifasciella* Tr. – 278: male genitalia (ETO 1286); 279: distal end of aedeagus (ETO 1286); 280: digitate process, juxta, aedeagus (ETO 1286).
Figs. 281–283. *Elachista nobilella* Zell. – 281: male genitalia (ETO 1121); 282: distal end of aedeagus (ETO 1121); 283: digitate process, juxta, aedeagus (ETO 1121).

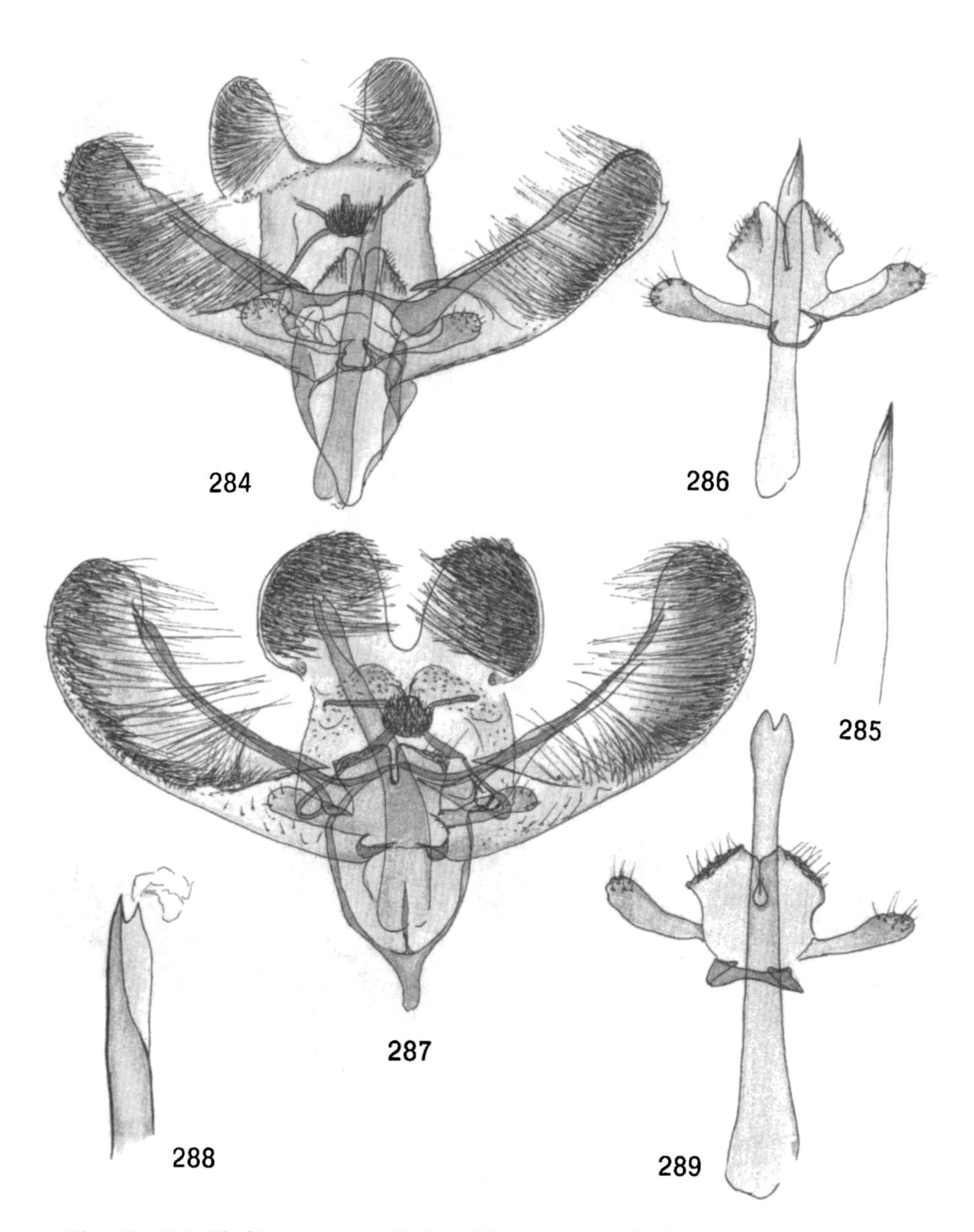

Figs. 284–286. *Elachista apicipunctella* Stt. – 284: male genitalia (ETO 1454); 285: distal end of aedeagus (ETO 1454); 286: digitate process (ETO 1454).
Figs. 287–289. *Elachista subnigrella* Dougl. – 287: male genitalia (ETO 1659); 288: distal end of aedeagus (ETO 1659); 289: digitate process, juxta, aedeagus (ETO 1343).

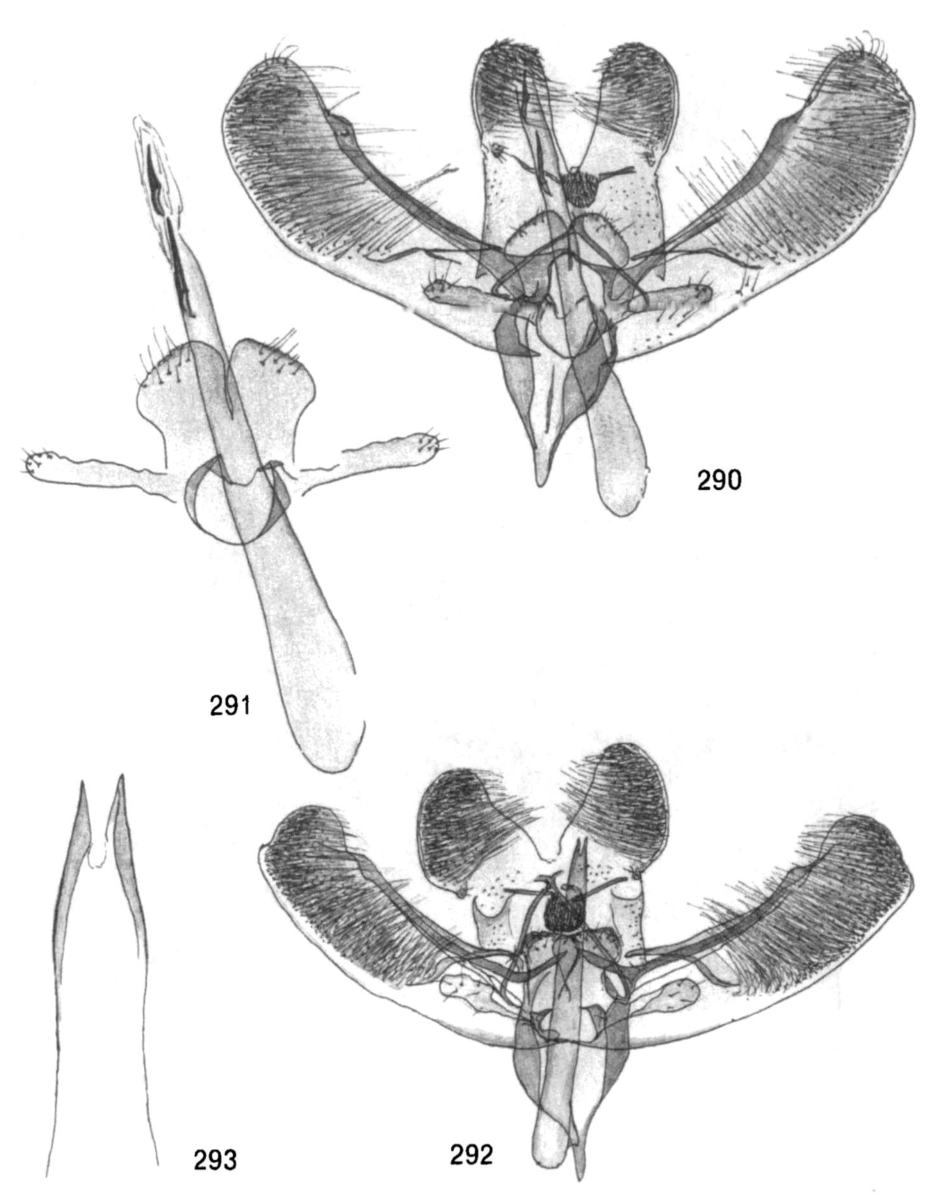

Figs. 290, 291. *Elachista reuttiana* Frey. – 290: male genitalia (ETO 5171); 291: digitate process, juxta, lobes (ETO 5171).
Figs. 292, 293. *Elachista orstadii* Palm. – 292: male genitalia (ETO S1); 293: distal end of aedeagus (ETO S1).

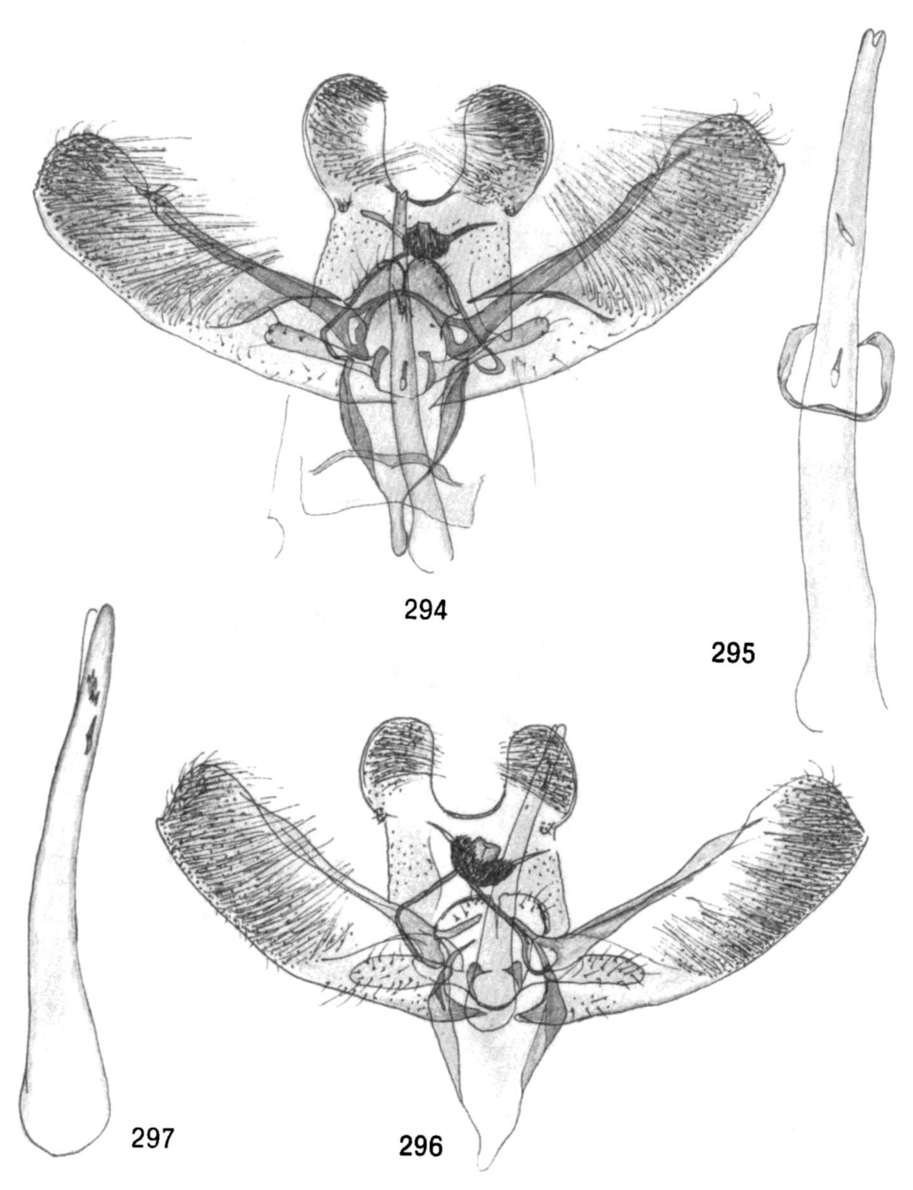

Figs. 294, 295. *Elachista ingvarella* Tr.–O. – 294: male genitalia (IS 3863); 295: aedeagus with anellus ring (IS 3863).
Figs. 296, 297. *Elachista krogeri* Svens. – 296: male gentalia (IS 5604); 297: aedeagus (IS 5604).

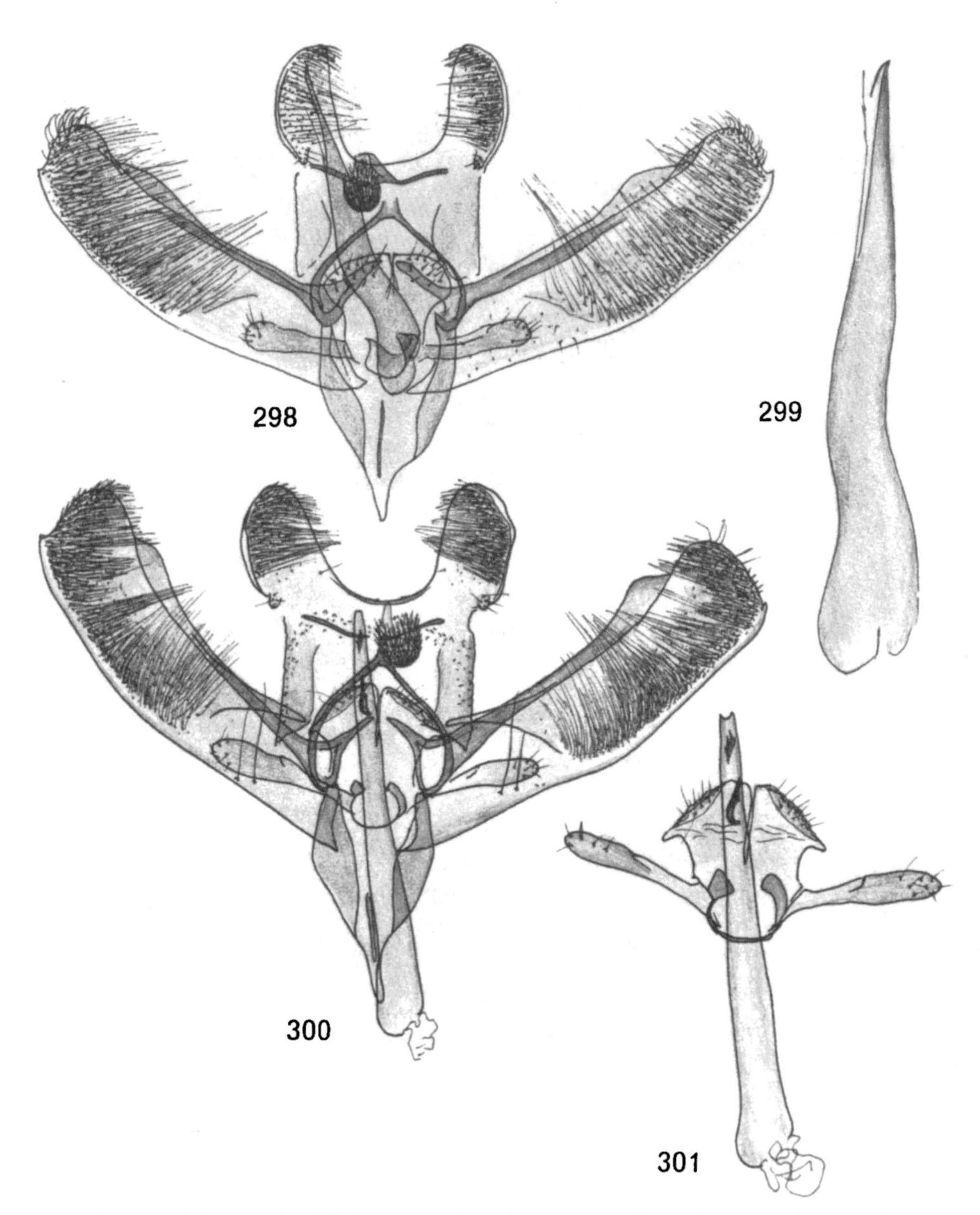

298, 299. *Elachista nielswolffi* Svens. – 298: male genitalia (IS 5641); 299: aedeagus (IS 5641).
Figs. 300, 301. *Elachista pomerana* Frey. – 300: male genitalia (ETO 2037); 301: digitate process, juxta, aedeagus (ETO 2037).

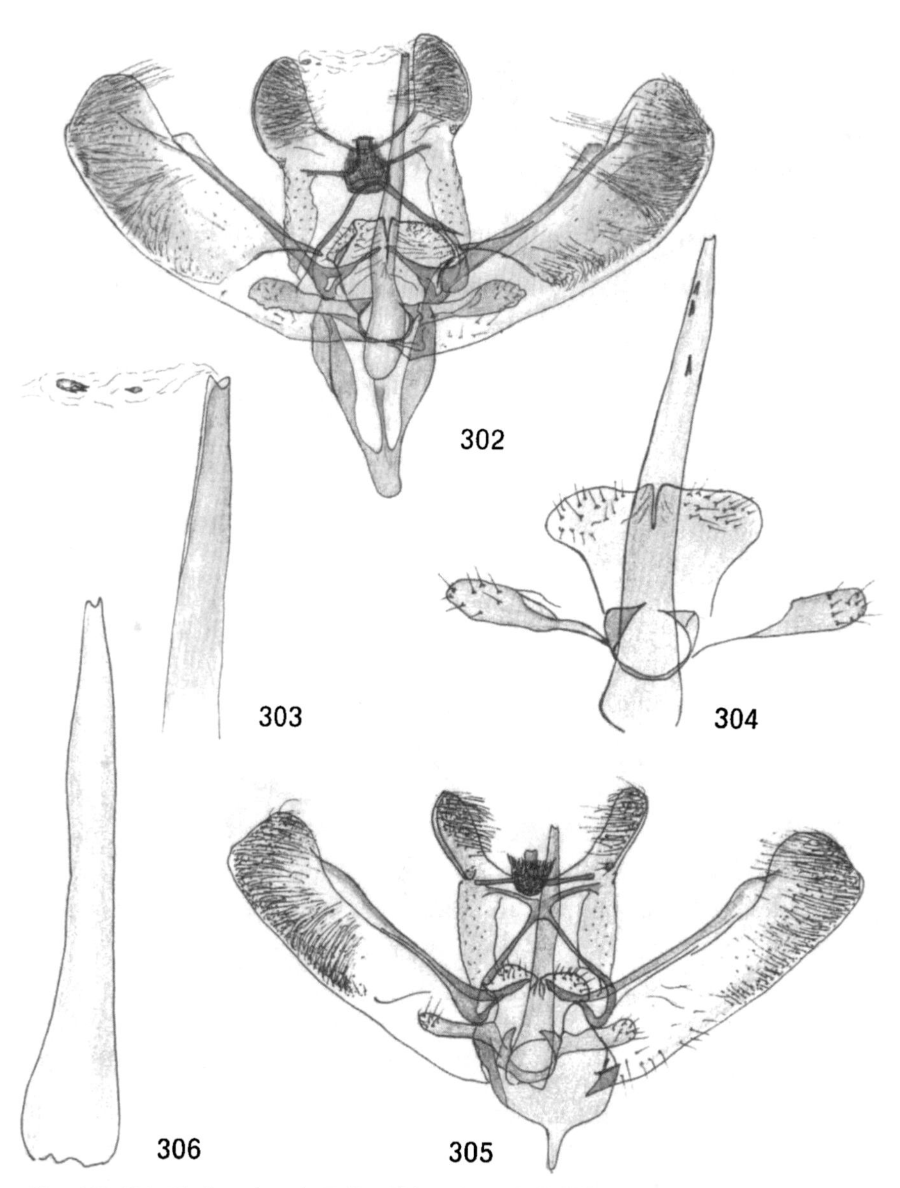

Figs. 302–304. *Elachista humilis* Zell. – 302: male genitalia (ETO 1290); 303: distal end of aedeagus (ETO 1290); 304: digitate process, juxta, aedeagus (OK 2421).
Figs. 305, 306. *Elachista vonschantzi* Svens. – 305: male genitalia (IS 5767); 306: aedeagus (IS 5767).

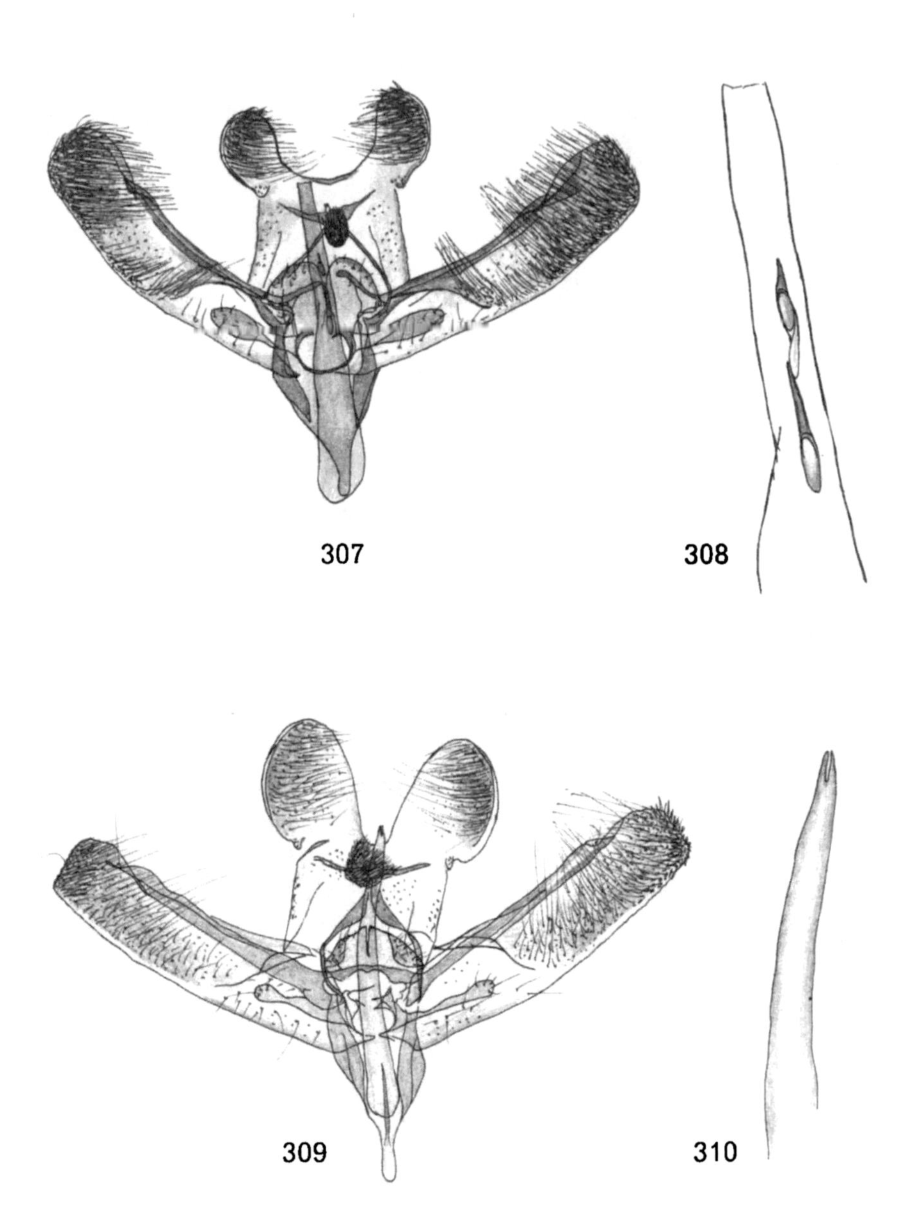

Figs. 307, 308. *Elachista pulchella* (Hw.). – 307: male genitalia (ETO 1357); 308: distal end of aedeagus (ETO 1357).
Figs. 309, 310. *Elachista anserinella* Zell. – 309: male genitalia (BMNH 11375); 310: distal end of aedeagus (BMNH 11375).

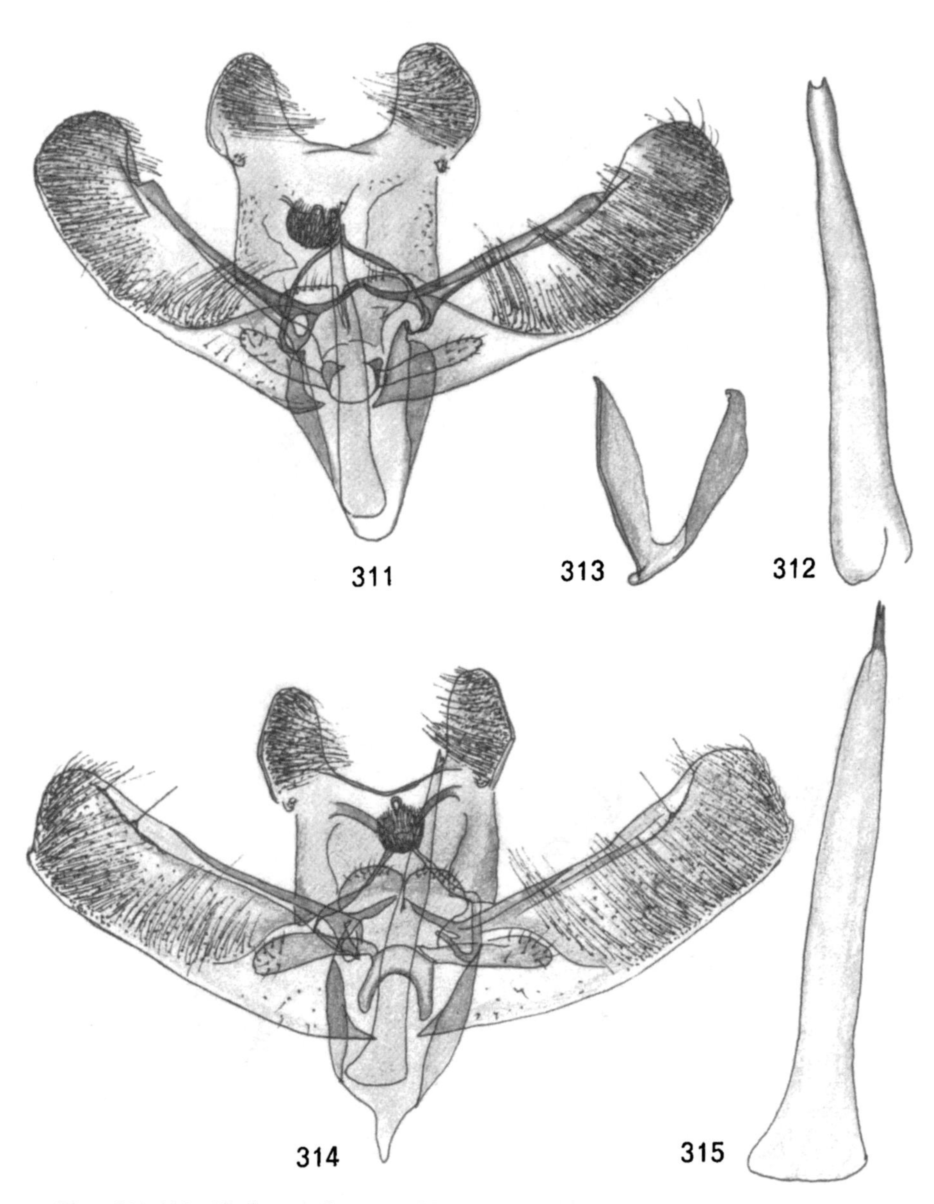

Figs. 311–313. *Elachista rufocinerea* (Hw.). – 311: male genitalia (ETO 1482); 312: aedeagus (ETO 1482); 313: vinculum with saccus (ETO 1277).
Figs. 314, 315. *Elachista lastrella* Chrét. – 314: male genitalia (ETO 5288); 315: aedeagus (ETO 5288).

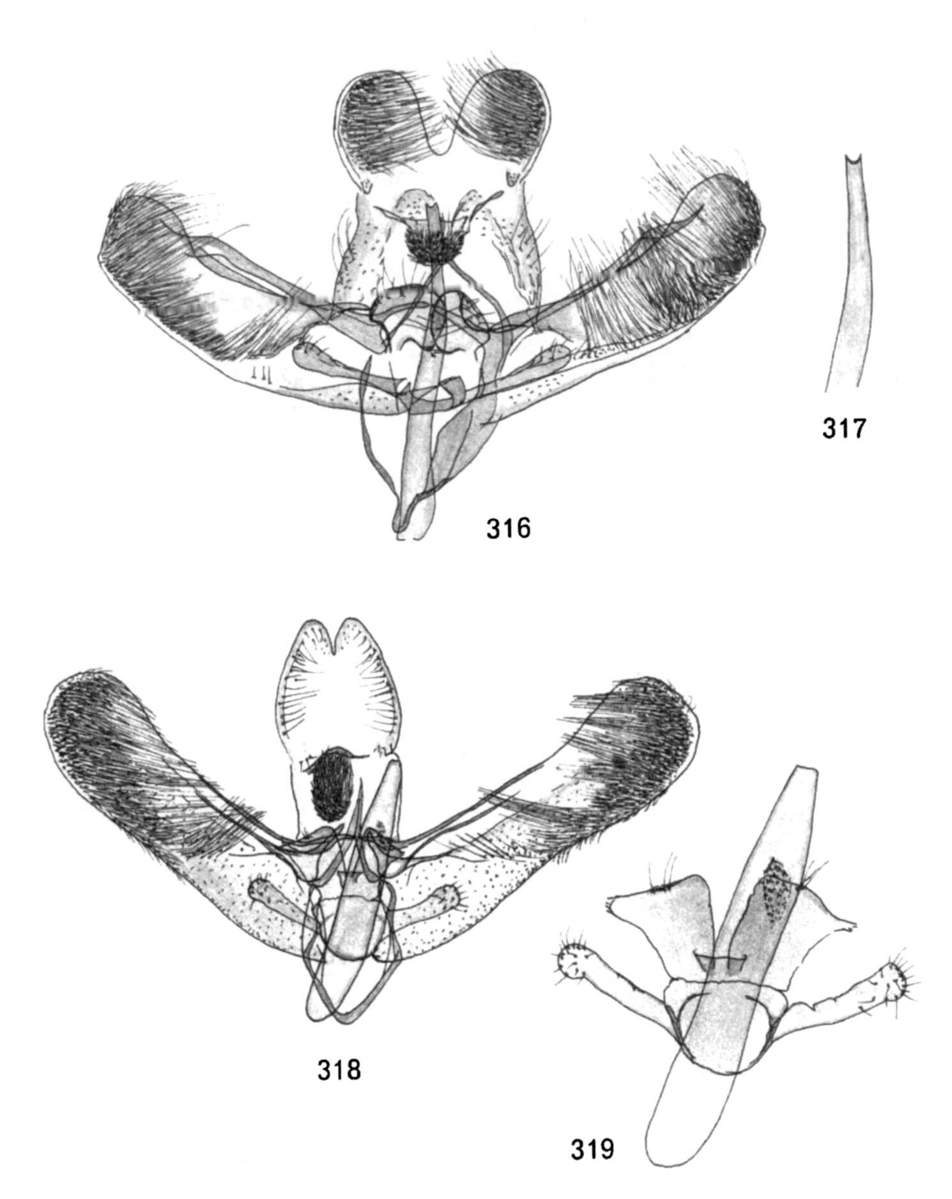

Figs. 316, 317. *Elachista cerusella* (Hb.). – 316: male genitalia (ETO 1414); 317: distal end of aedeagus (ETO 1414).
Figs. 318, 319. *Elachista argentella* (Cl.). – 318: male genitalia (ETO 1012); 319: digitate process, juxta, aedeagus (ETO 1012).

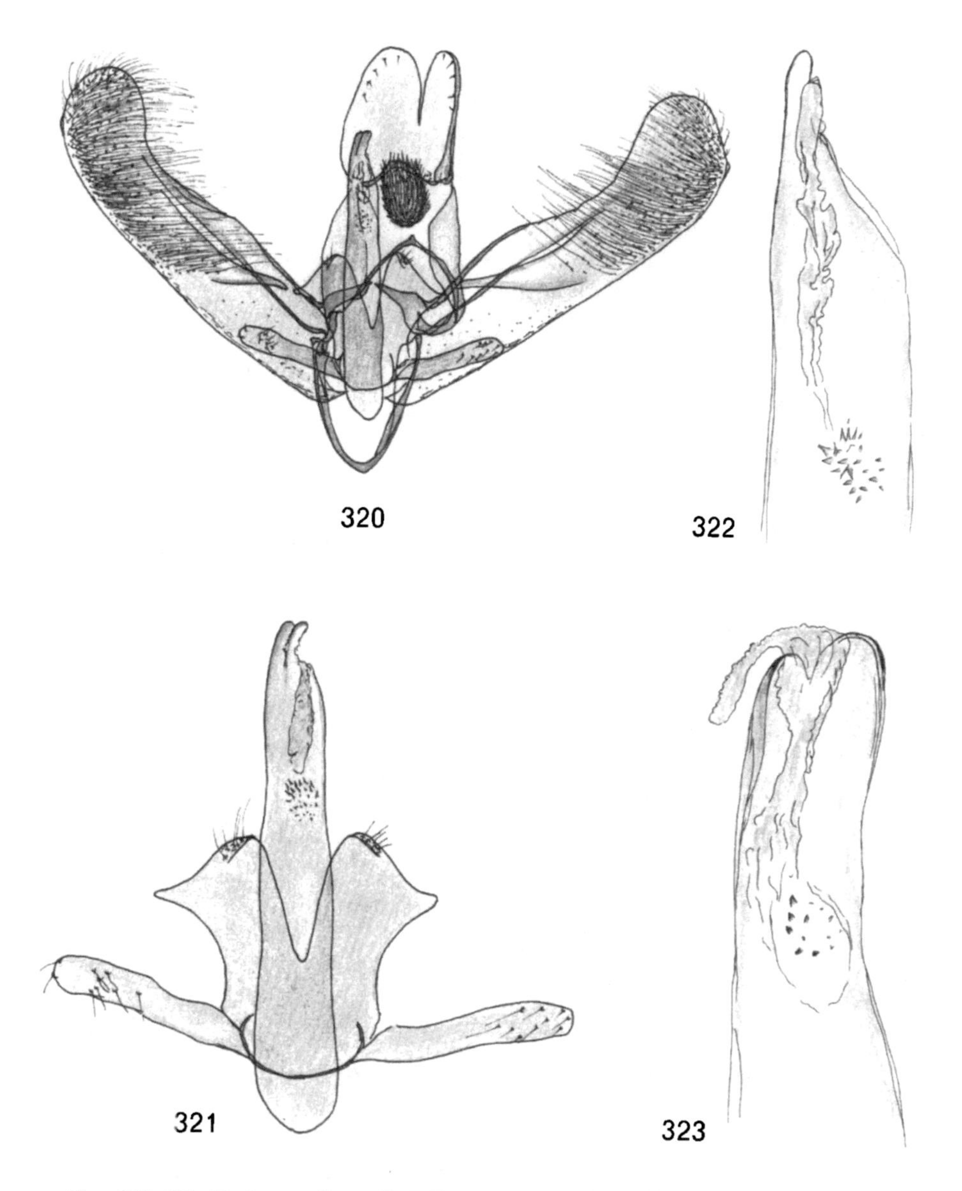

Figs. 320–323. *Elachista pollinariella* Zell. – 320: male genitalia (NLW 658); 321: digitate process, juxta, aedeagus (NLW 658); 322: distal end of aedeagus (ETO 1123); 323: distal end of aedeagus (ETO 1599).

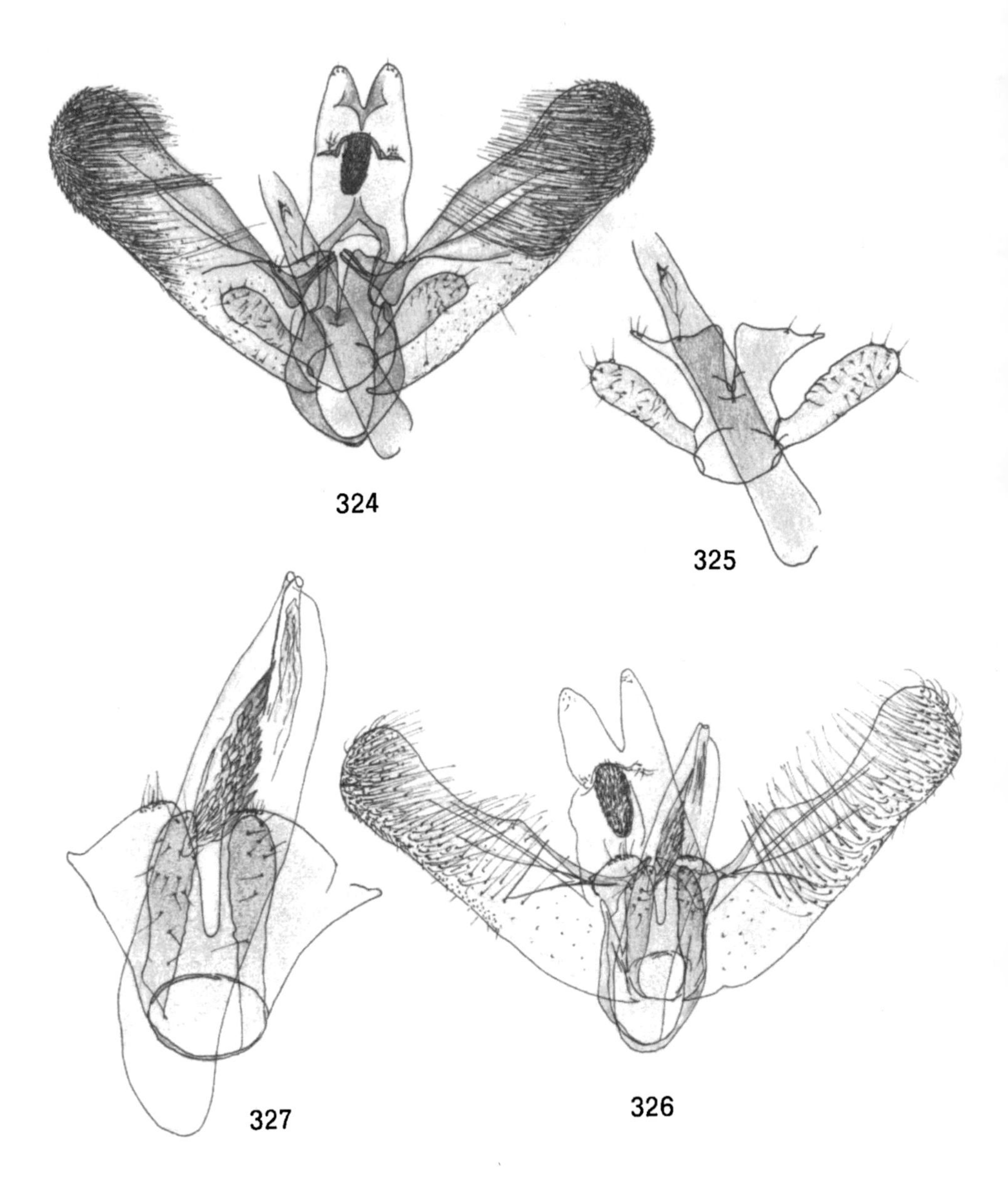

Figs. 324, 325. *Elachista triatomea* (Hw.). – 324: male genitalia (ETO 1033); 325: digitate process, juxta, aedeagus (ETO 1033).
Figs. 326, 327. *Elachista collitella* (Dup.). – 326: male genitalia (ETO 5174); 327: digitate process, juxta, aedeagus (ETO 5174).

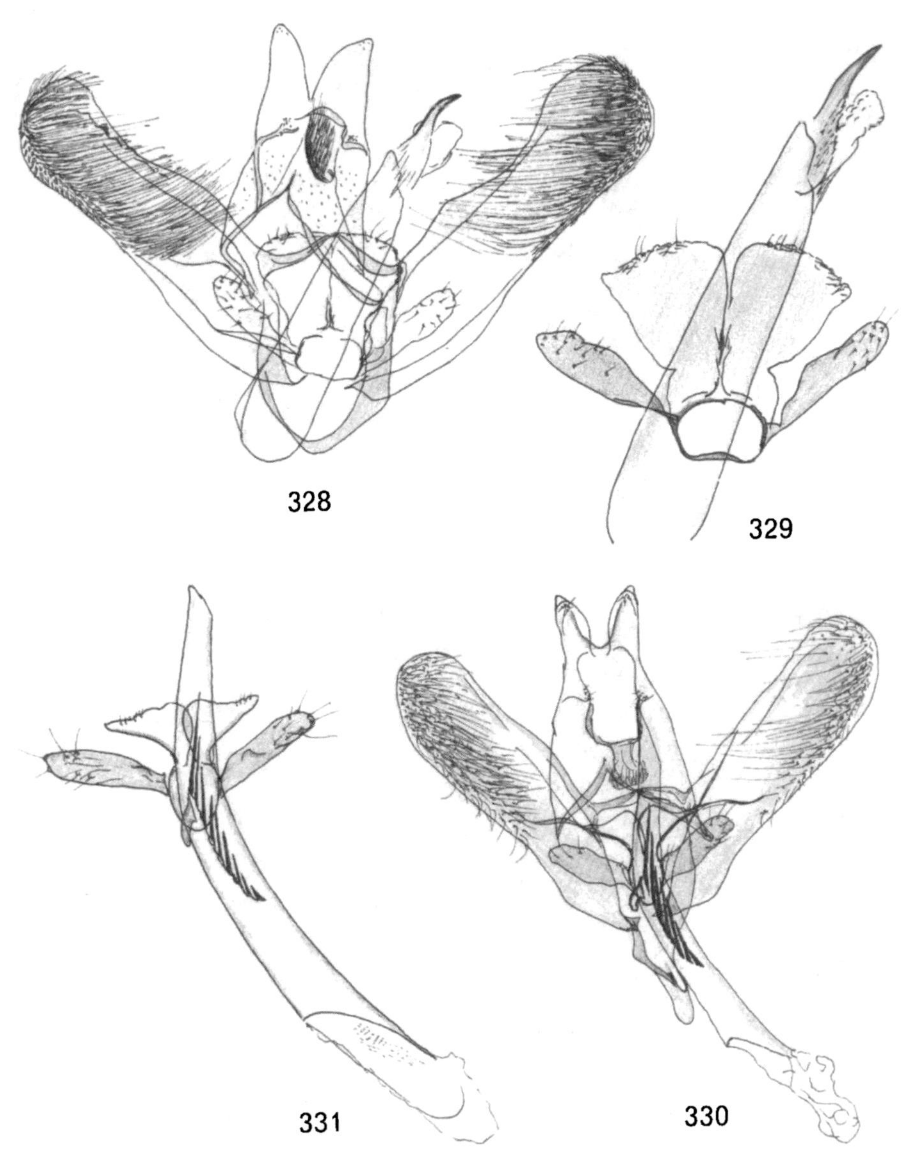

Figs. 328, 329. *Elachista subocellea* (Stph.). – 328: male genitalia (ETO 1816); 329: digitate process, juxta, aedeagus (ETO 1816).
Figs. 330, 331. *Elachista festucicolella* Zell. – 330: male genitalia (ETO 1783); 331: digitate process, juxta, aedeagus (ETO 1783).

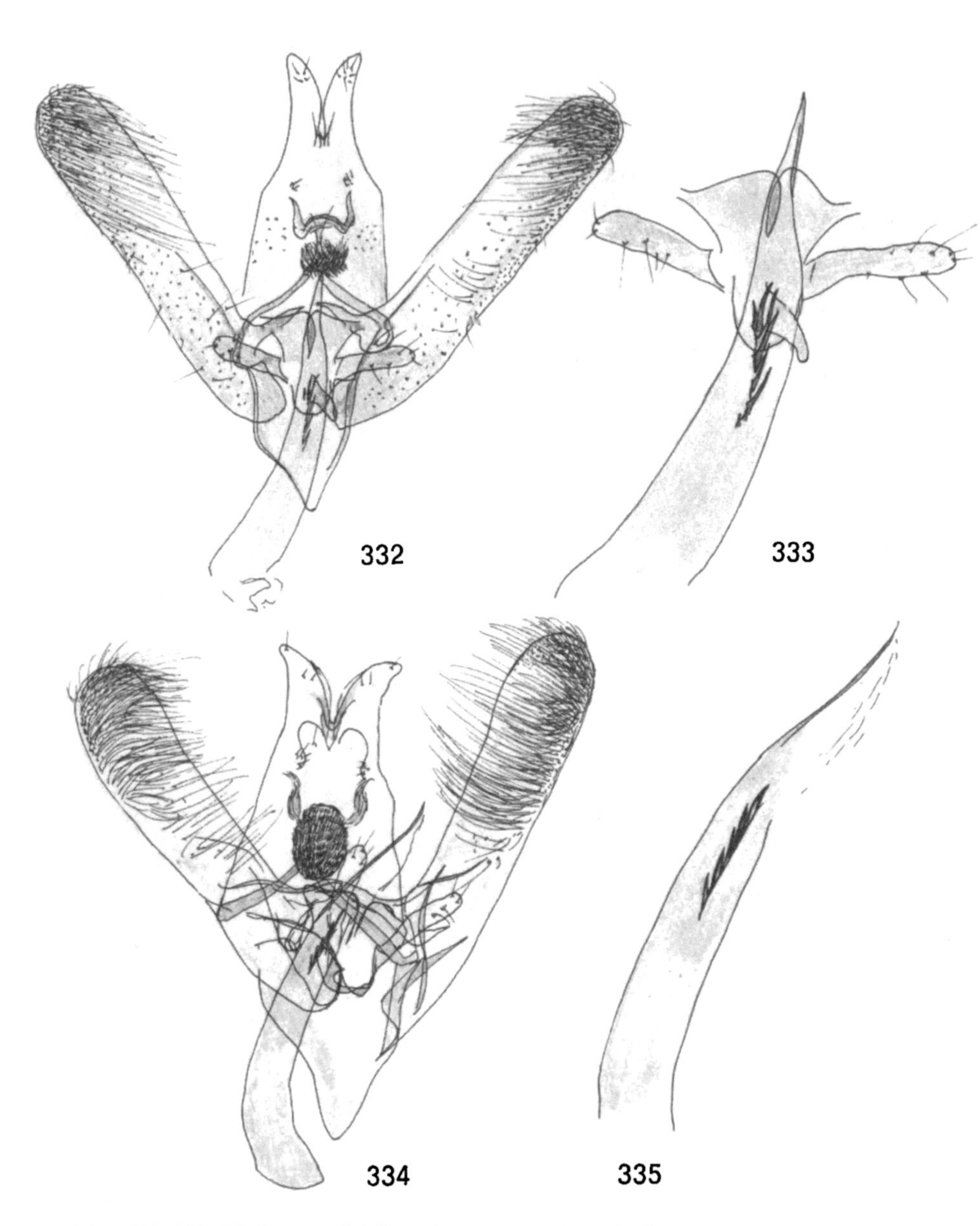

Figs. 332, 333. *Elachista nitidulella* (HS.). – 332: male genitalia (ETO B.25.10.74.); 333: digitate process, juxta, aedeagus (ETO B. 25.10.74).
Figs. 334, 335. *Elachista dispilella* Zell. – 334: male genitalia (NLW 3094); 335: distal end of aedeagus (NLW 3094).

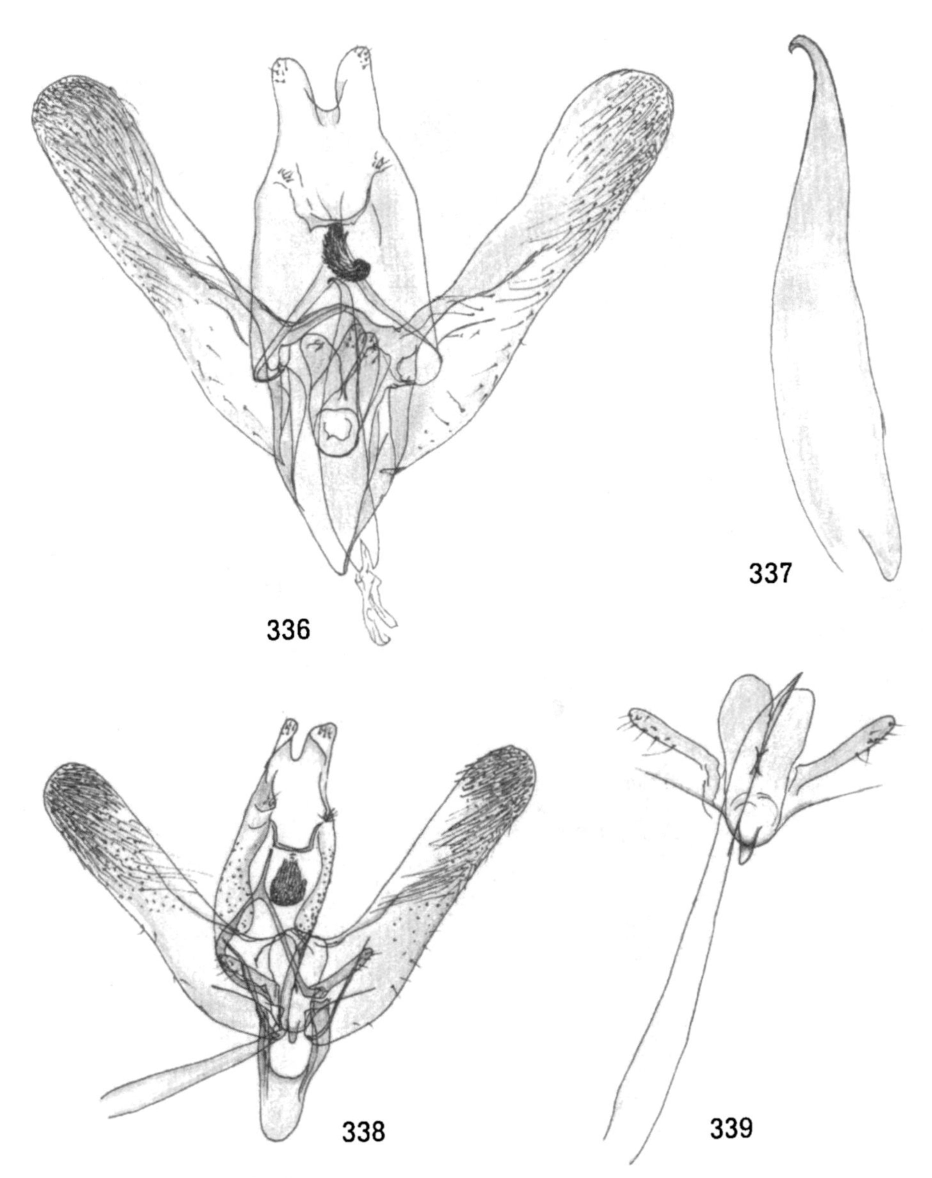

Figs. 336, 337. *Elachista triseriatella* Stt. - 336: male genitalia (ETO 1876); 337: aedeagus (ETO 1876).
Figs. 338, 339. *Elachista dispunctella* (Dup.). - 338. male genitalia (ETO C. 18.8.75); 339: digitate process, juxta, aedeagus (ETO C. 18.8.75).

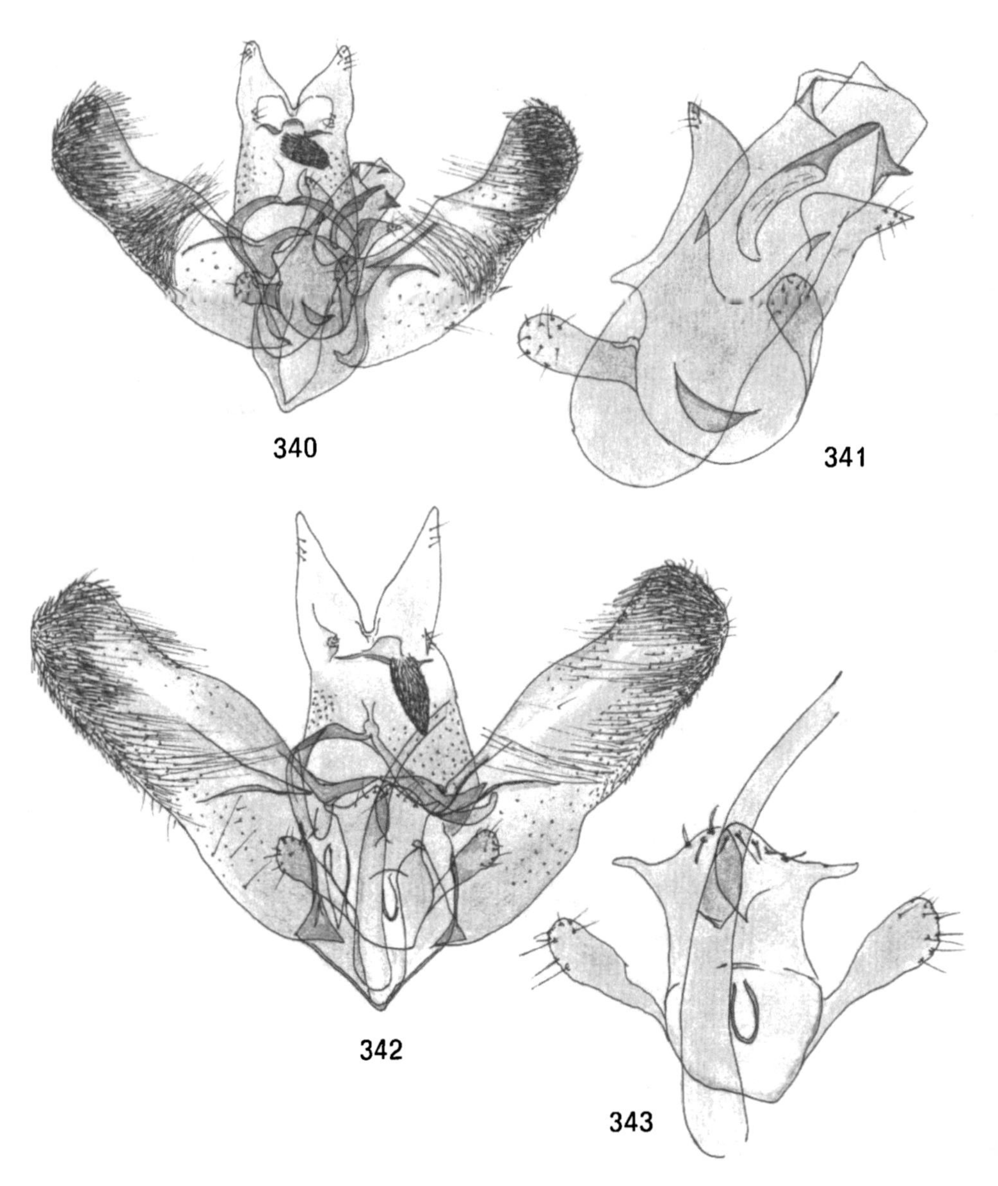

Figs. 340, 341. *Elachista rudectella* (Stt.). – 340: male genitalia (ETO H. 10.11.74); 341: digitate process, juxta, aedeagus (ETO H. 10.11.74).
Figs. 342, 343. *Elachista squamosella* (HS.). – 342: male genitalia (ETO A. 10.11.74); 343: digitate process, juxta, aedeagus (ETO A. 10.11.74).

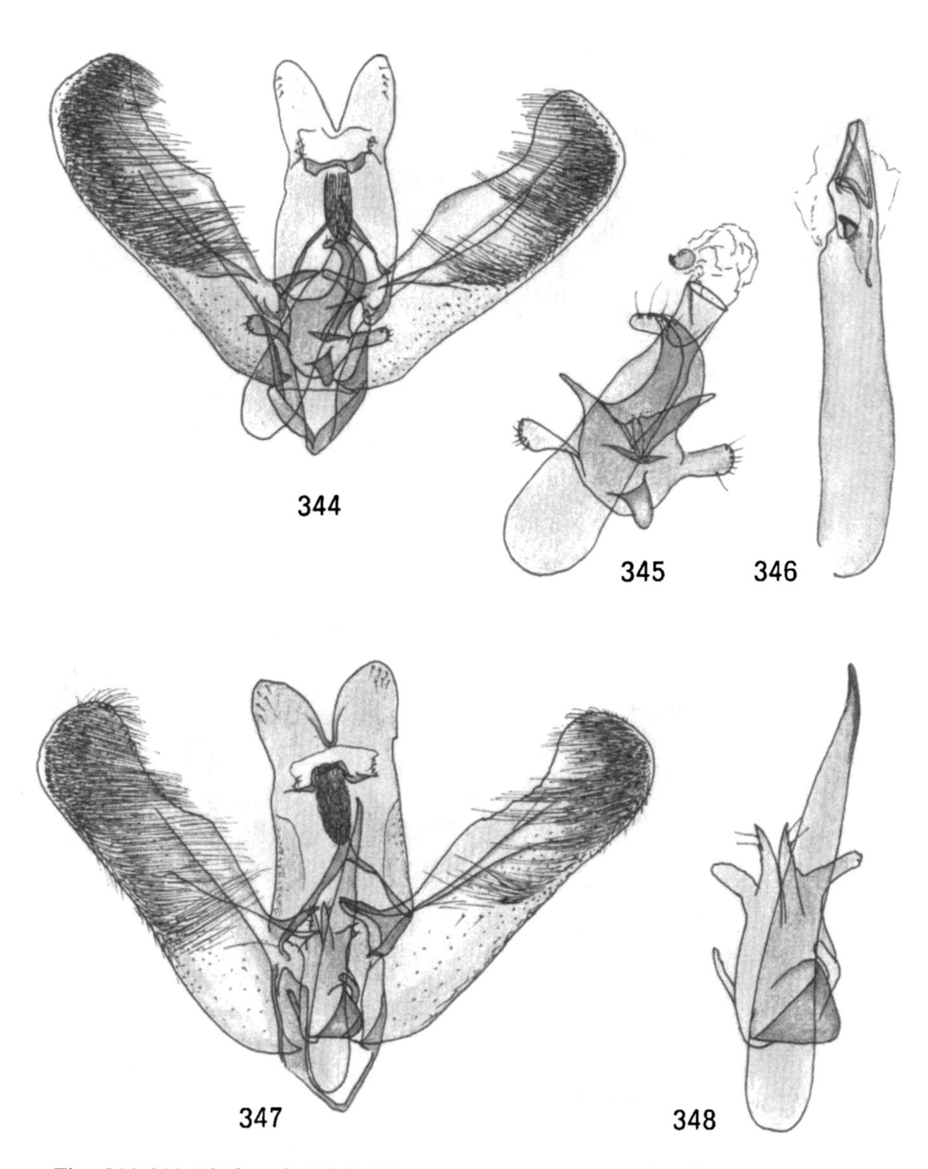

Figs. 344–346. *Elachista bedellella* (Sirc.). – 344: male genitalia (ETO 2024); 345: digitate process, juxta, aedeagus (ETO 2024); 346: aedeagus (ETO B. 16.12.76).
Figs. 347, 348. *Elachista pullicomella* Zell. – 347: male genitalia (ETO 425); 348: digitate process, juxta, aedeagus (ETO 425).

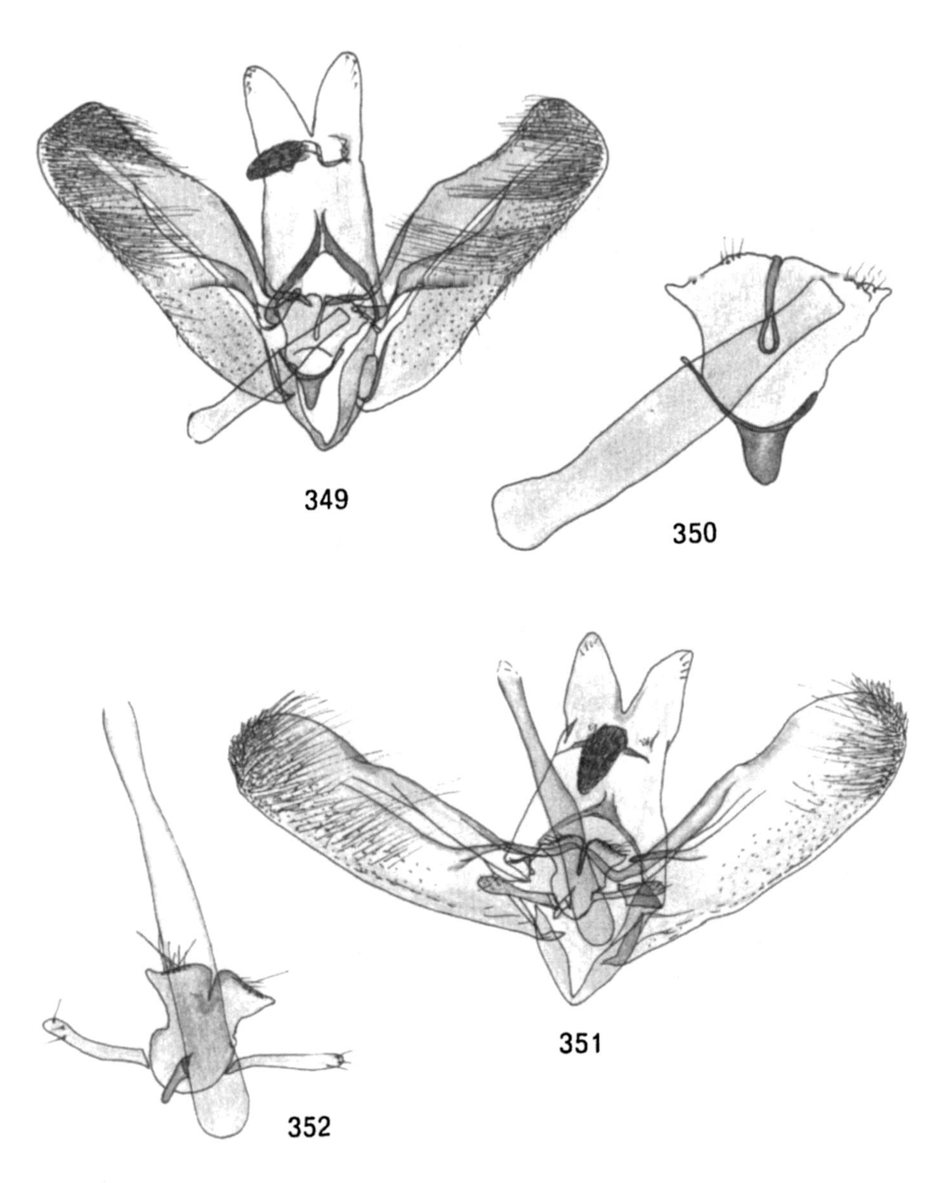

Figs. 349, 350. *Elachista littoricola* Le March. – 349: male genitalia (ETO A.26.9.76); 350: juxta and aedeagus (ETO A.26.9.76).
Figs. 351, 352. *Elachista chrysodesmella* Zell. – 351: male genitalia (IS 5294); 352: digitate process, juxta, aedeagus (IS 5294).

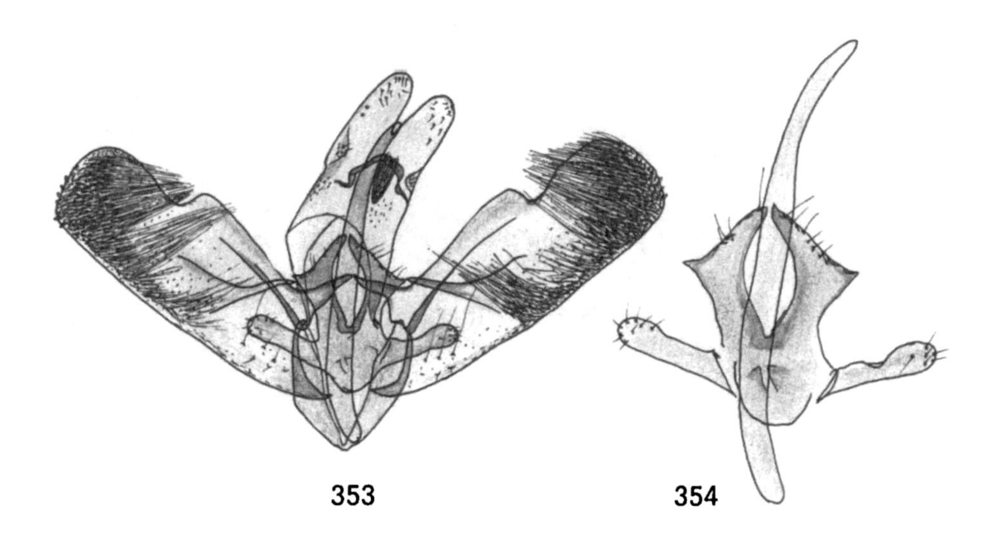

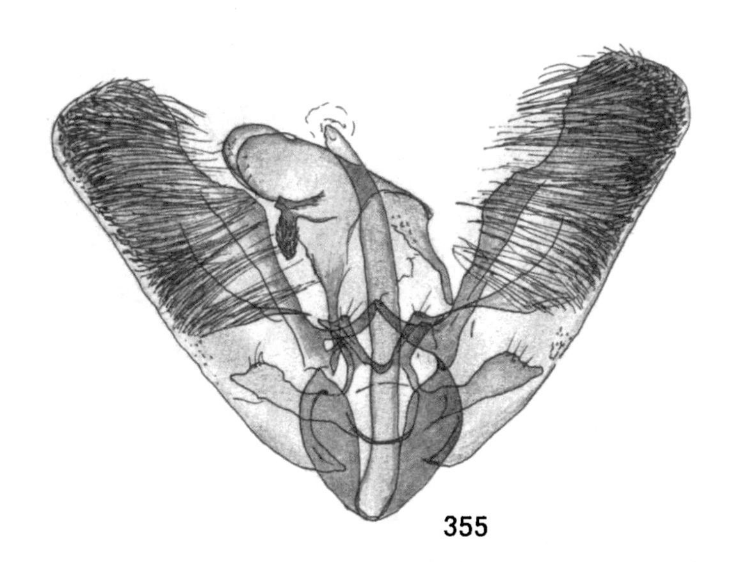

Figs. 353, 354. *Elachista megerlella* (Hb.). – 353: male genitalia (ETO 2019); 354: digitate process, juxta, aedeagus (ETO 2019).
Fig. 355. *Elachista cingillella* (HS.), male genitalia (IS 4652).

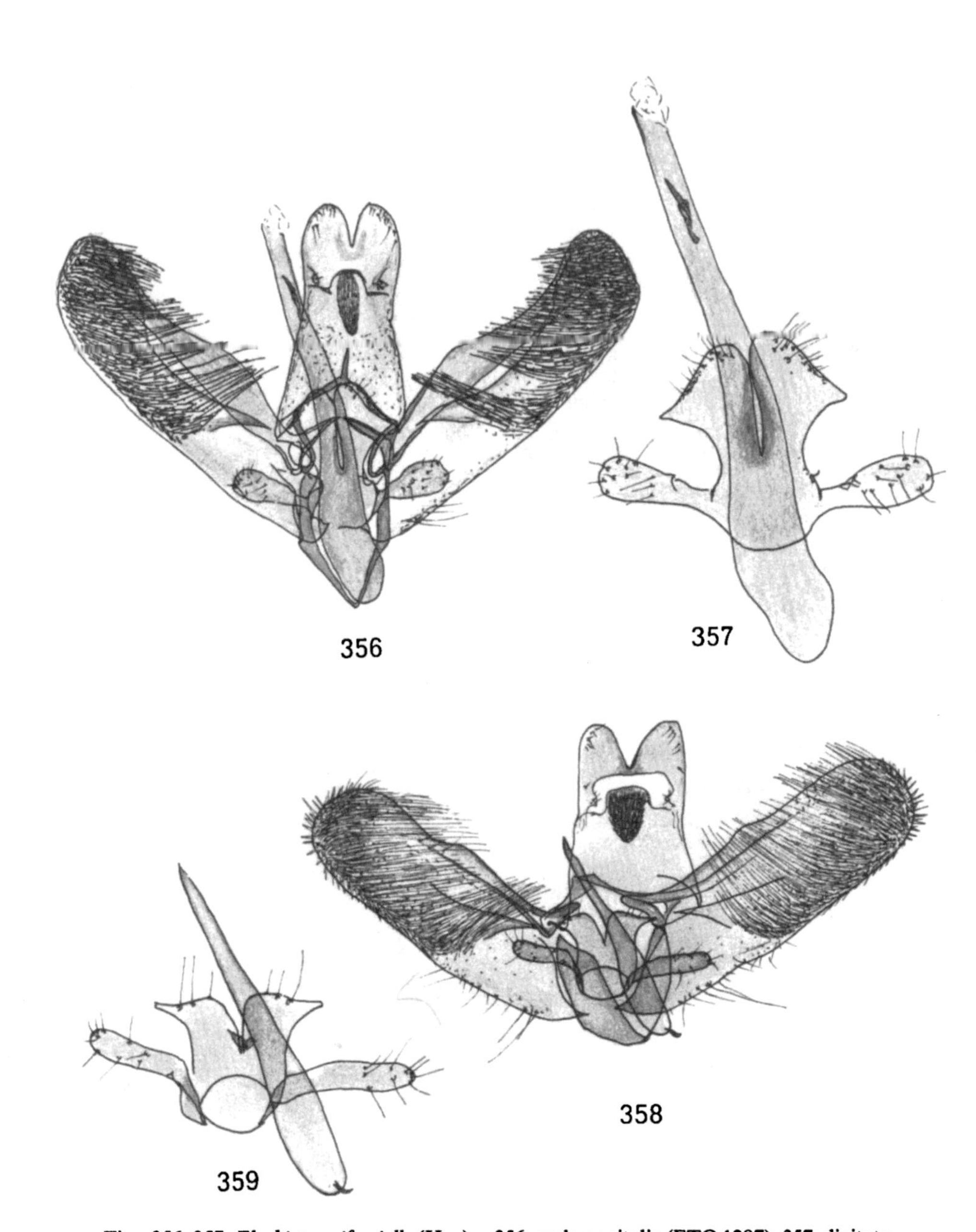

Figs. 356, 357. *Elachista unifasciella* (Hw.). – 356: male genitalia (ETO 1287); 357: digitate process, juxta, aedeagus (ETO 1287).
Figs. 358, 359. *Elachista gangaballa* Zell. – 358: male genitalia (OK 2320); 359: digitate process, juxta, aedeagus (OK 2320).

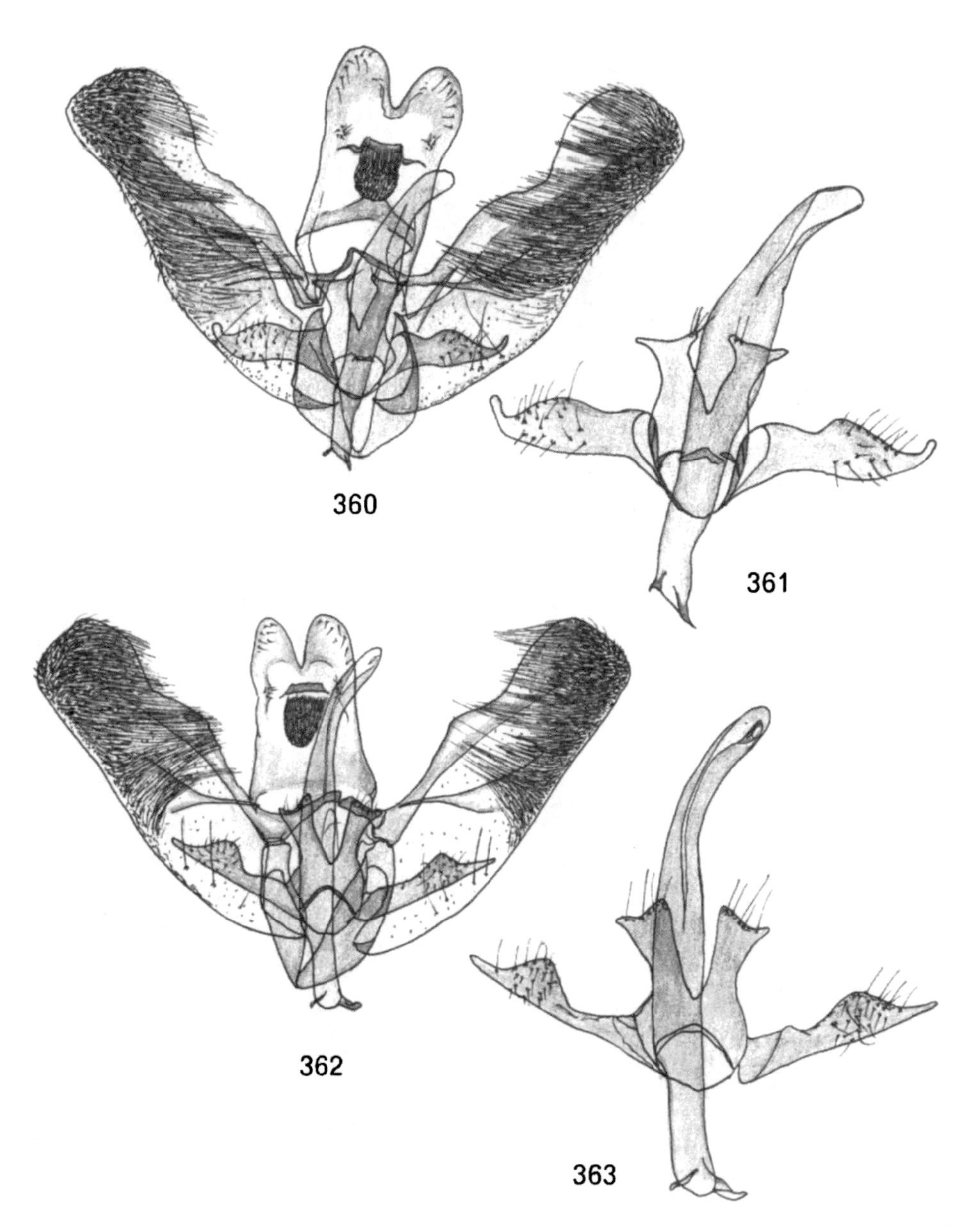

Figs. 360, 361. *Elachista subalbidella* Schl. – 360: male genitalia (ETO 1359); 361: digitate process (ETO 1359).
Figs. 362, 363. *Elachista revinctella* Zell. – 362: male genitalia (ETO 1259); 363: digitate process, juxta, aedeagus (ETO 1259).

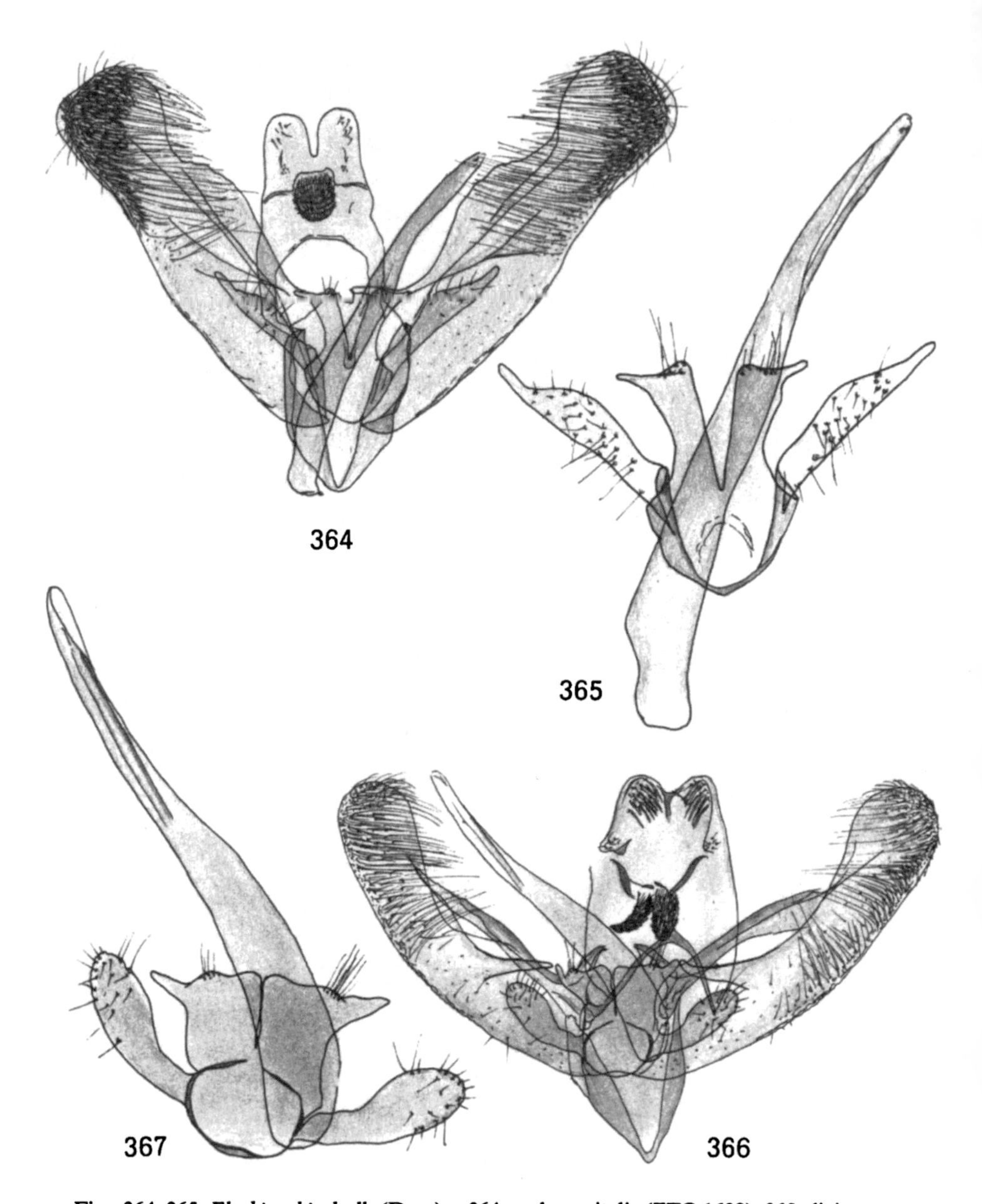

Figs. 364, 365. *Elachista bisulcella* (Dup.). – 364: male genitalia (ETO 1632); 365: digitate process, juxta, aedeagus (ETO 1632).
Figs. 366, 367. *Biselachista trapeziella* (Stt.). – 366: male genitalia (IS 4640); 367: digitate process, juxta, aedeagus (IS 4640).

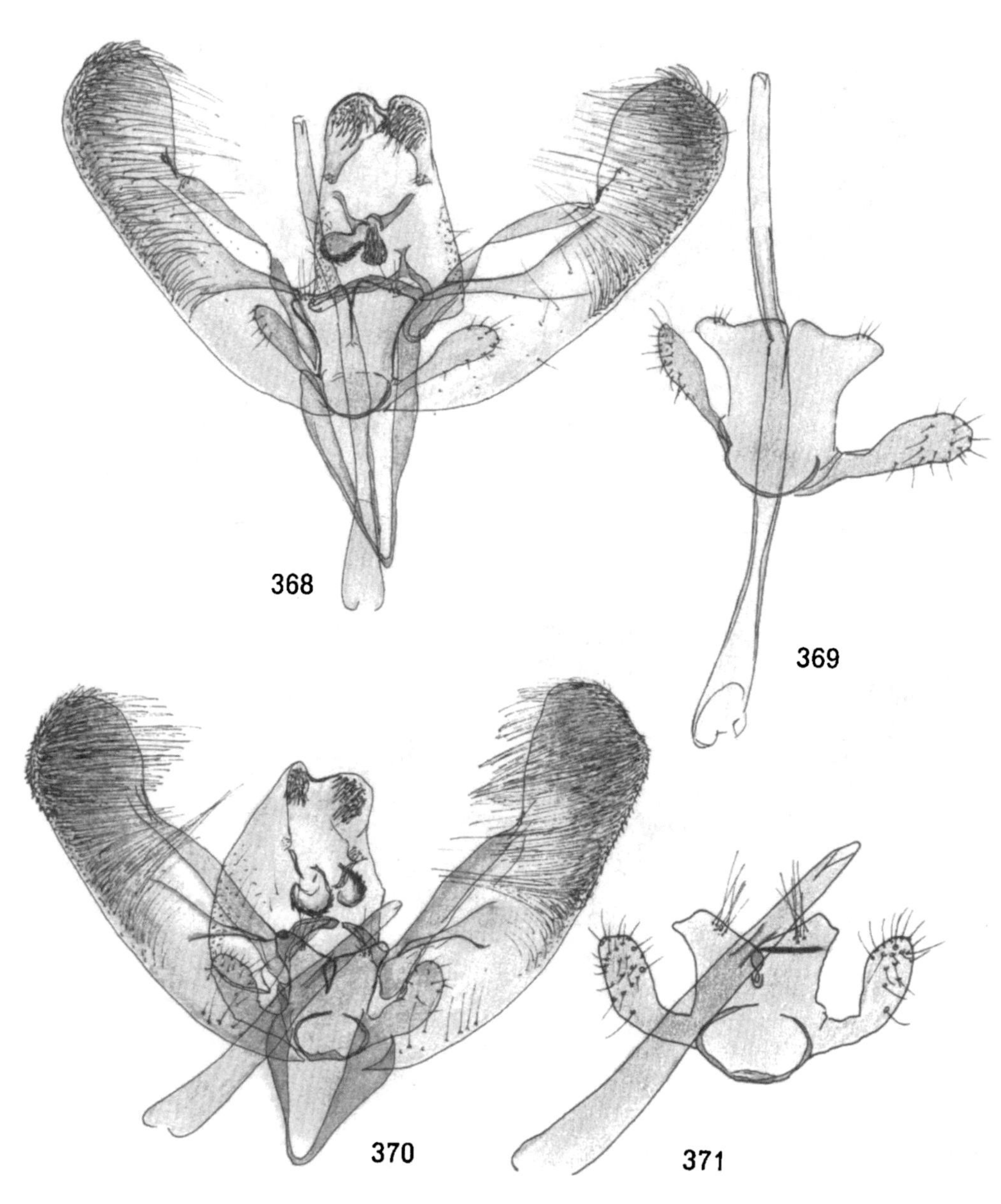

Figs. 368, 369. *Biselachista imatrella* (Schantz). – 368: male genitalia (MvS 532); 369: digitate process, juxta, aedeagus (MvS 532).
Figs. 370, 371. *Biselachista cinereopunctella* (Hw.). – 370: male genitalia (ETO 1869); 371: digitate process, juxta, aedeagus (ETO 1869).

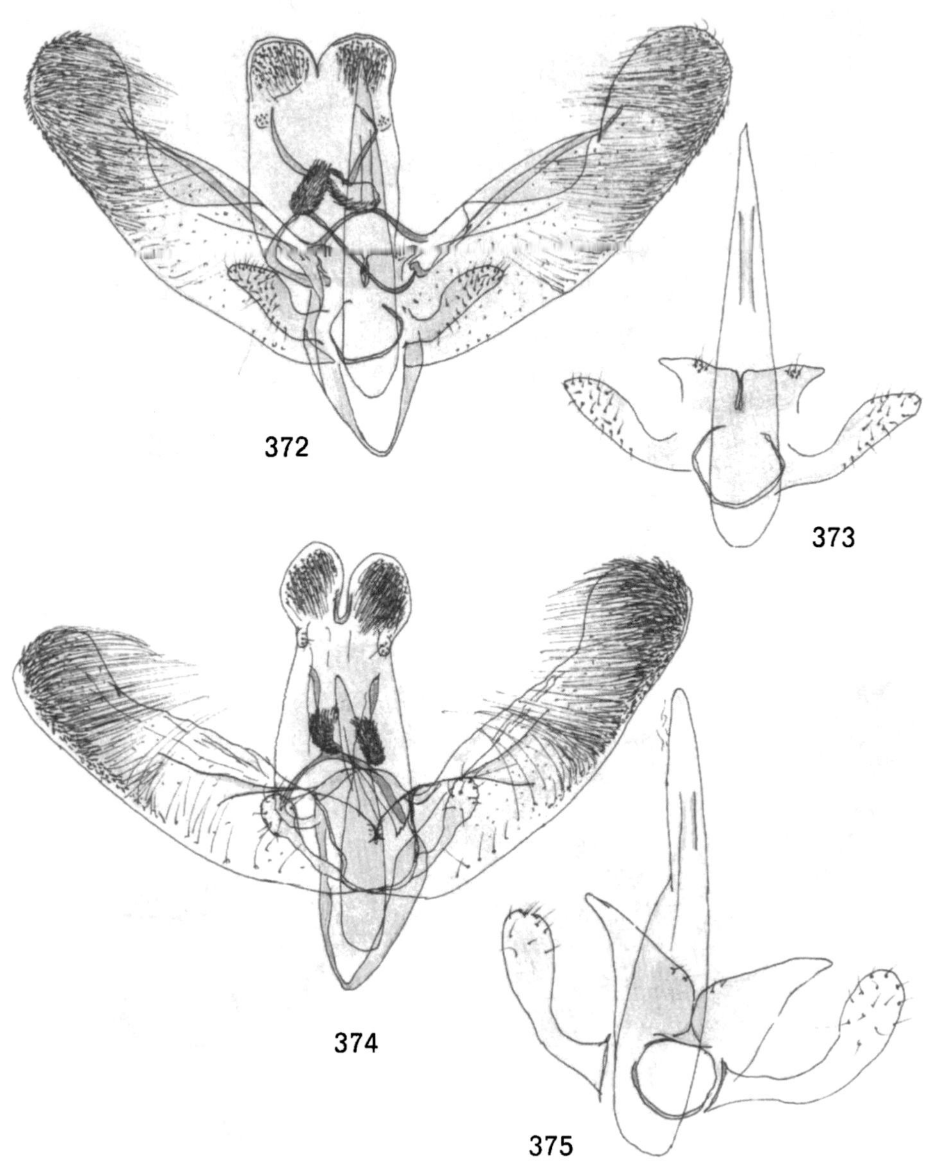

Figs. 372, 373. ***Biselachista kebneella*** sp. n. – 372: male genitalia (IS 5541); 373: digitate process, juxta, aedeagus (IS 5541).
Figs. 374, 375. ***Biselachista ornithopodella*** (Frey). – 374: male genitalia (BMNH 19285); 375: digitate process, juxta, aedeagus (BMNH 19285).

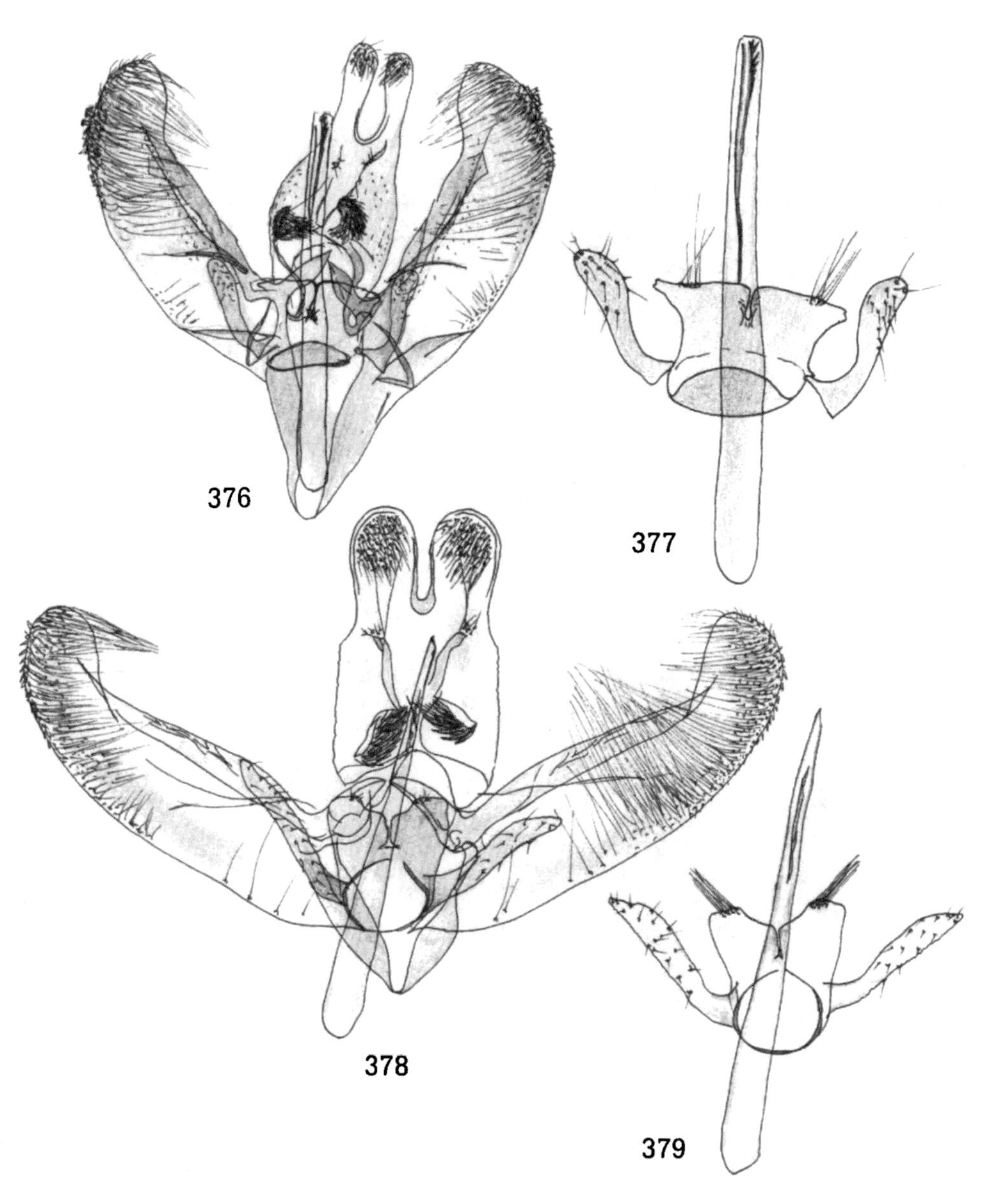

Figs. 376, 377. *Biselachista serricornis* (Stt.). – 376: male genitalia (ETO 1351); 377: digitate process, juxta, aedeagus (ETO 1351).
Figs. 378, 379. *Biselachista freyi* (Stgr.). – 378: male genitalia (IS 4788); 379: digitate process, juxta, aedeagus (IS 4788).

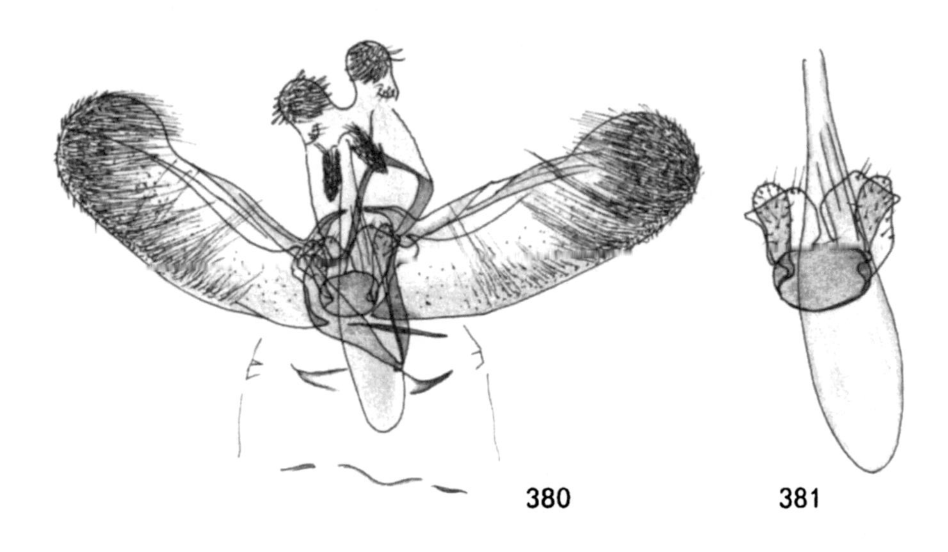

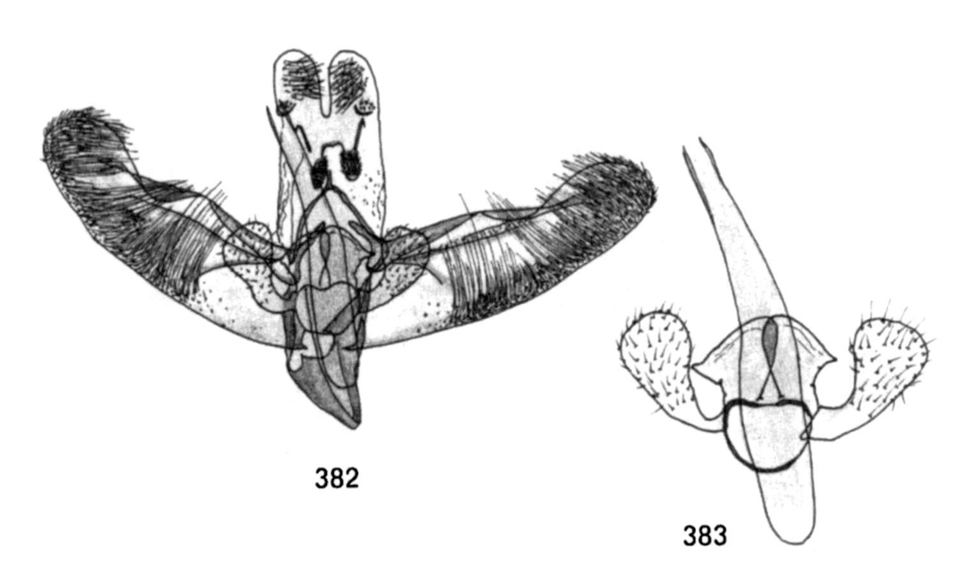

Figs. 380, 381. *Biselachista scirpi* (Stt.). – 380: male genitalia (ETO 1705); 381: digitate process, juxta, aedeagus (ETO 1705).
Figs. 382, 383. *Biselachista eleochariella* (Stt.). – 382: male genitalia (ETO 1317); 383: digitate process, juxta, aedeagus (ETO 1317).

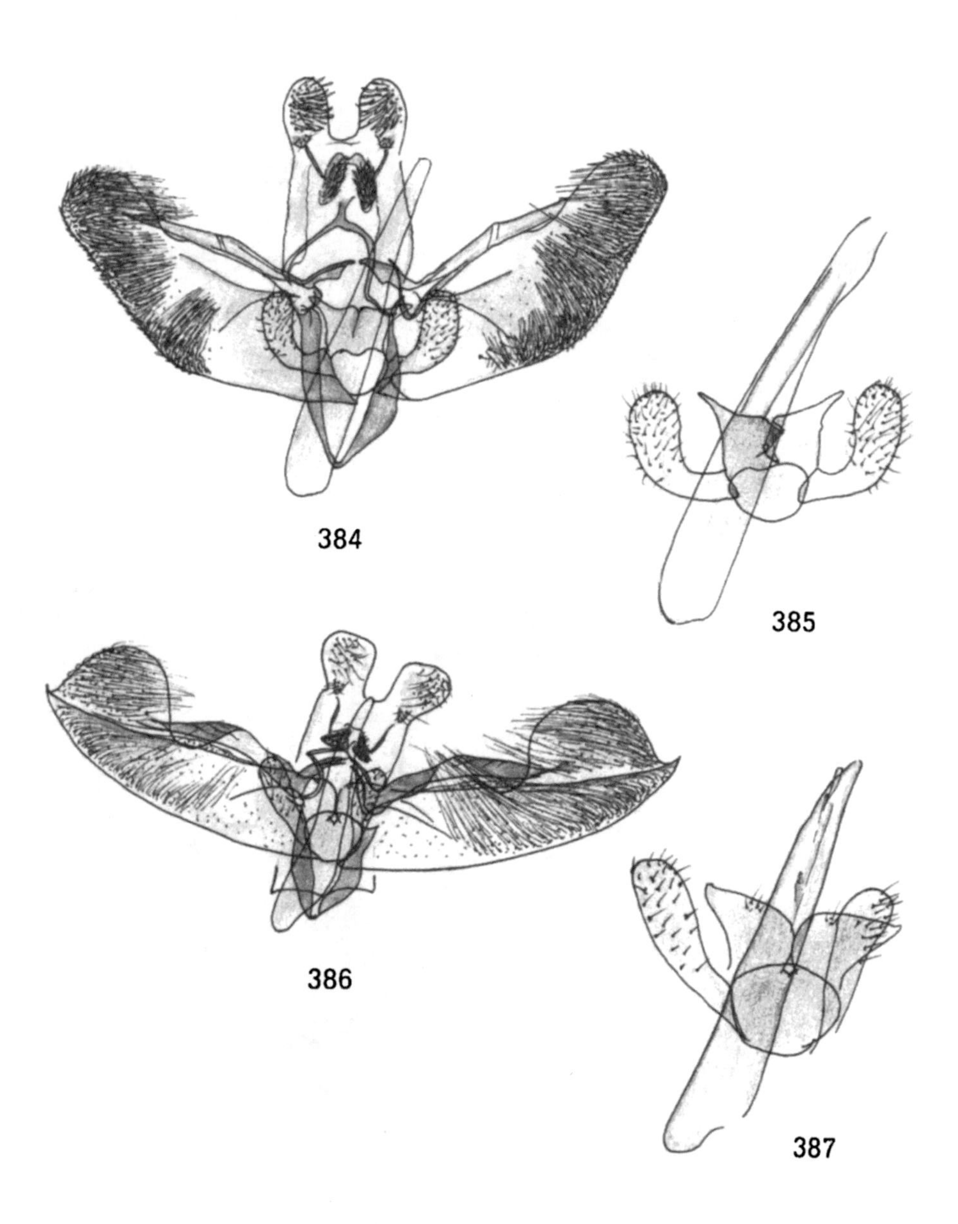

Figs. 384, 385. *Biselachista utonella* (Frey). – 384: male genitalia (ETO 1318); 385: digitate process, juxta, aedeagus (ETO 1447).
Figs. 386, 387. *Biselachista albidella* (Nyl.). – 386: male genitalia (ETO 1305); 387: digitate process, juxta, aedeagus (ETO 1305).

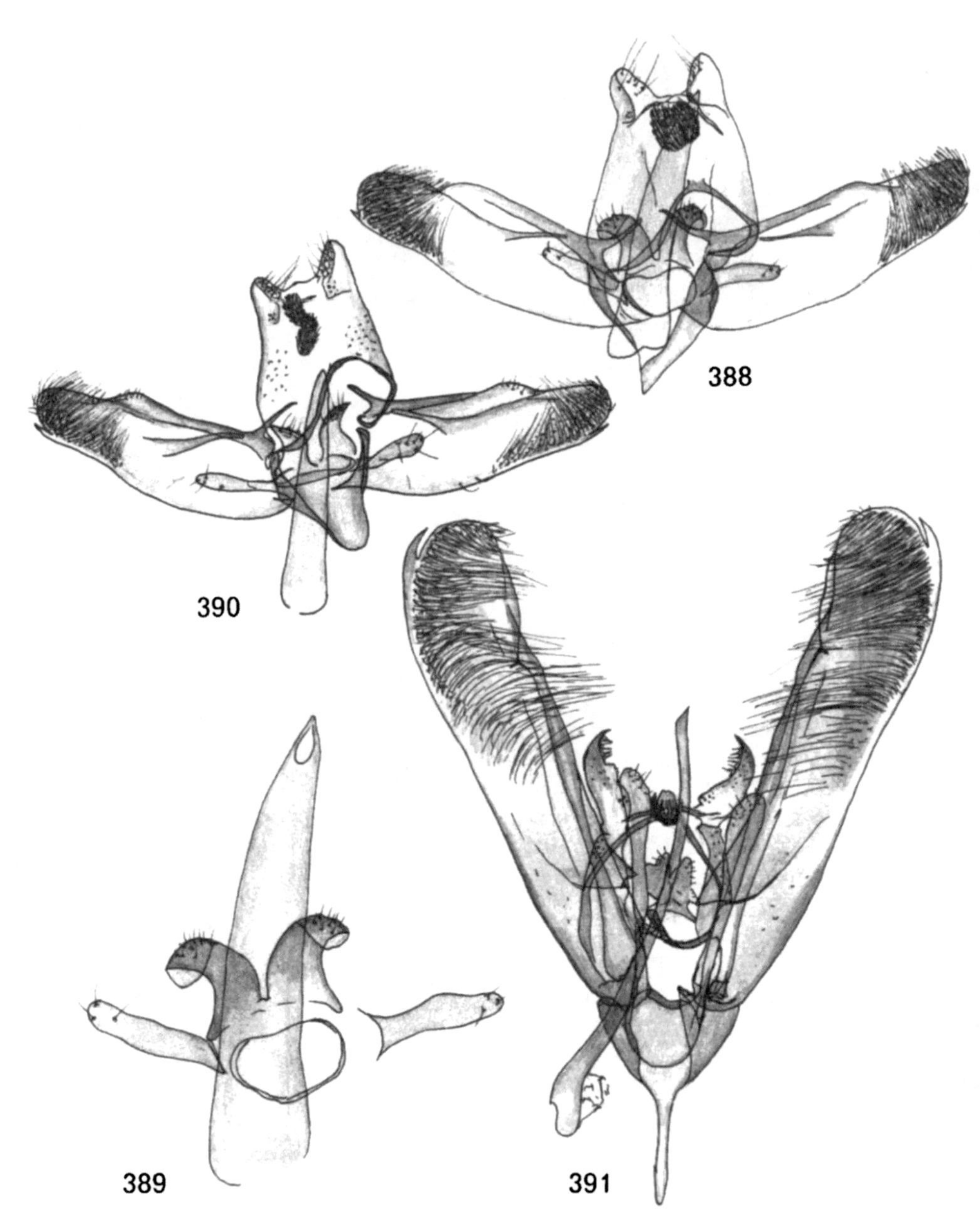

Figs. 388–390. *Biselachista abiskoella* (Bengts.). – 388: male genitalia (BÅB 507–15027); 389: digitate process, juxta, aedeagus (BÅB 507–15027); 390: male genitalia (OP 80). Fig. 391. *Cosmiotes freyerella* (Hb.), male genitalia (ETO 1368).

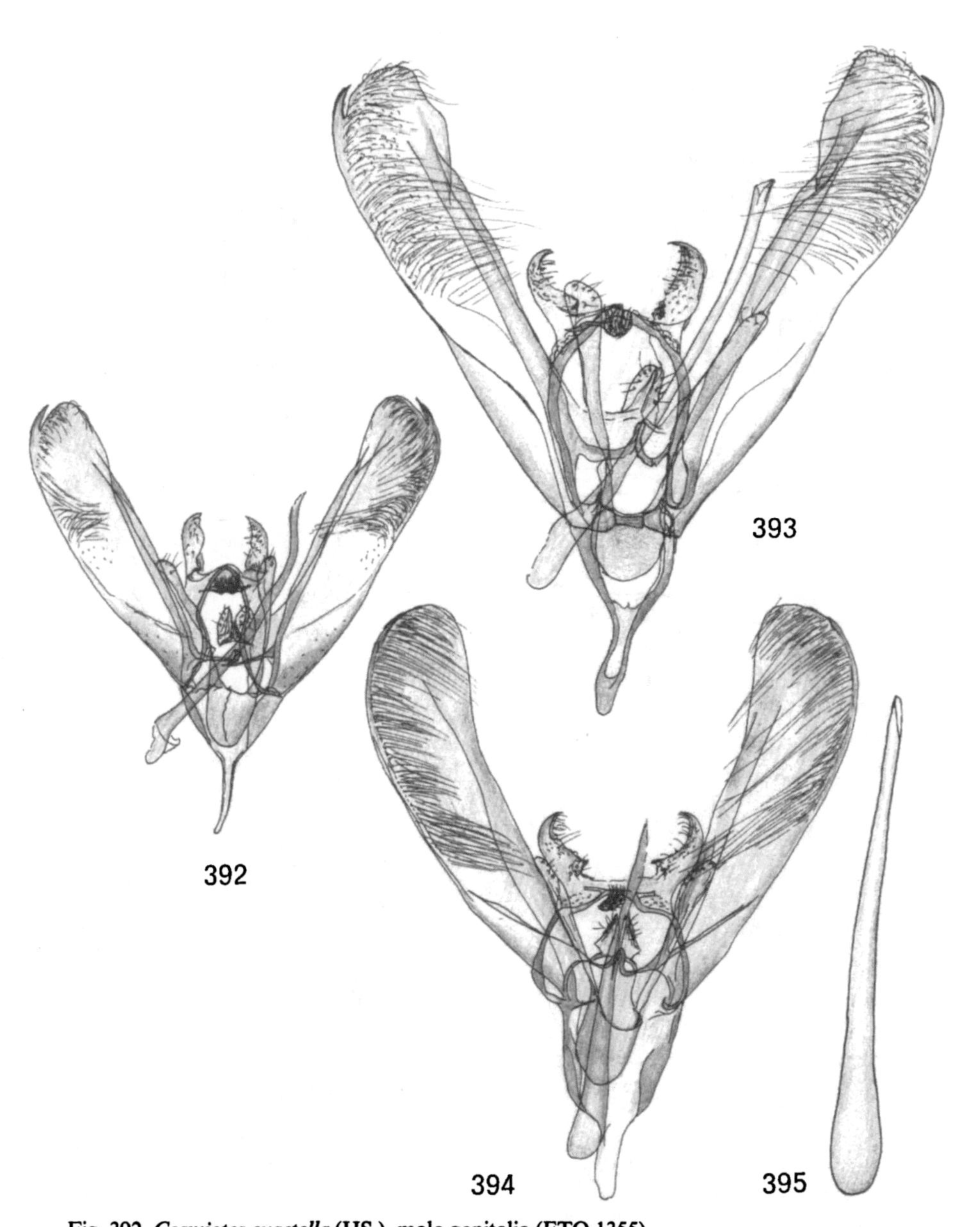

Fig. 392. *Cosmiotes exactella* (HS.), male genitalia (ETO 1355).
Fig. 393. *Cosmiotes stabilella* (Stt.), male genitalia (ETO 937).
Figs. 394, 395. *Cosmiotes consortella* (Stt.). – 394: male genitalia (ETO 1363); 395: aedeagus (ETO 346).

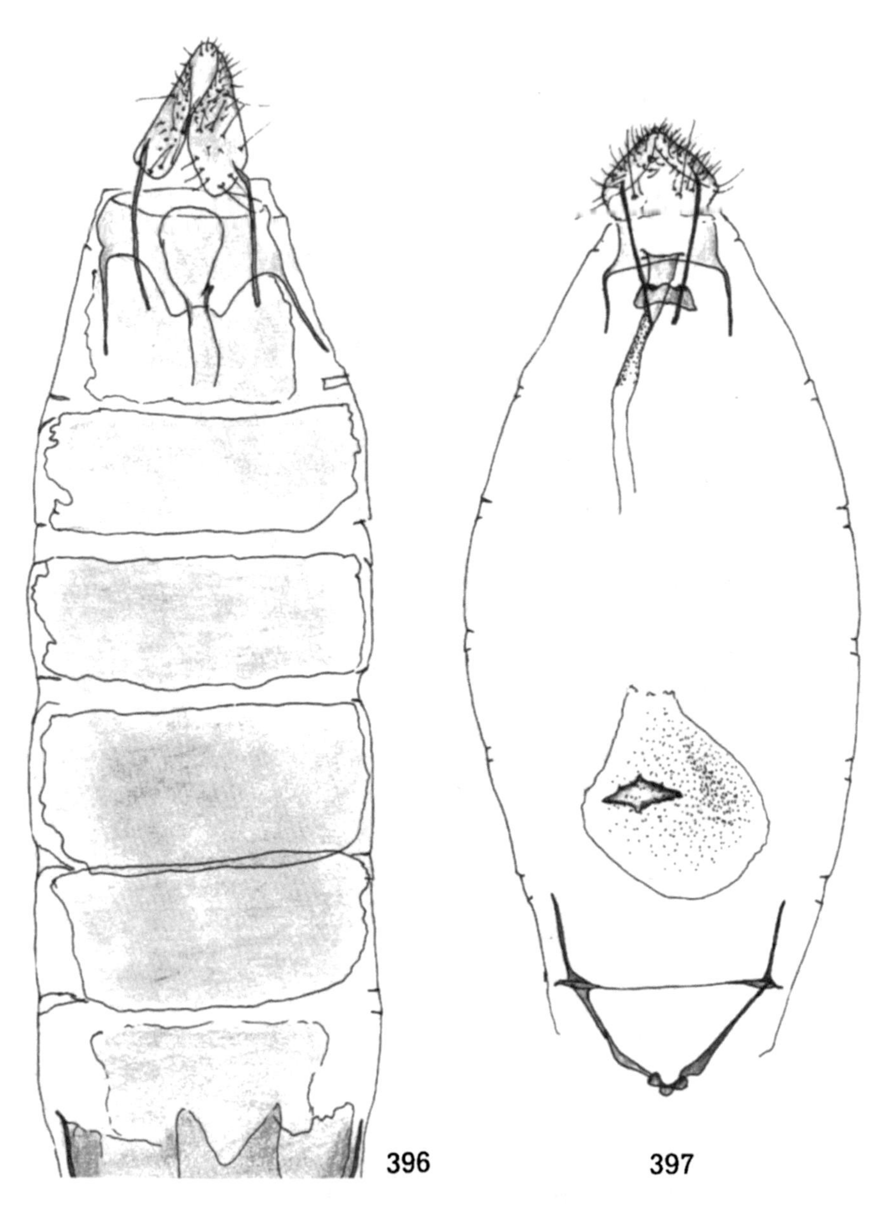

Fig. 396. *Mendesia farinella* (Thnbg.), female genitalia (ETO 1637).
Fig. 397. *Perittia herrichiella* (HS.), female genitalia (ETO 1490).

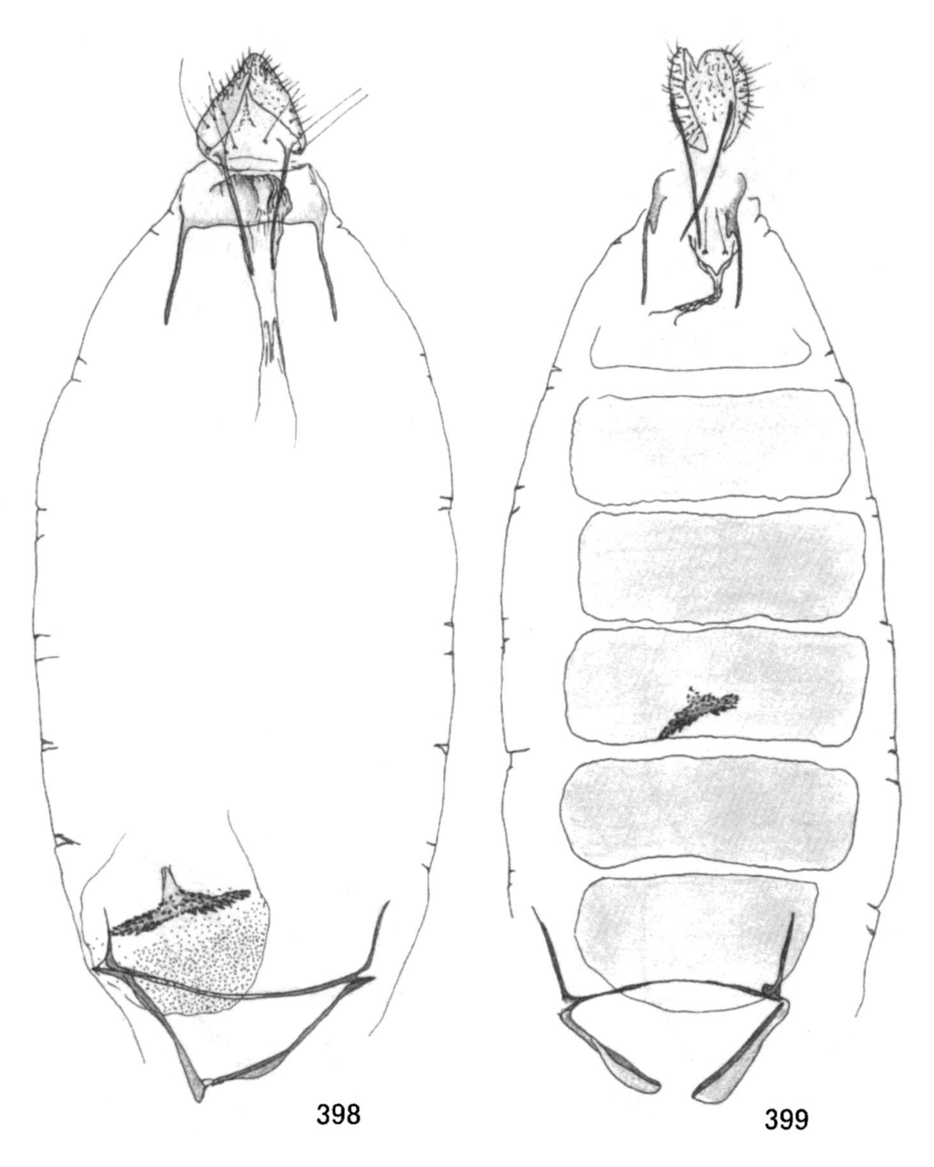

Fig. 398. *Perittia obscurepunctella* (Stt.), female genitalia (ETO 1491).
Fig. 399. *Stephensia brunnichella* (L.), female genitalia (ETO 1633).

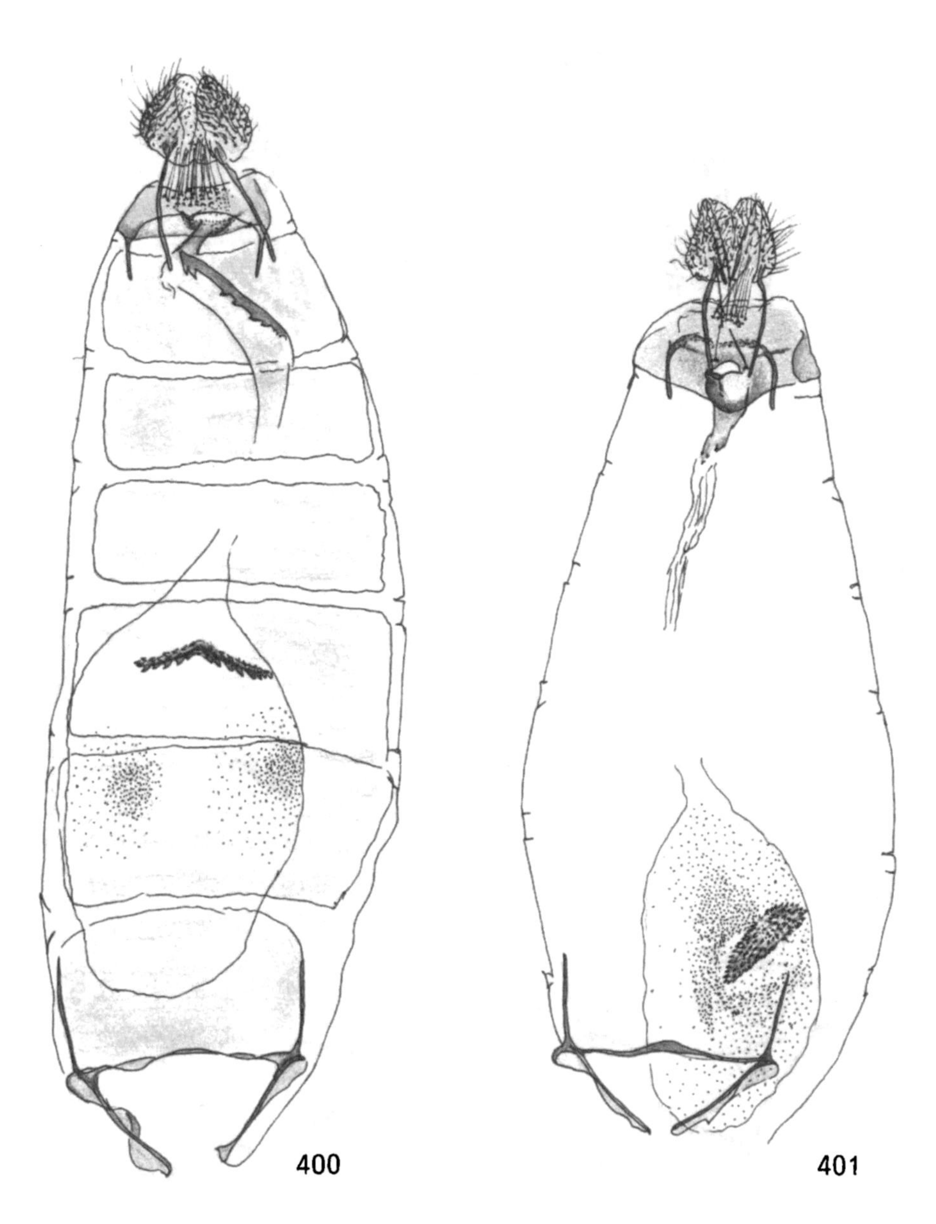

Fig. 400. *Elachista regificella* Sirc., female genitalia (ETO 1313).
Fig. 401. *Elachista gleichenella* (F.), female genitalia (ETO 1345).

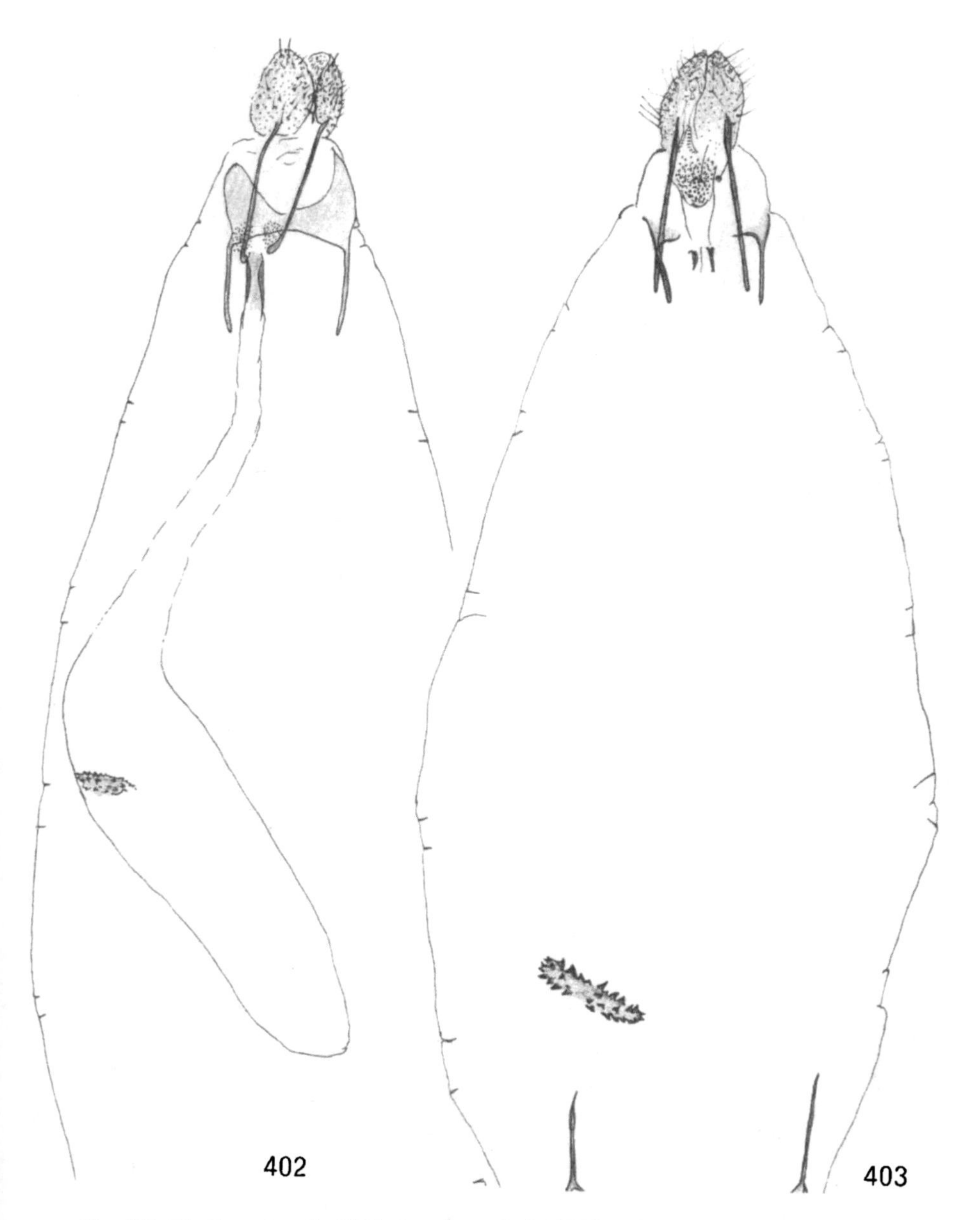

Fig. 402. *Elachista pigerella* (HS.), female genitalia (ETO 2050).
Fig. 403. *Elachista quadripunctella* (Hb.), female genitalia (ETO 1865).

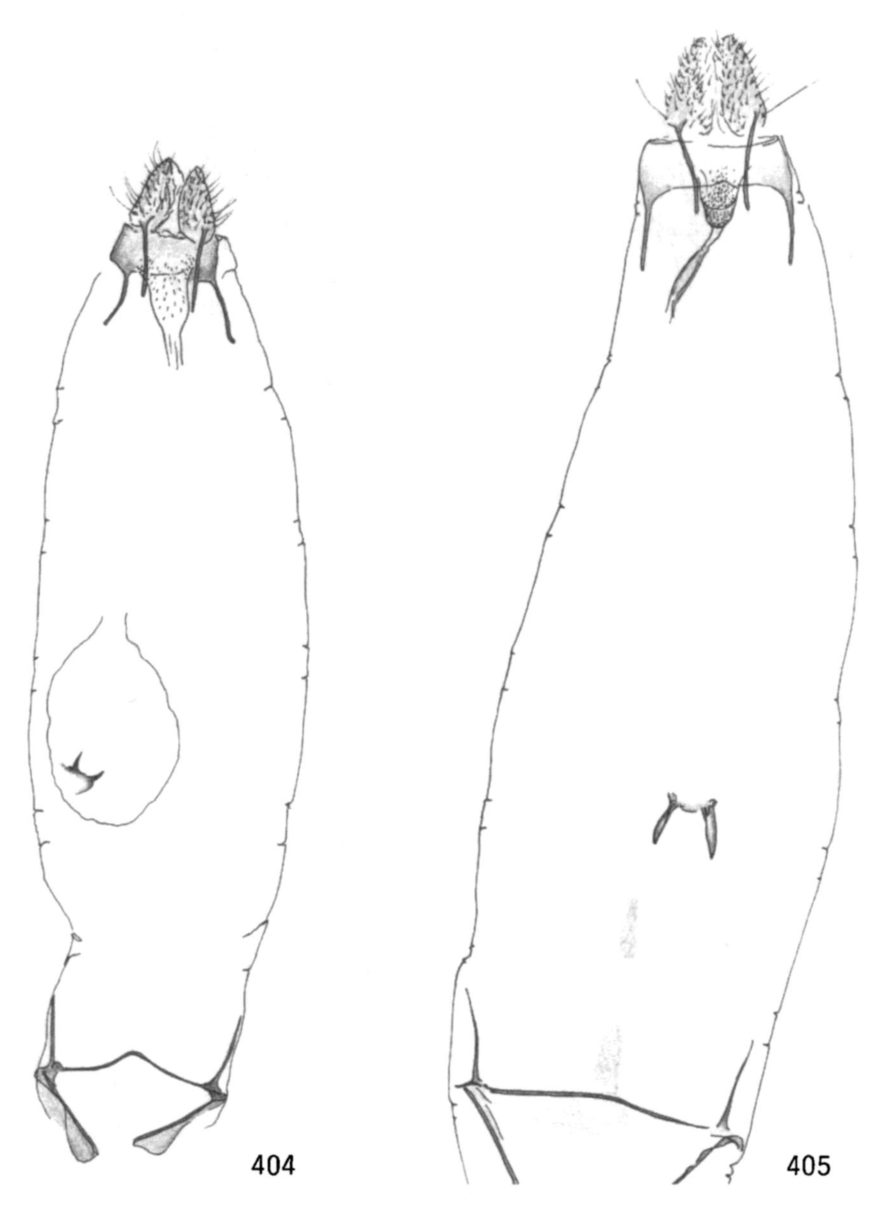

Fig. 404. *Elachista tetragonella* (HS.), female genitalia (ETO 1867).
Fig. 405. *Elachista biatomella* (Stt.), female genitalia (BMNH 19185).

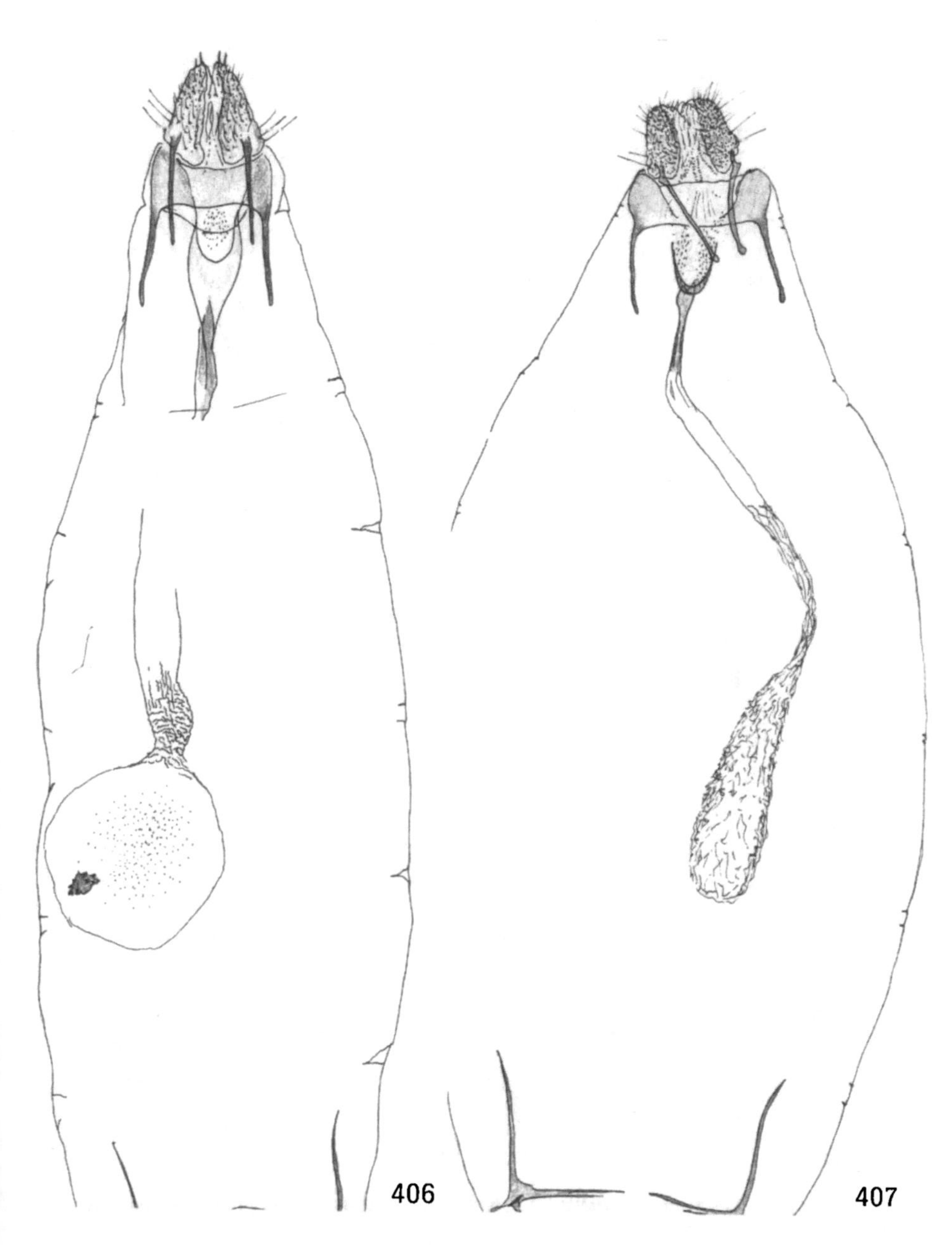

Fig. 406. *Elachista martinii* Hofm., female genitalia (ETO 2054).
Fig. 407. *Elachista poae* Stt., female genitalia (ETO C.7.2.77).

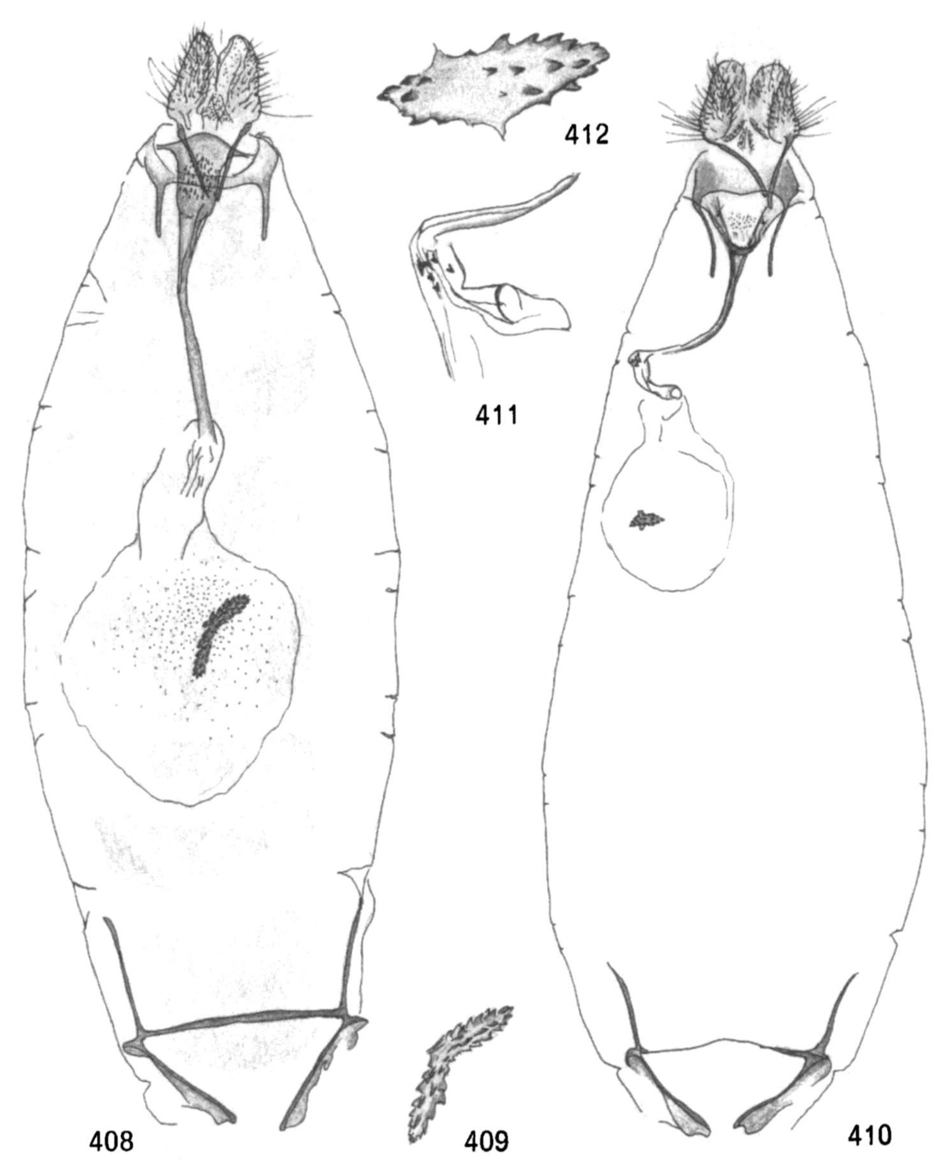

Figs. 408, 409. *Elachista atricomella* Stt. – 408: female genitalia (ETO 1298); 409: signum (ETO 1298).
Figs. 410–412. *Elachista kilmunella* Stt. – 410: female genitalia (IS 4411); 411: inception of ductus seminalis into ductus bursae (IS 4411); 412: signum (IS 4411).

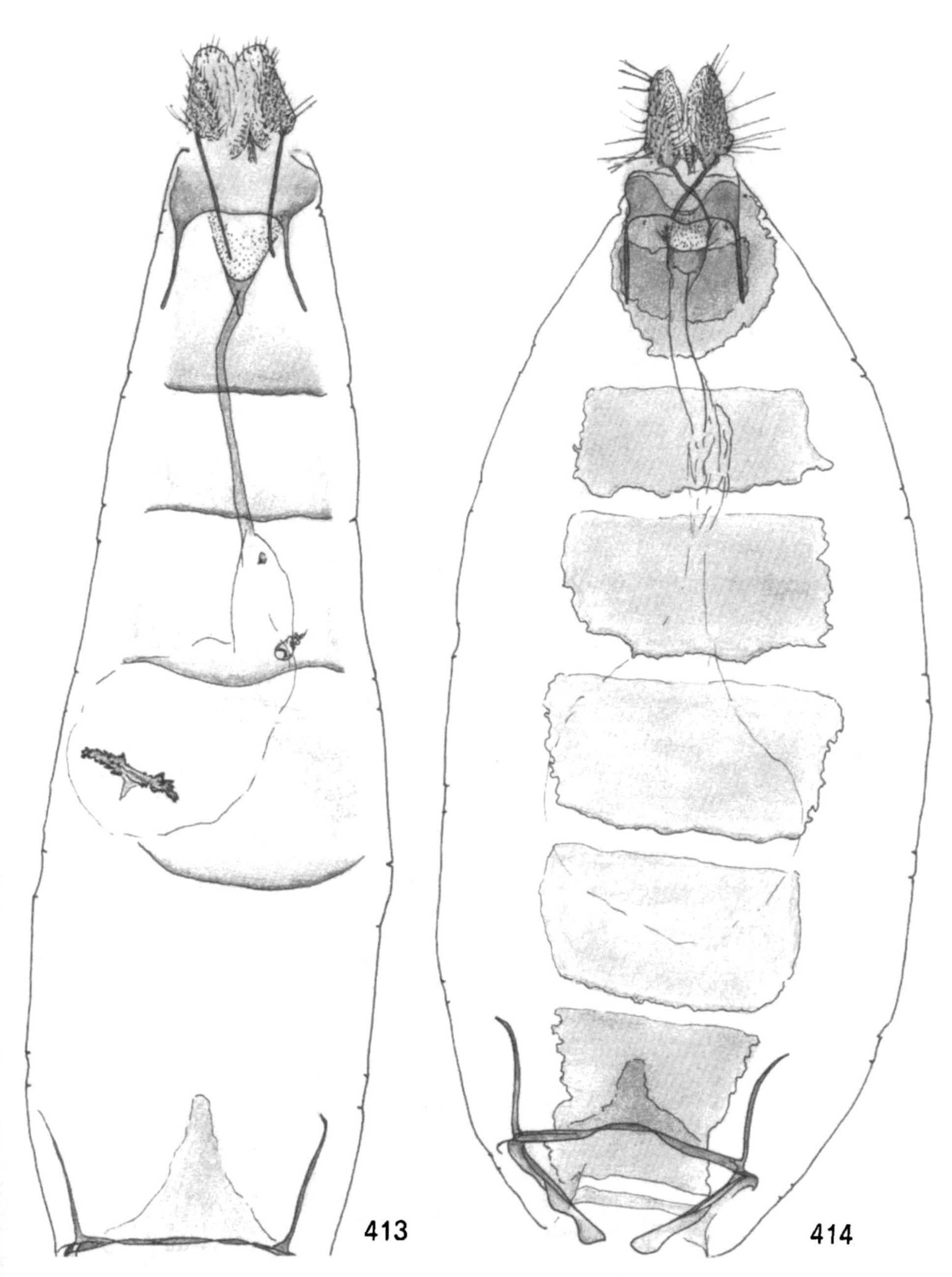

Fig. 413. *Elachista parasella* Tr.–O., female genitalia (IS 5330).
Fig. 414. *Elachista alpinella* Stt., female genitalia (ETO E.26.11.76).

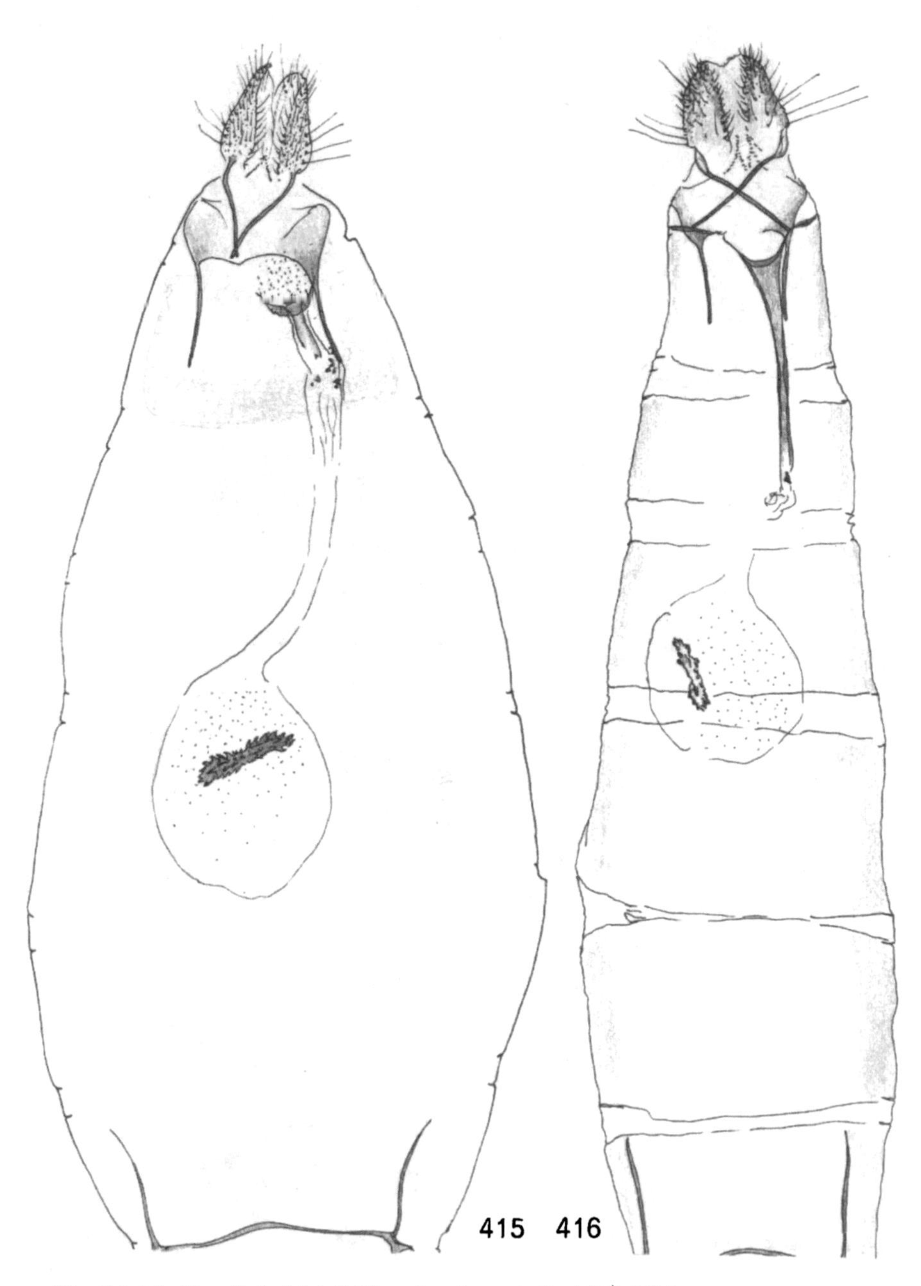

Fig. 415. *Elachista diederichsiella* Her., female genitalia (ETO 1836).
Fig. 416. *Elachista compsa* Tr.-O., female genitalia (ETO 1785).

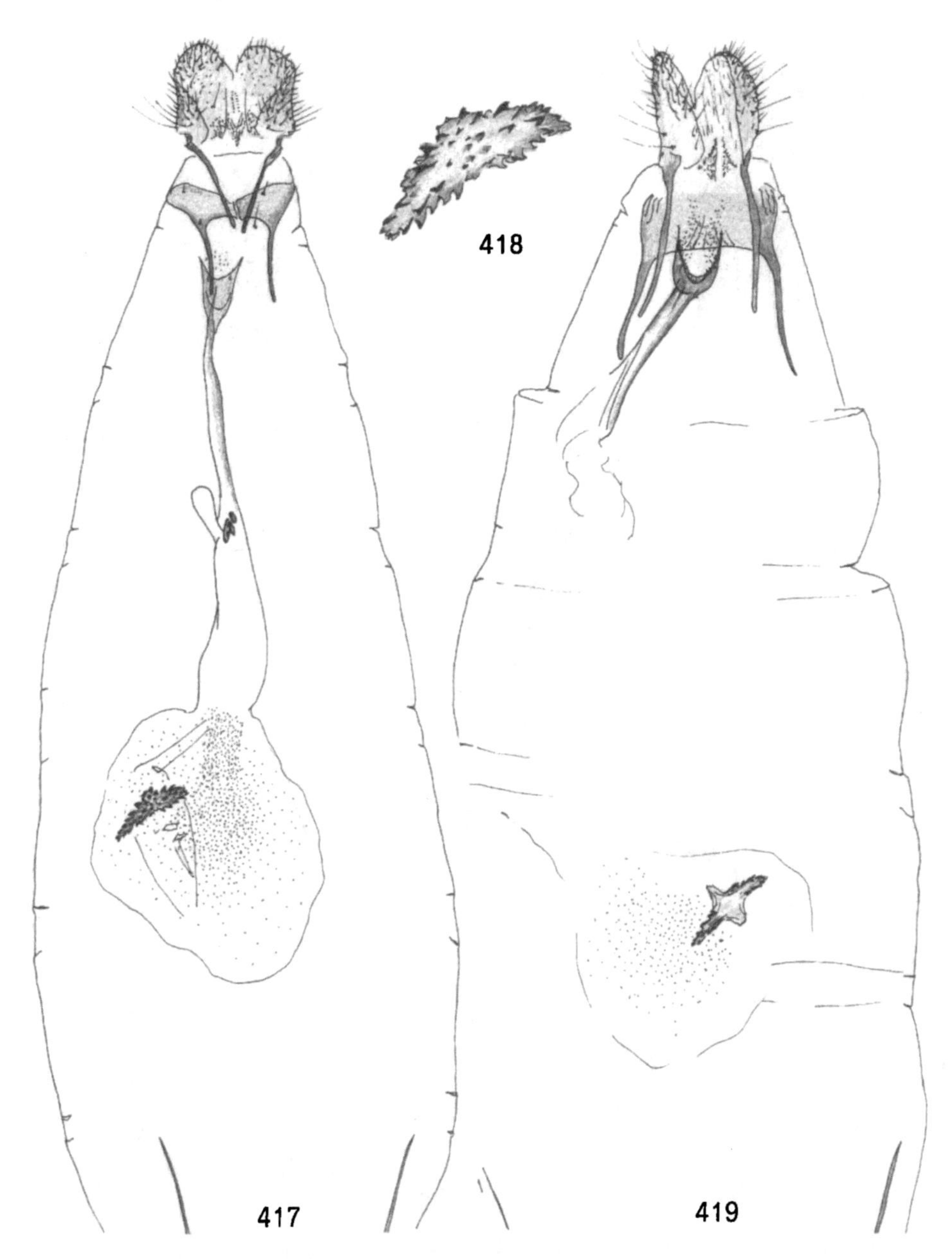

Figs. 417, 418.*Elachista elegans* Frey. – 417: female genitalia (IS 4780); 418: signum (IS 4780).
Fig. 419. *Elachista luticomella* Zell., female genitalia (ETO 1344).

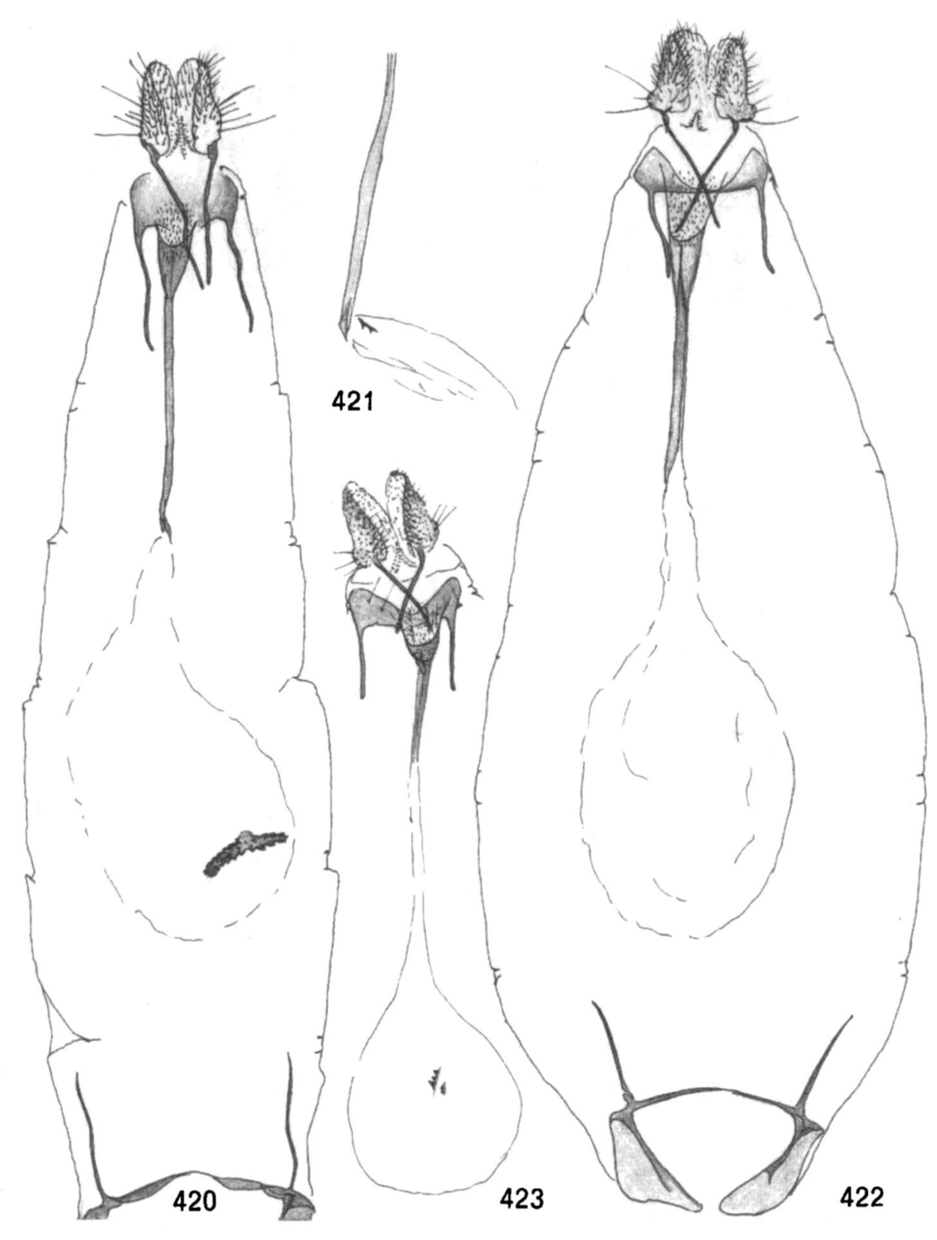

Figs. 420, 421. *Elachista albifrontella* (Hb.). – 420: female genitalia (ETO 505); 421: ductus bursae (GP 111).
Figs. 422, 423. *Elachista bifasciella* Tr. – 422: female genitalia (ETO 1746); 423: female genitalia (KL 144).

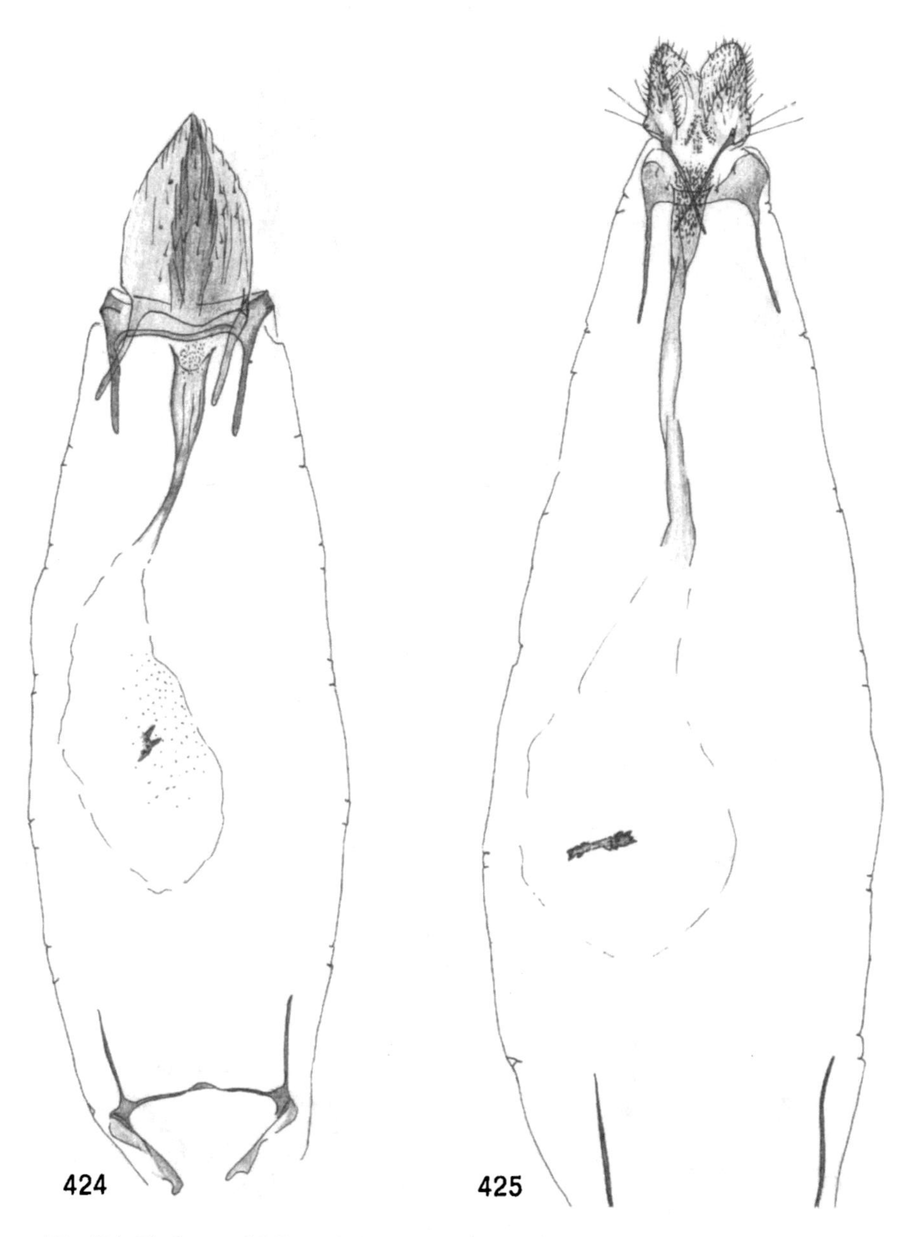

Fig. 424. *Elachista nobilella* Zell., female genitalia (ETO 1362).
Fig. 425. *Elachista apicipunctella* Stt., female genitalia (ETO 1480).

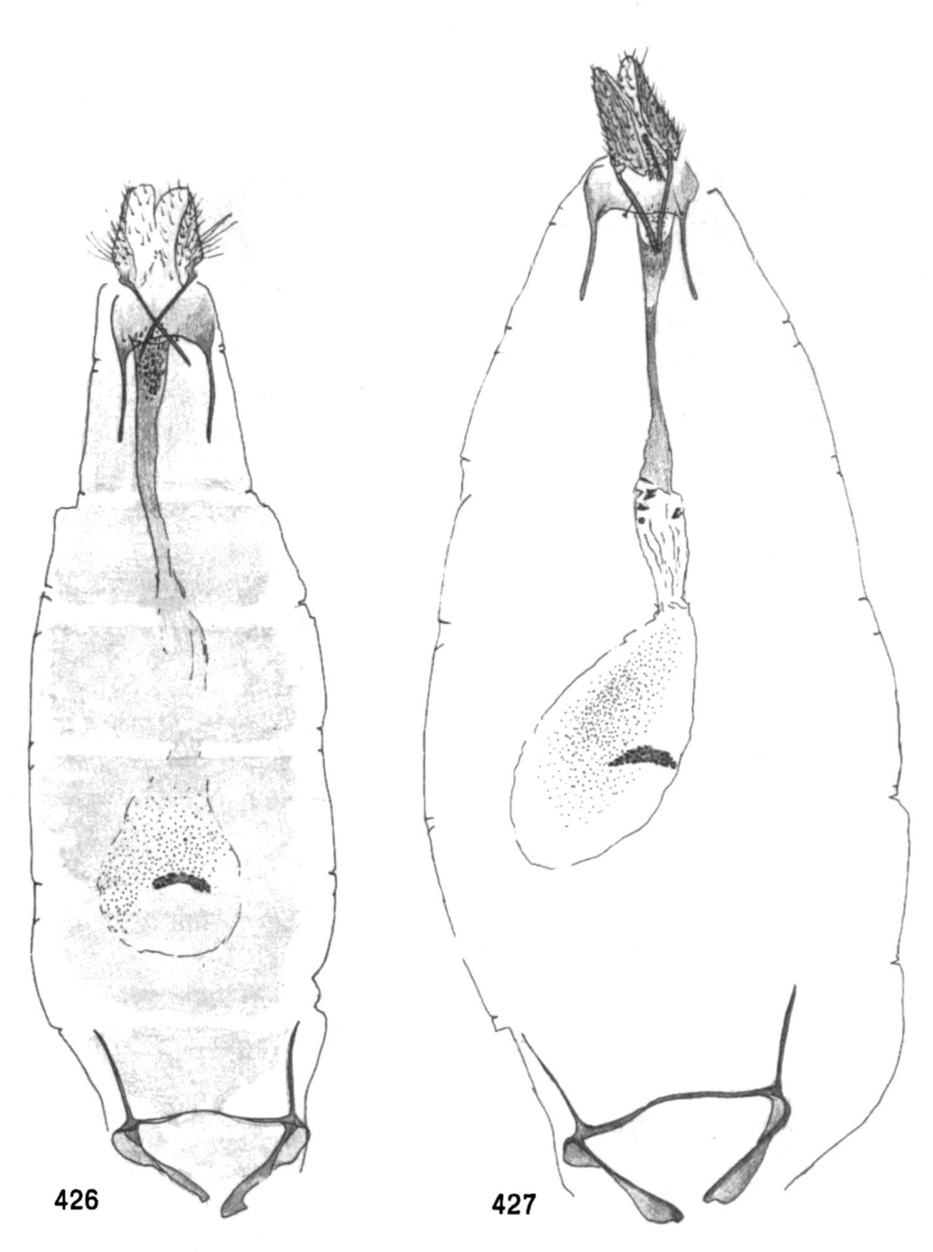

Fig. 426. *Elachista subnigrella* Dougl., female genitalia (ETO 1347).
Fig. 427. *Elachista reuttiana* Frey, female genitalia (ETO 5172).

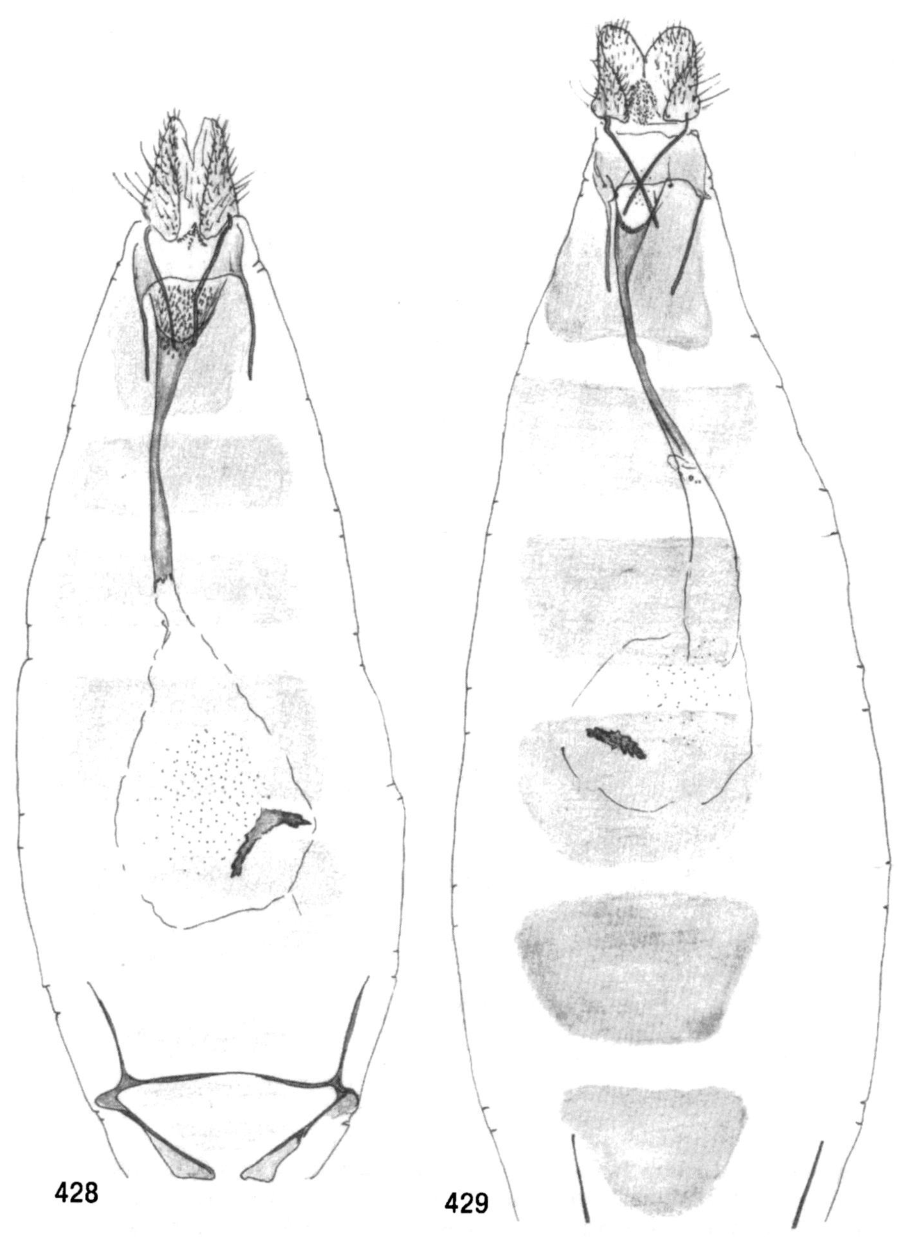

Fig. 428. *Elachista orstadii* Palm, female genitalia (ETO 1781).
Fig. 429. *Elachista ingvarella* Tr.–O., female genitalia (IS 3827).

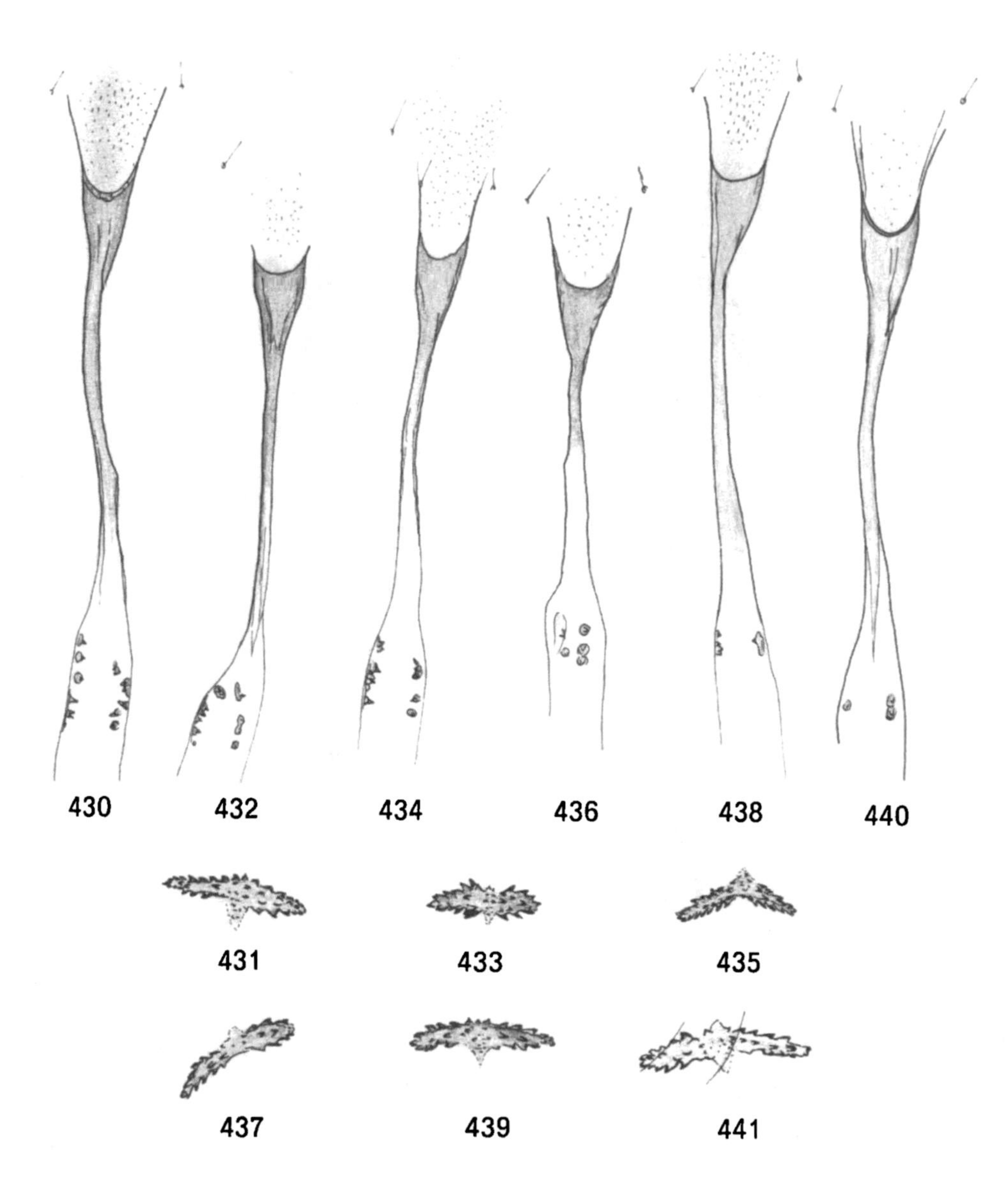

Figs. 430–441. *Elachista ingvarella* Tr.–O. – 430: ostium and ductus bursae (IS 5650); 431: signum (IS 5650); 432: ostium and ductus bursae (IS 5646); 433: signum (IS 5646); 434: ostium and ductus bursae (IS 5353); 435: signum (IS 5353); 436: ostium and ductus bursae (IS 4713); 437: signum (IS 4713); 438: ostium and ductus bursae (IS 3833); 439: signum (IS 3833); 440: ostium and ductus bursae (IS 3861); 441: signum (IS 3861).

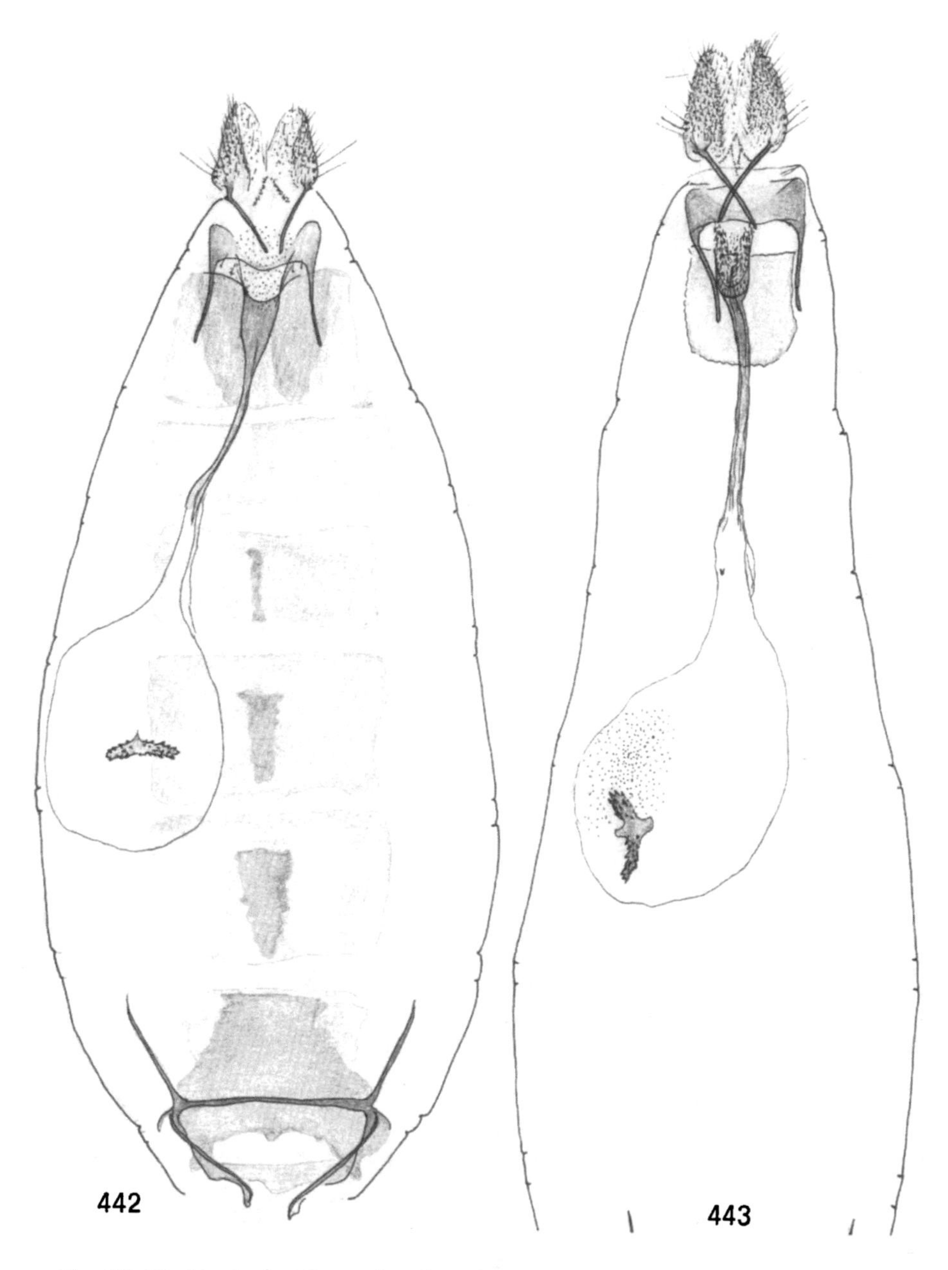

Fig. 442. *Elachista krogeri* Svens., female genitalia (IS 5642).
Fig. 443. *Elachista pomerana* Frey, female genitalia (ETO 2012).

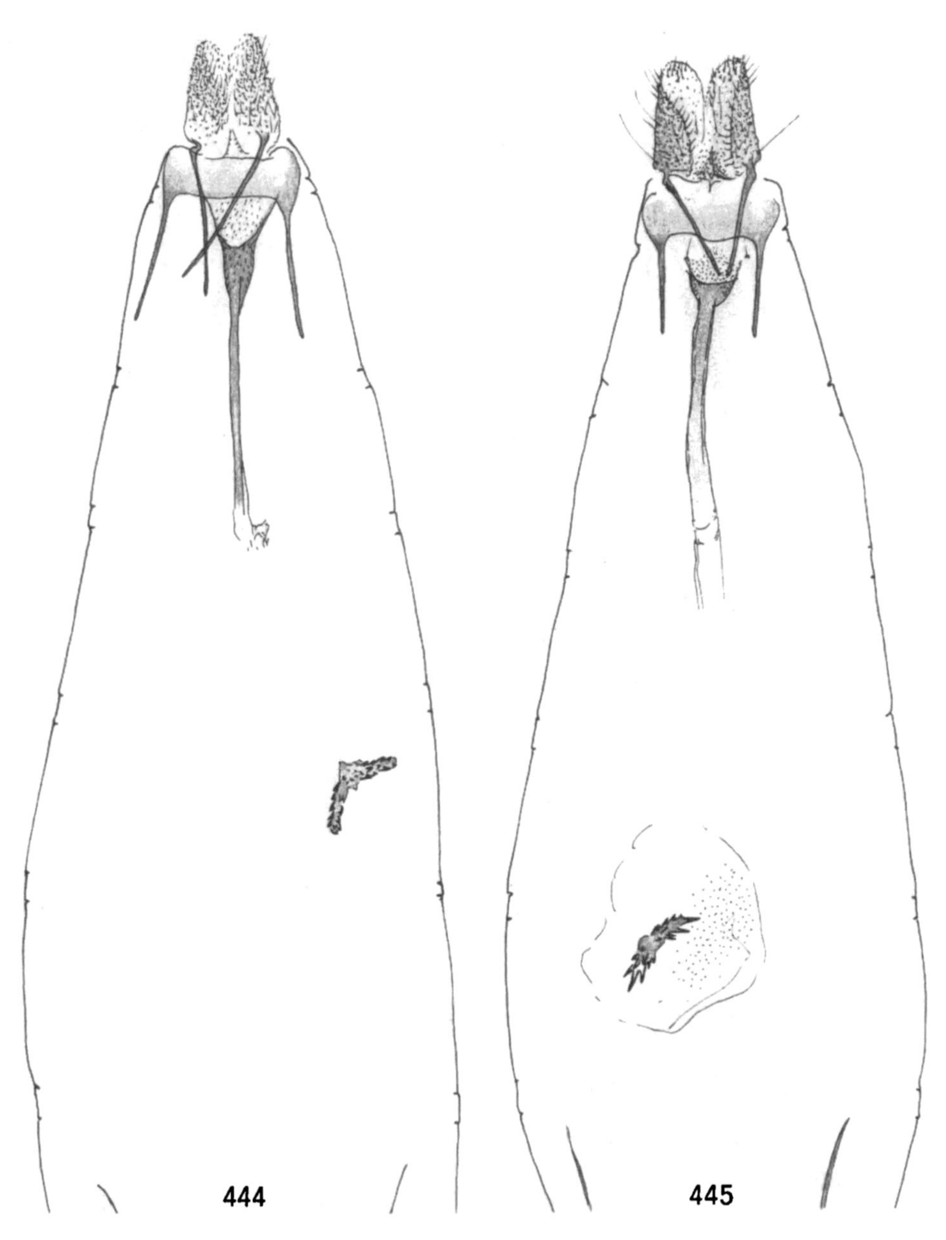

Fig. 444. *Elachista humilis* Zell., female genitalia (BMNH 19280).
Fig. 445. *Elachista vonschantzi* Svens., female genitalia (IS 5620).

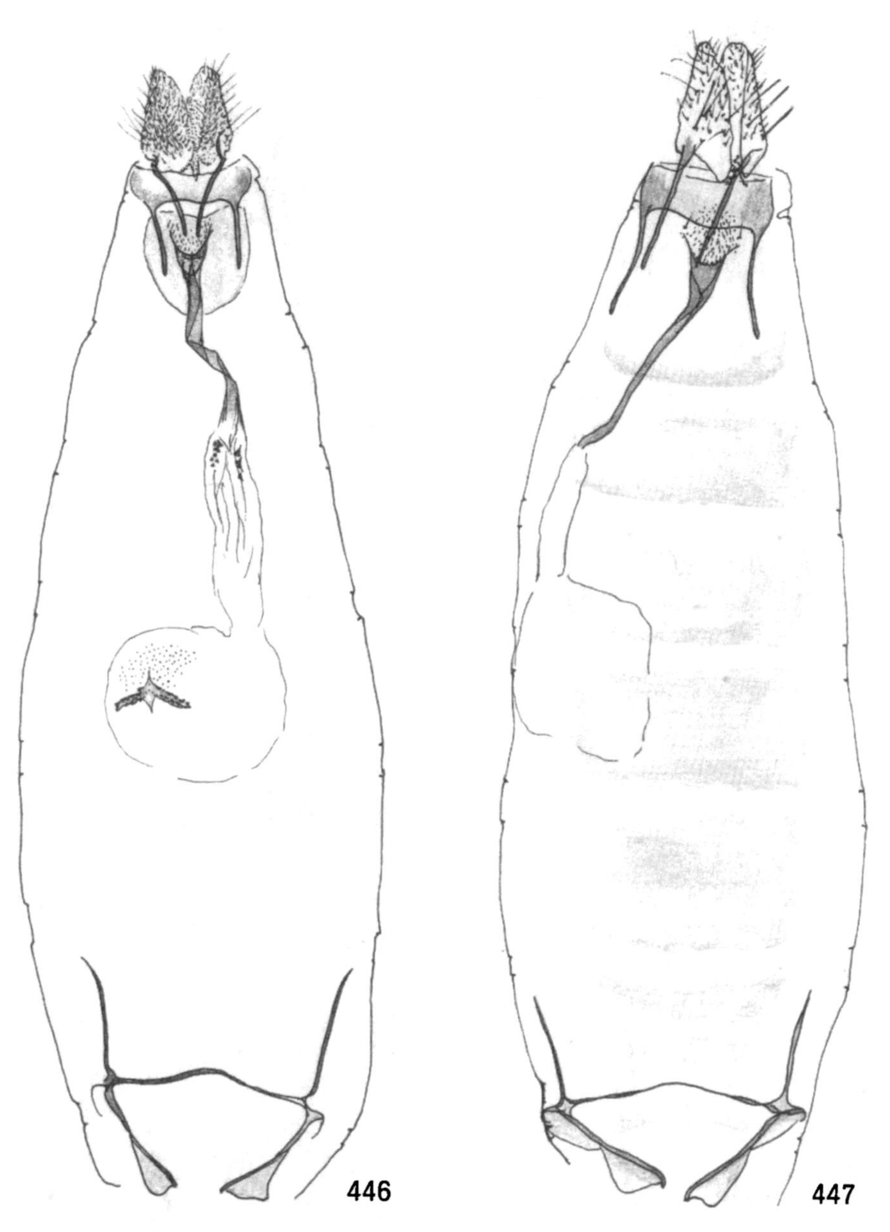

Fig. 446. *Elachista pulchella* (Hw.), female genitalia (ETO 2009).
Fig. 447. *Elachista anserinella* Zell., female genitalia (ETO 1871).

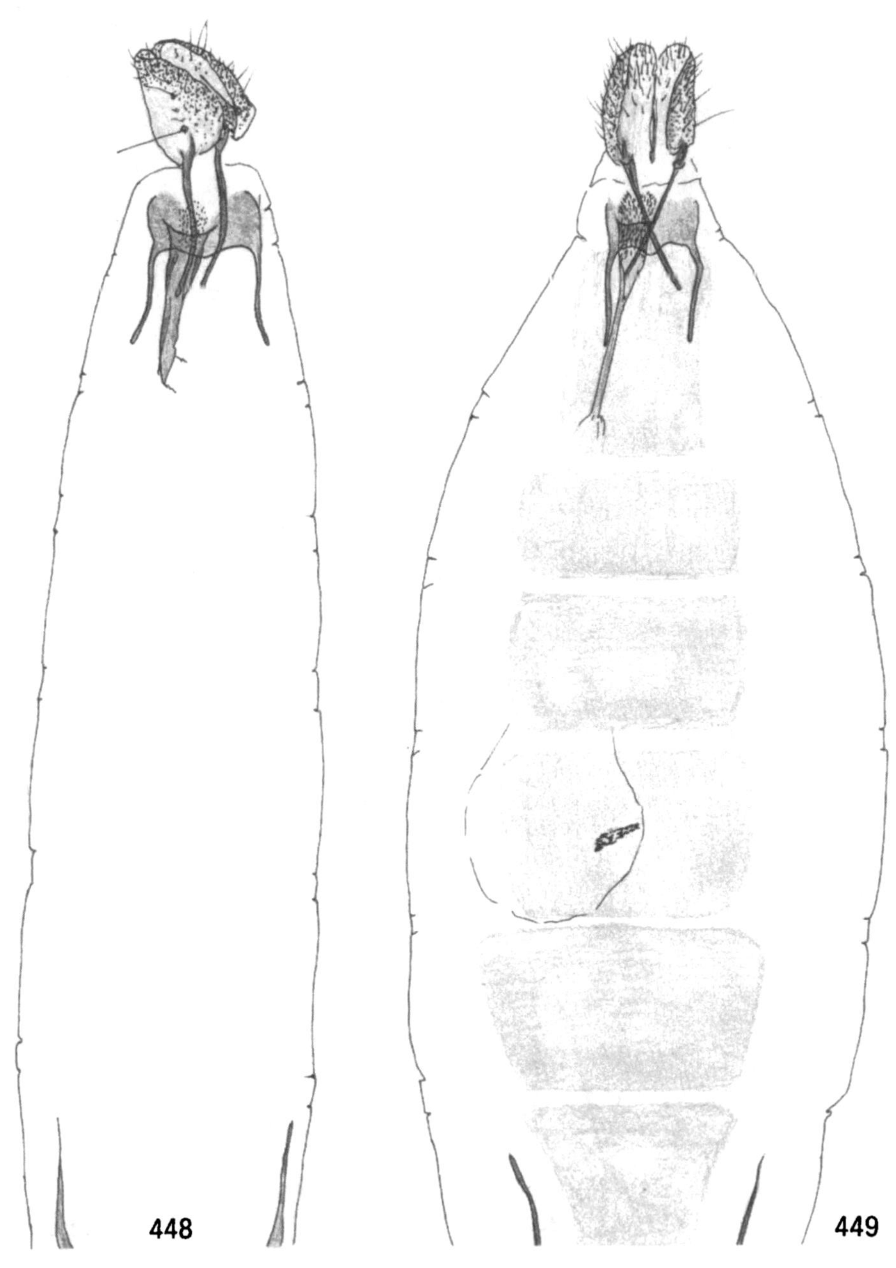

Fig. 448. *Elachista rufocinerea* (Hw.), female genitalia (EJ 247).
Fig. 449. *Elachista lastrella* Chrét., female genitalia (ETO 5279).

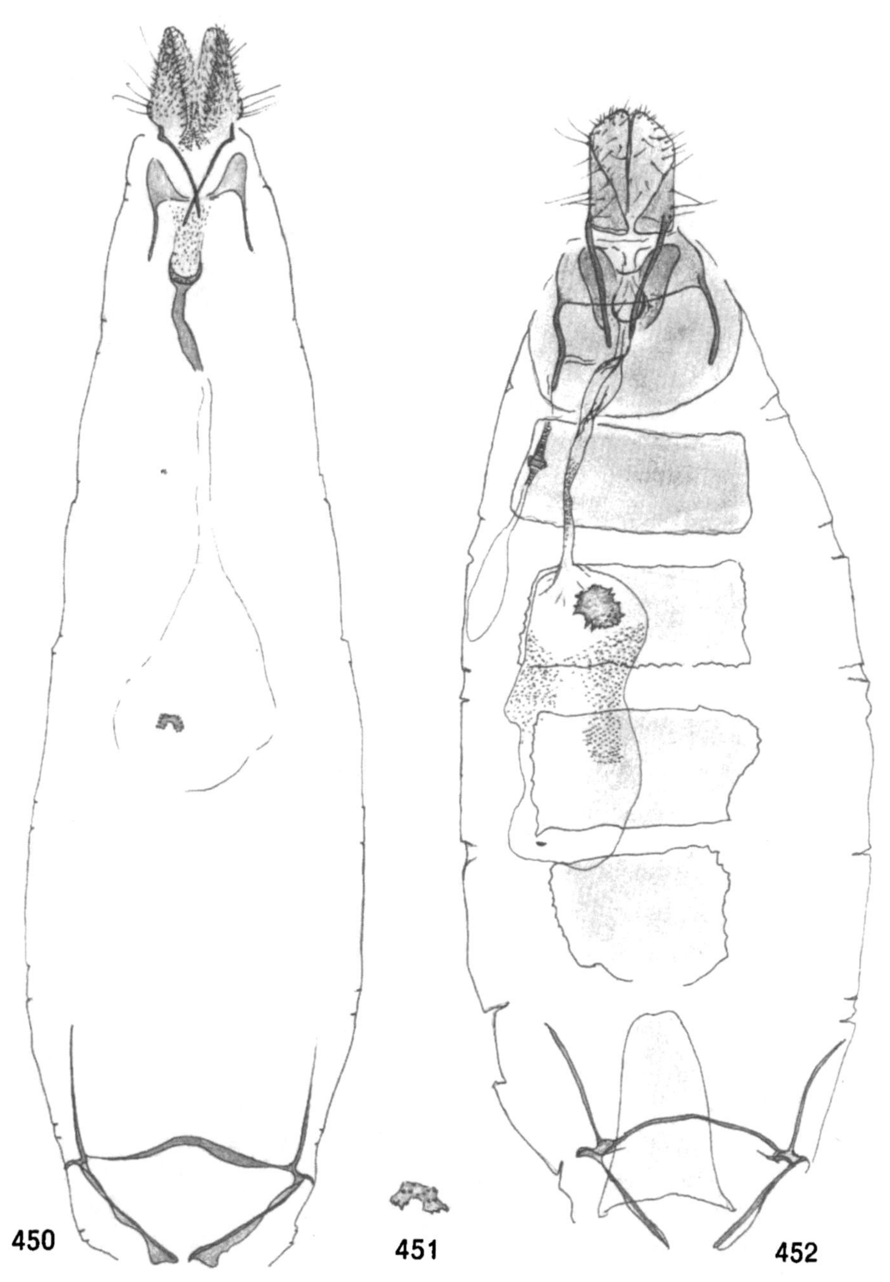

Figs. 450, 451. *Elachista cerusella* (Hb.). – 450: female genitalia (ETO 1311); 451: signum (ETO 1311).
Fig. 452. *Elachista argentella* (Cl.), female genitalia (ETO 1992).

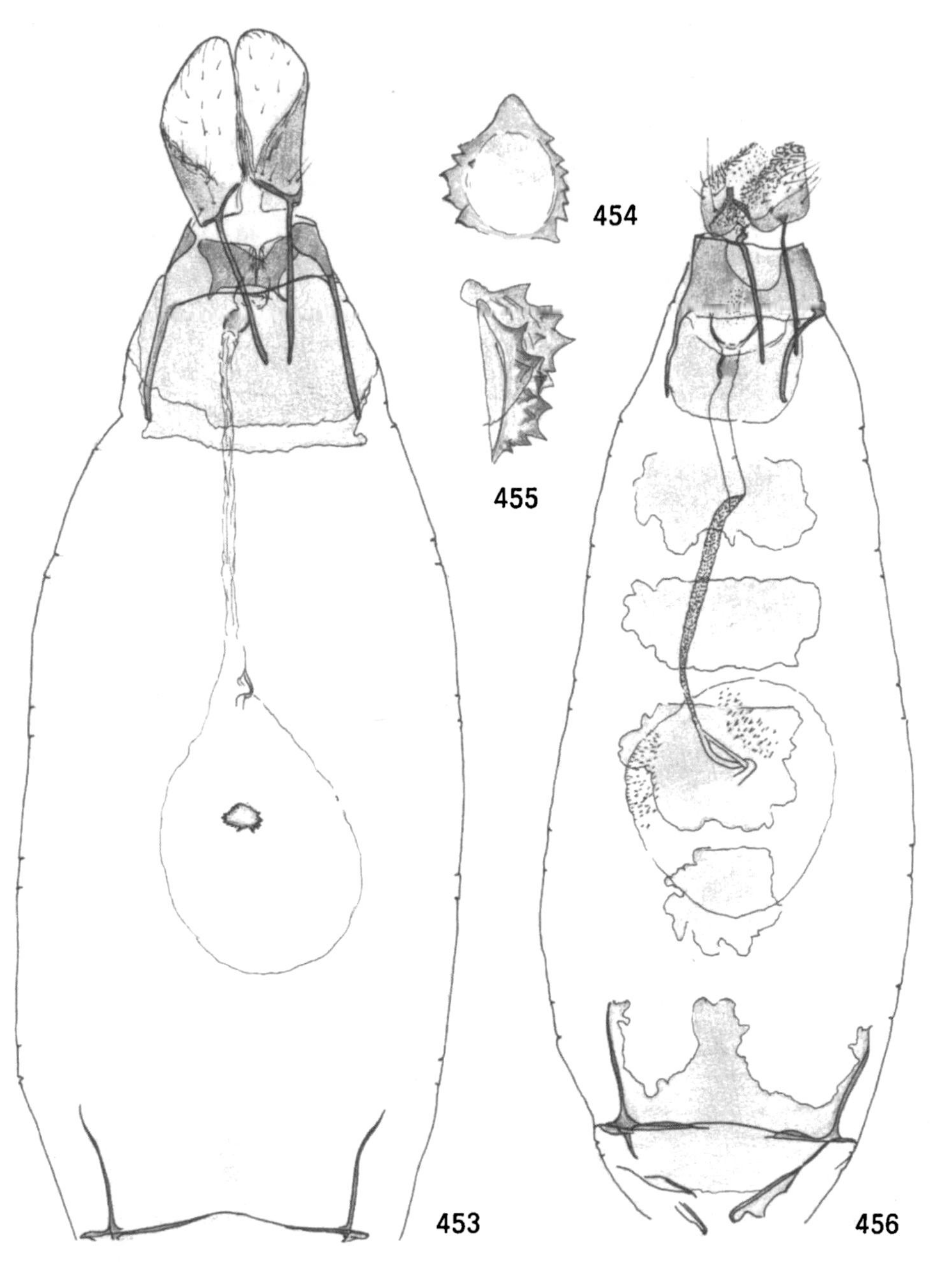

Figs. 453–455. *Elachista pollinariella* Zell. - 453: female genitalia (ETO 5175); 454: signum (BMNH 19282); 455: signum, lateral view (ETO 1678).
Fig. 456. *Elachista triatomea* (Hw.), female genitalia (ETO 1031).

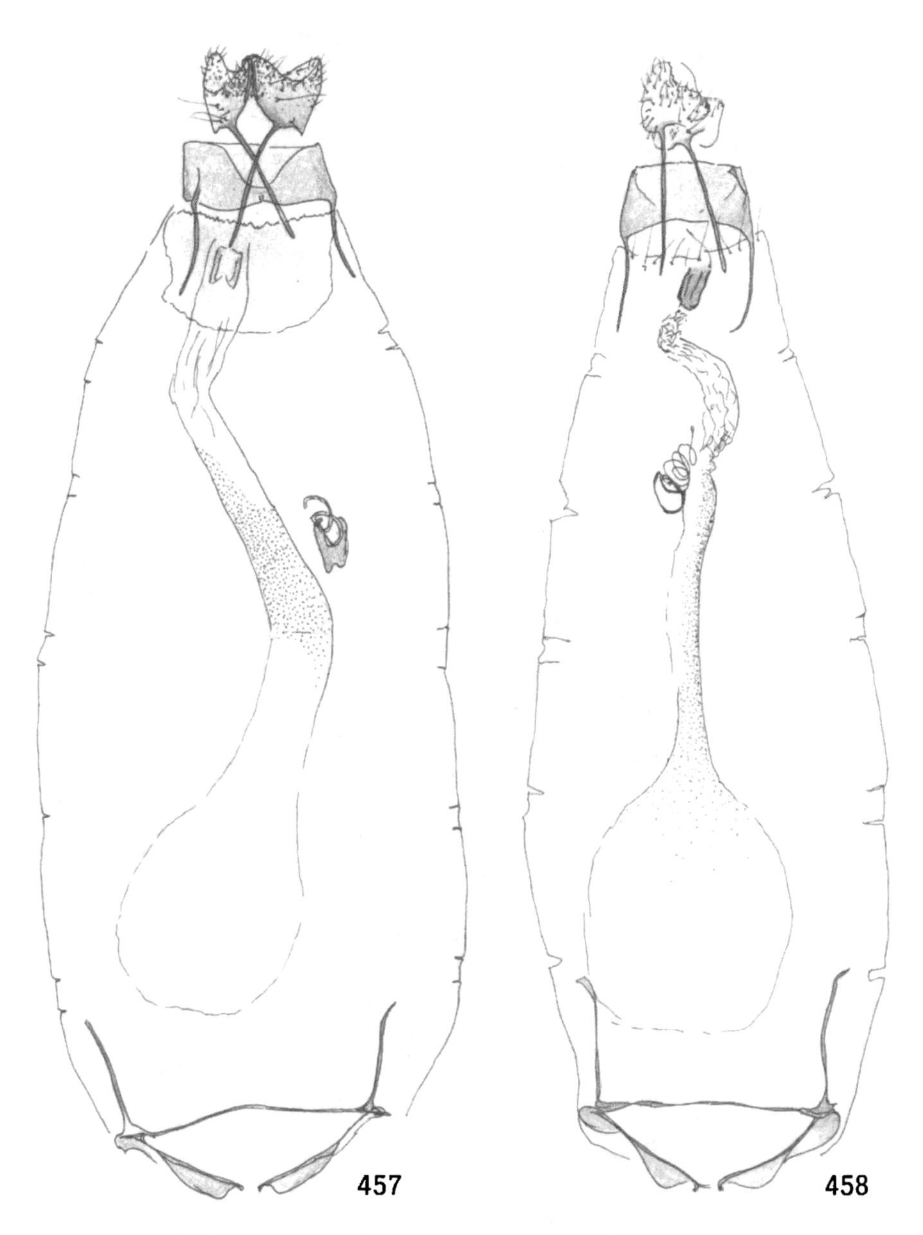

Fig. 457. *Elachista collitella* (Dup.), female genitalia (ETO 5190).
Fig. 458. *Elachista subocellea* (Stph.), female genitalia (ETO 1817).

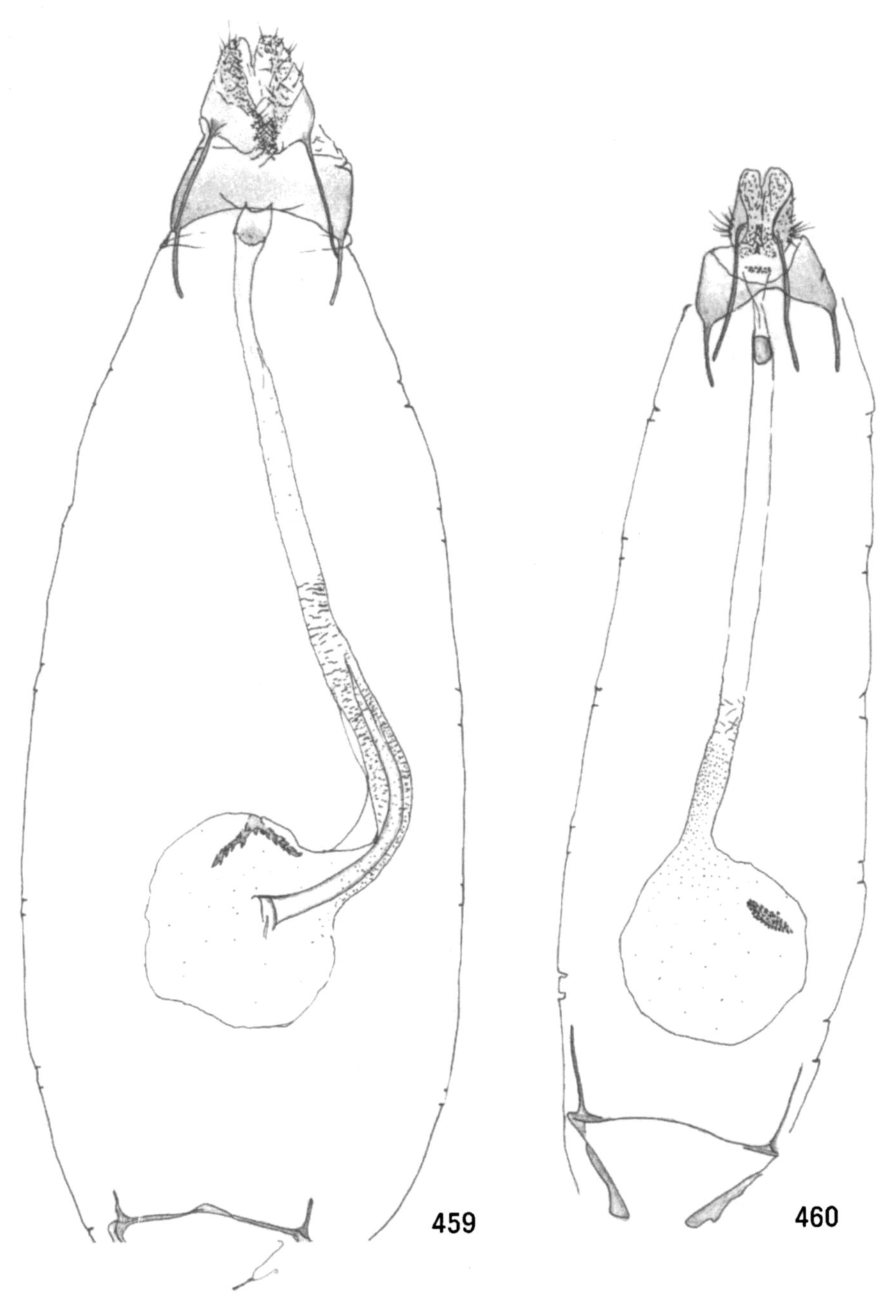

Fig. 459. *Elachista festucicolella* Zell., female genitalia (IS 5445).
Fig. 460. *Elachista nitidulella* (HS.), female genitalia (ETO C. 25.10.74).

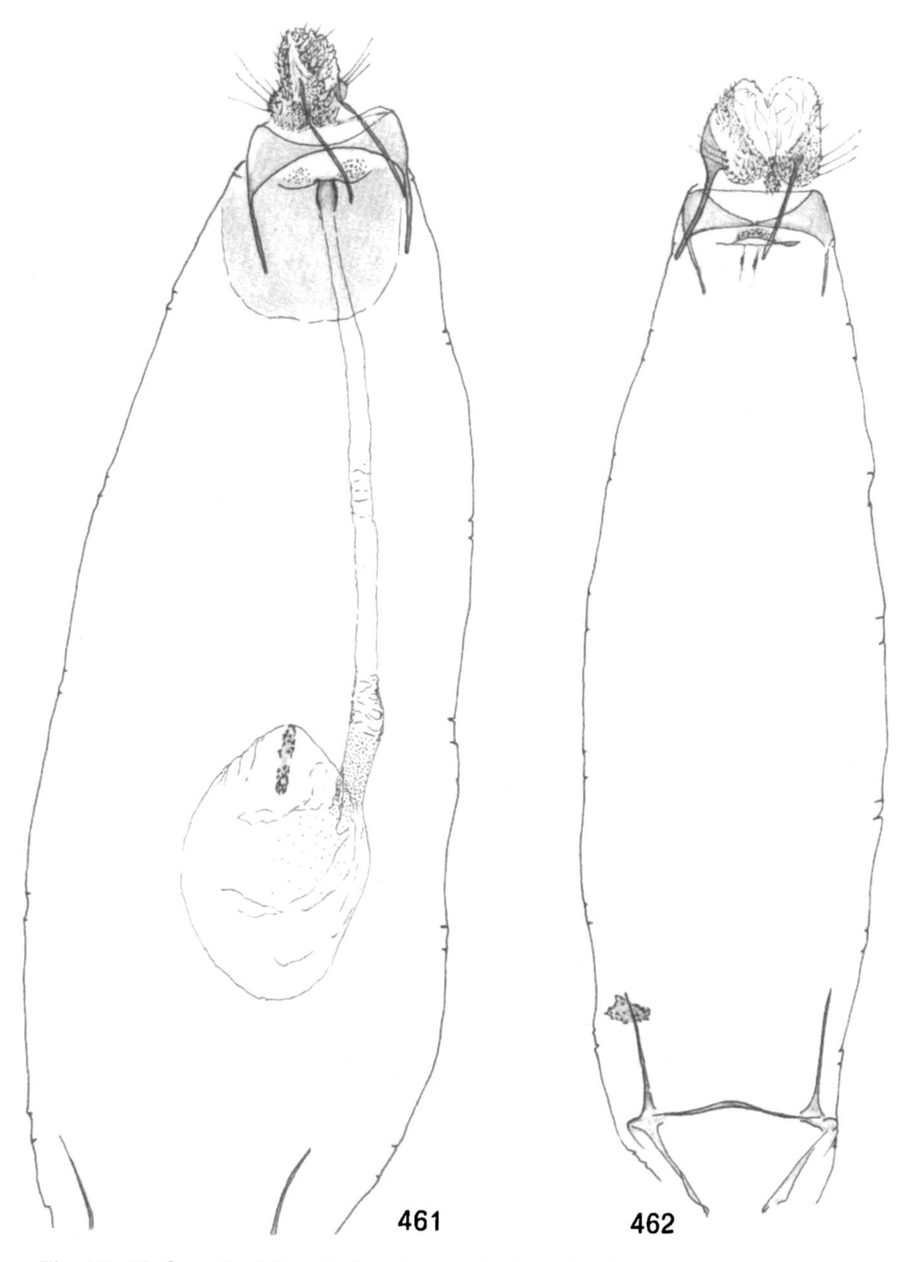

Fig. 461. *Elachista dispilella* Zell., female genitalia (ETO 1668).
Fig. 462. *Elachista triseriatella* Stt., female genitalia (IS 3911).

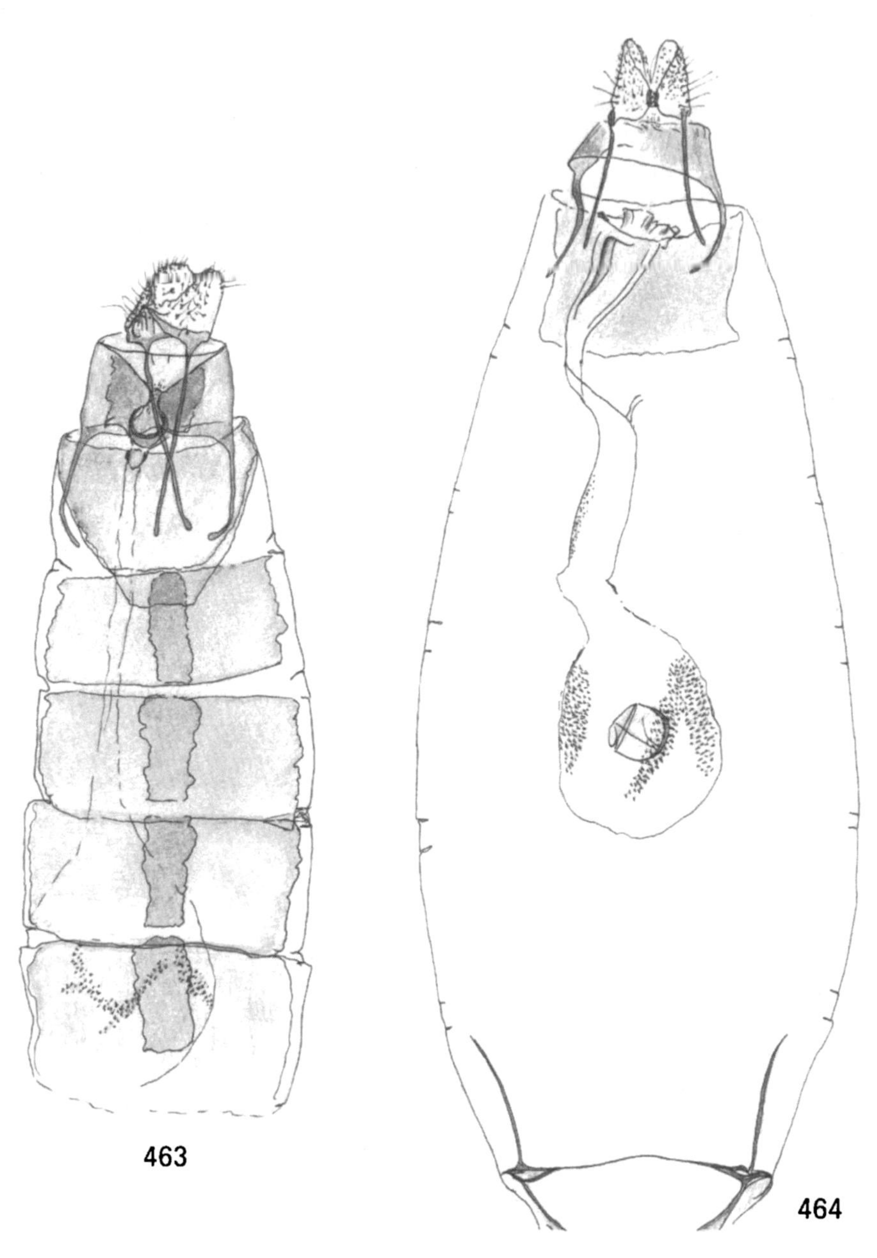

Fig. 463. *Elachista dispunctella* (Dup.), female genitalia (ETO B. 18.8.75).
Fig. 464. *Elachista rudectella* Stt., female genitalia (ETO 2052).

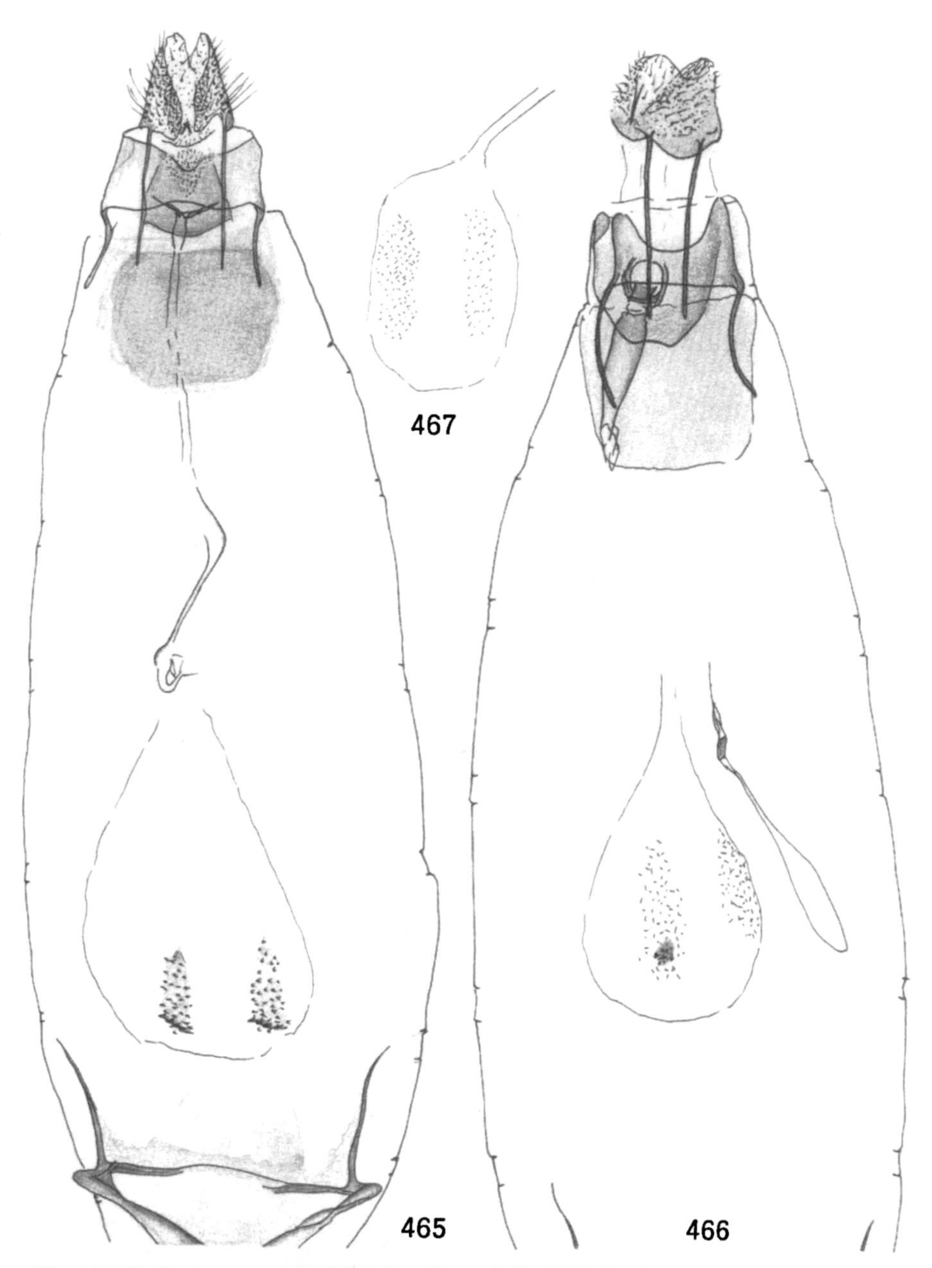

Fig. 465. *Elachista squamosella* (HS.), female genitalia (ETO B. 10.11.74).
Figs. 466, 467. *Elachista bedellella* (Sirc.). – 466: female genitalia (No. El 8a, F. N. Pierce); 467: corpus bursae (ETO 2013).

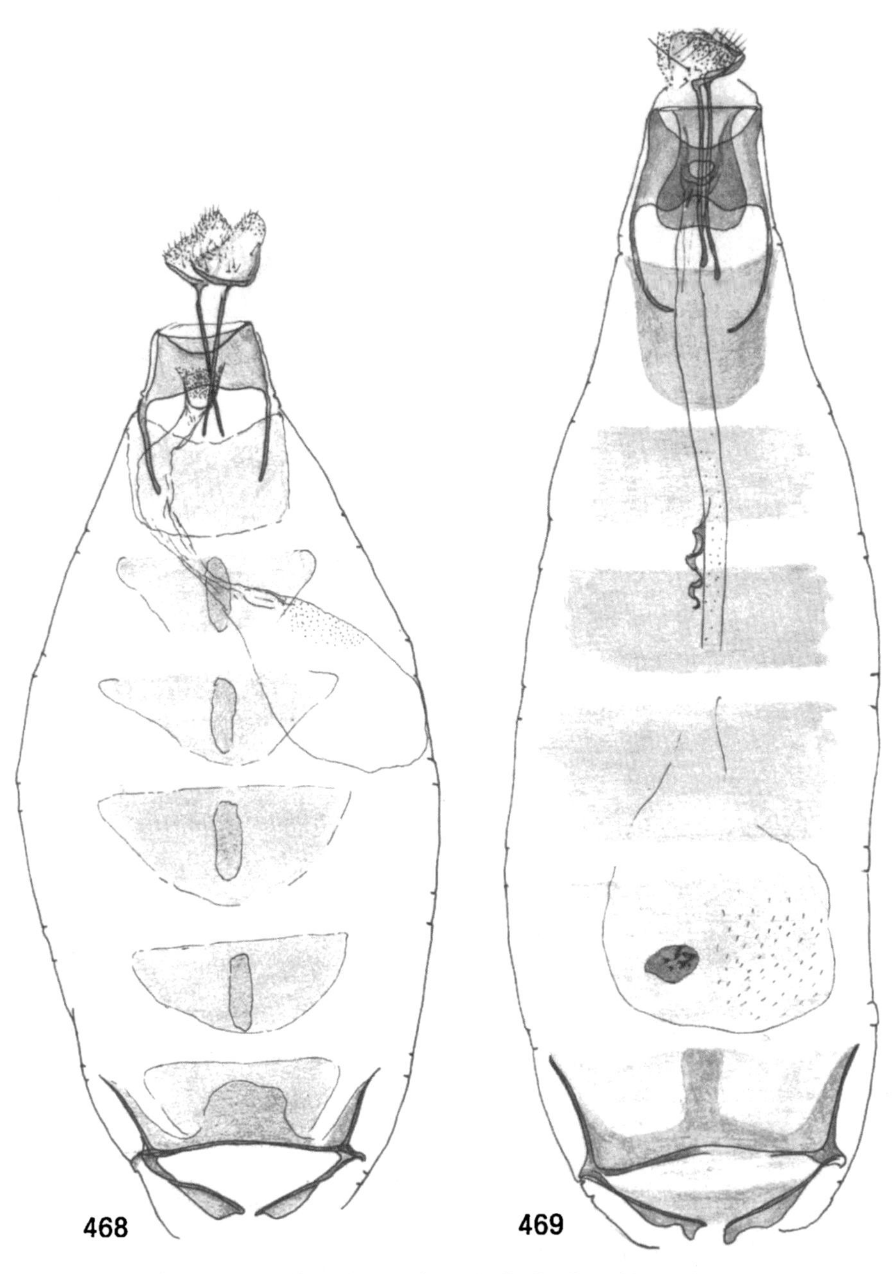

Fig. 468. *Elachista pullicomella* Zell., female genitalia (ETO A. 23.11.76).
Fig. 469. *Elachista littoricola* Le March., female genitalia (ETO B. 23.11.76).

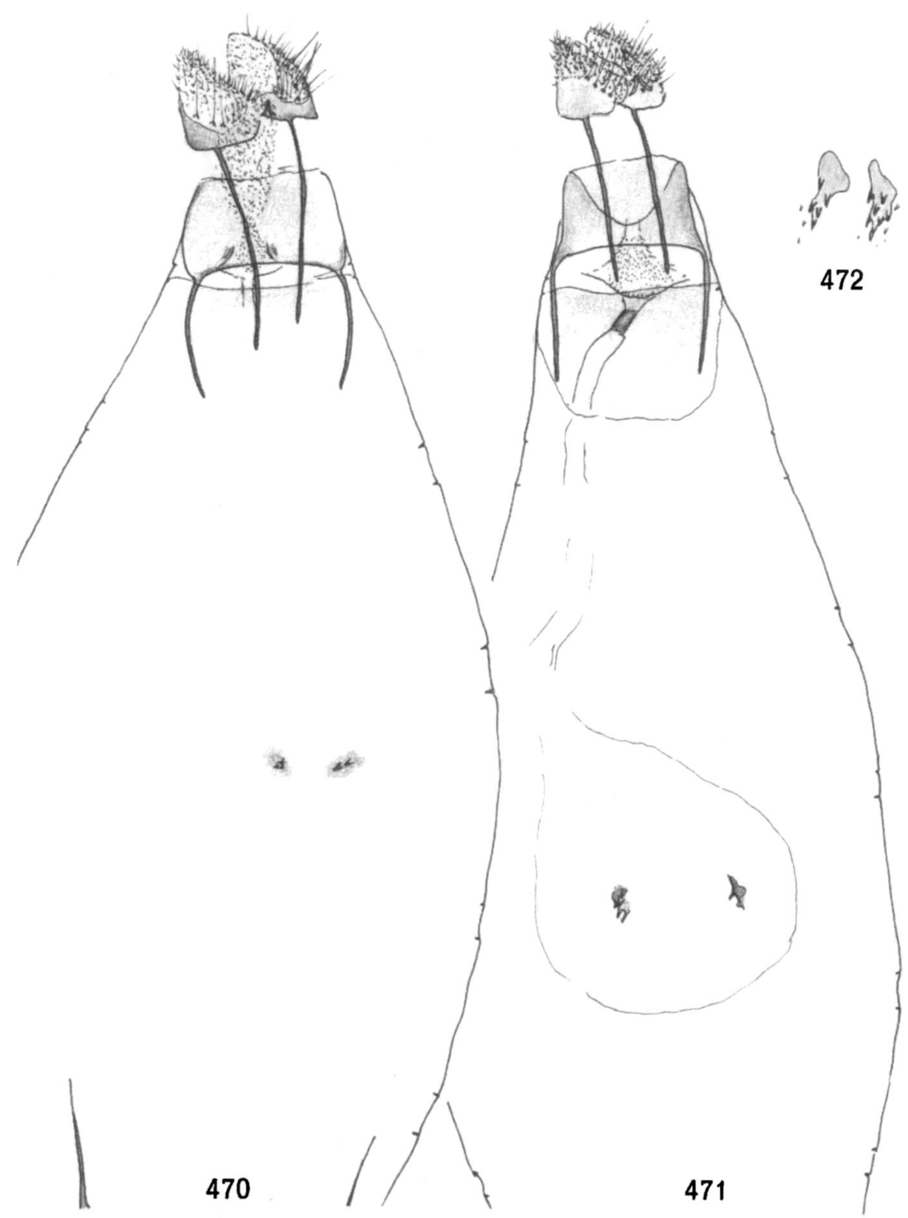

Fig. 470. *Elachista chrysodesmella* Zell., female genitalia (IS 5295).
Figs. 471, 472. *Elachista megerlella* (Hb.). – 471: female genitalia (OK 14.IX.1976); 472: signum (BMNH 18936).

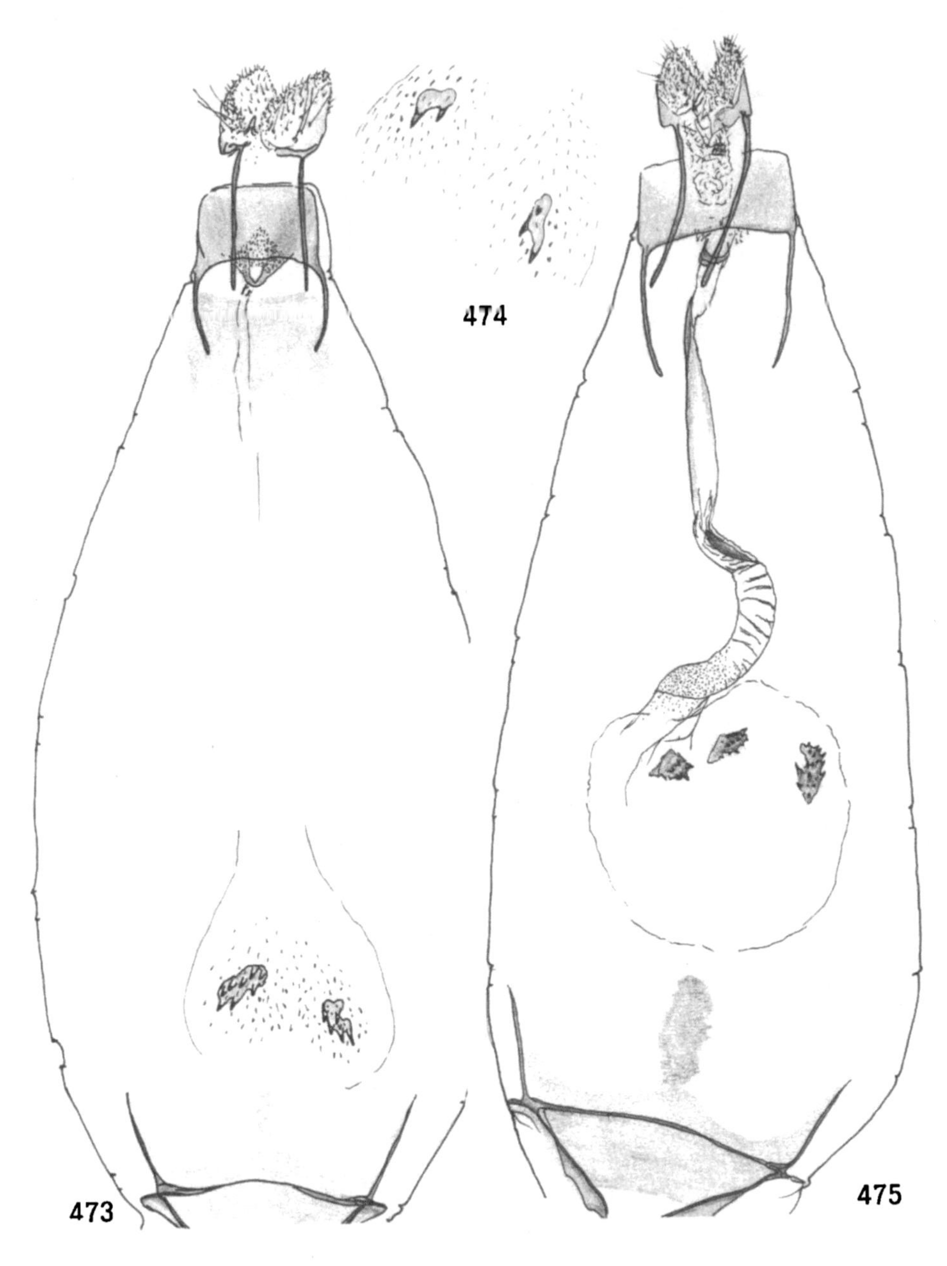

Figs. 473, 474. *Elachista cingillella* (HS.). – 473: female genitalia (IS 4414); 474: signum (IS 3195).
Fig. 475. *Elachista unifasciella* (Hw.), female genitalia (ETO 2027).

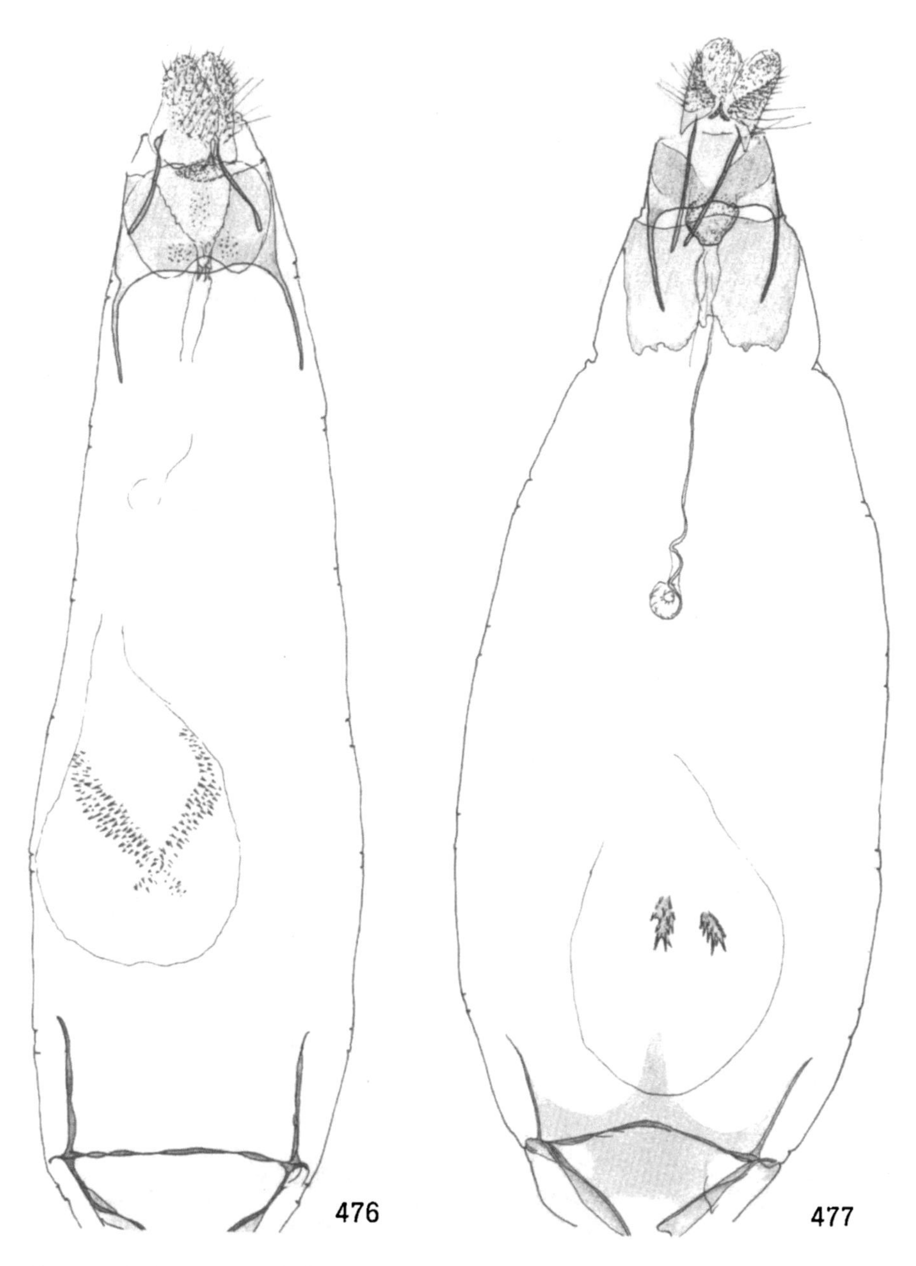

Fig. 476. *Elachista gangabella* Zell., female genitalia (ETO 1814).
Fig. 477. *Elachista subalbidella* Schl., female genitalia (ETO 2026).

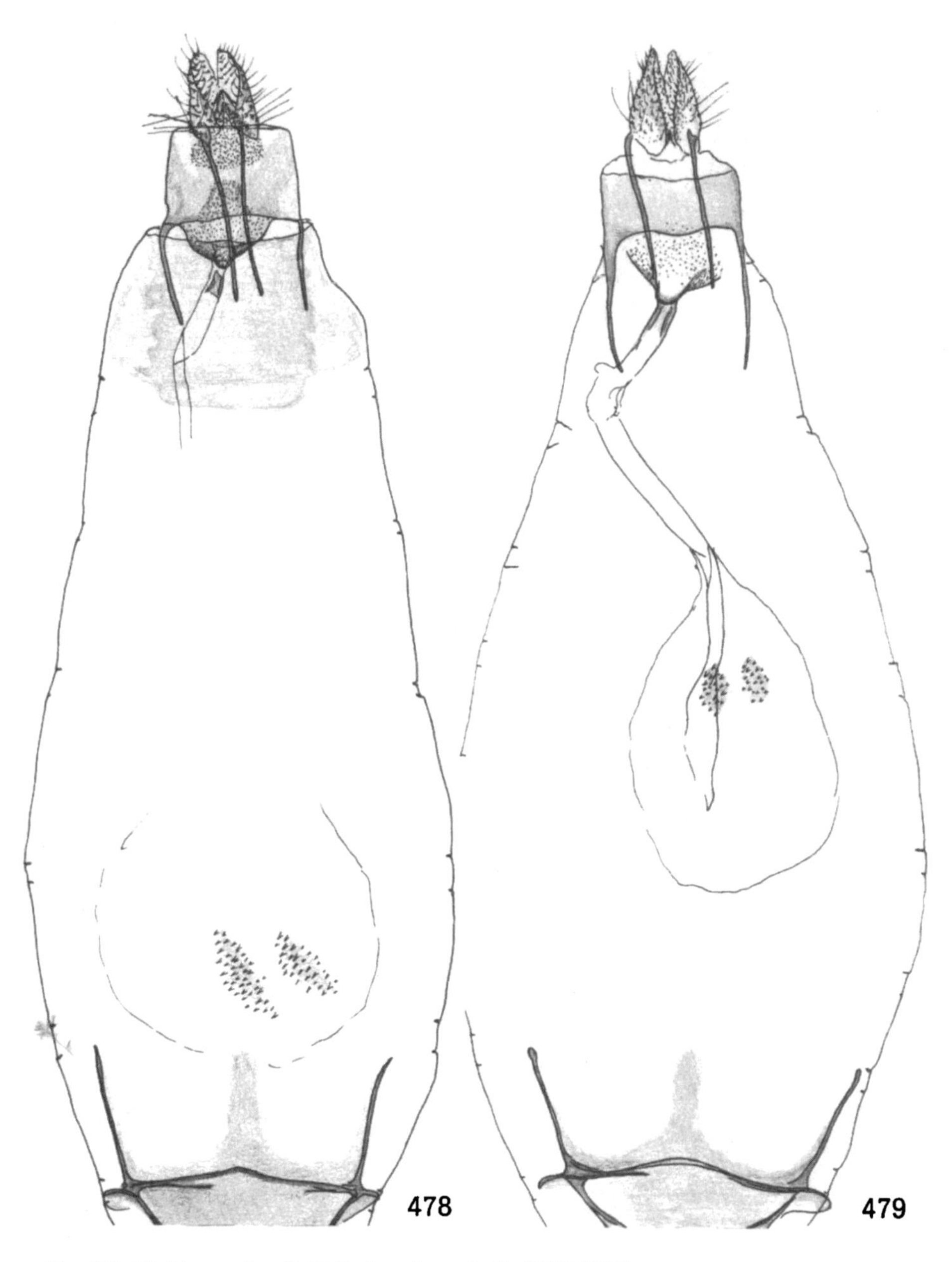

Fig. 478. *Elachista revinctella* Zell., female genitalia (ETO 1780).
Fig. 479. *Elachista bisulcella* (Dup.), female genitalia (ETO 1635).

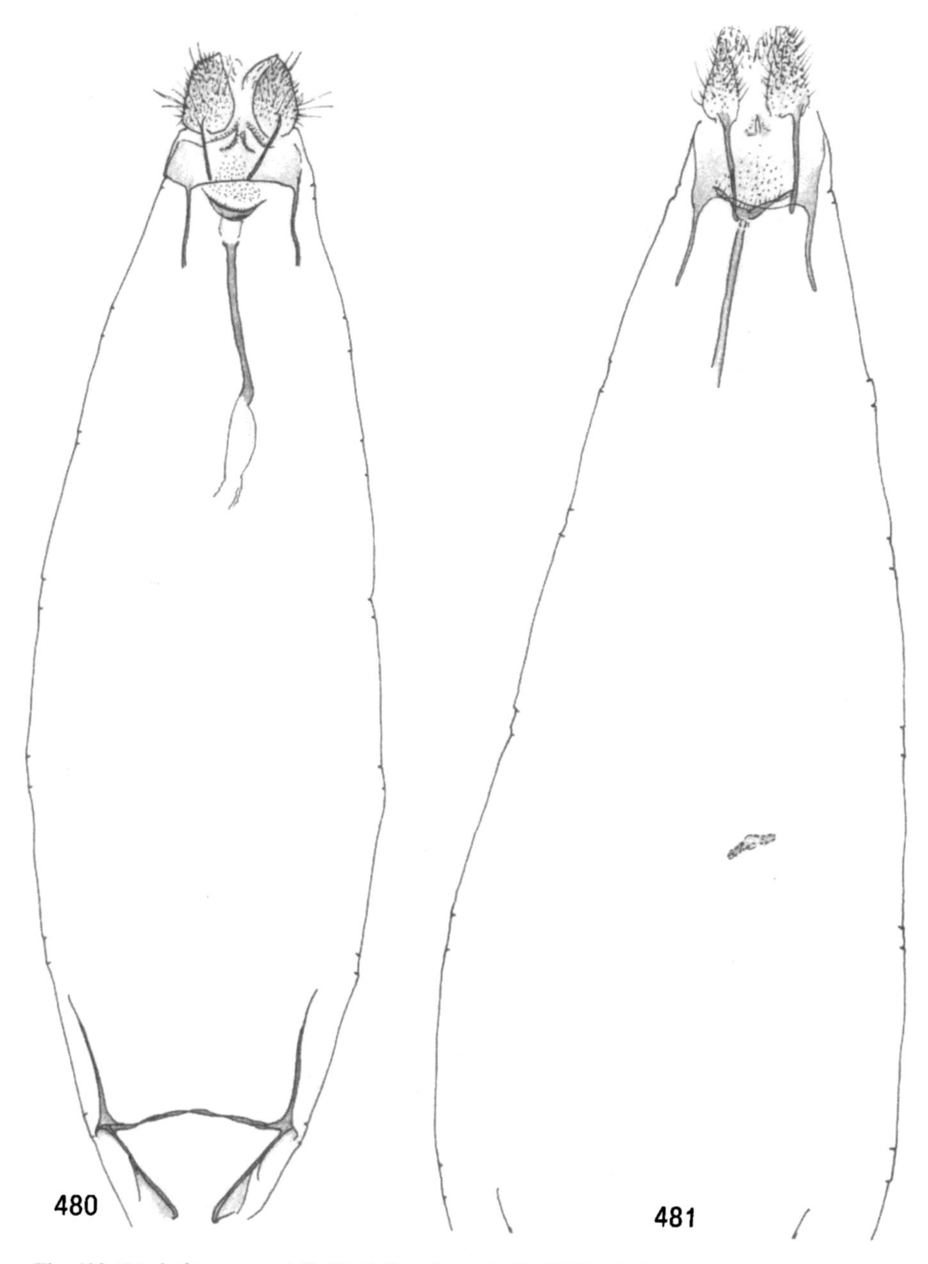

Fig. 480. *Biselachista trapeziella* (Stt.), female genitalia (ETO 1866).
Fig. 481. *Biselachista cinereopunctella* (Hw.), female genitalia (ETO 1877).

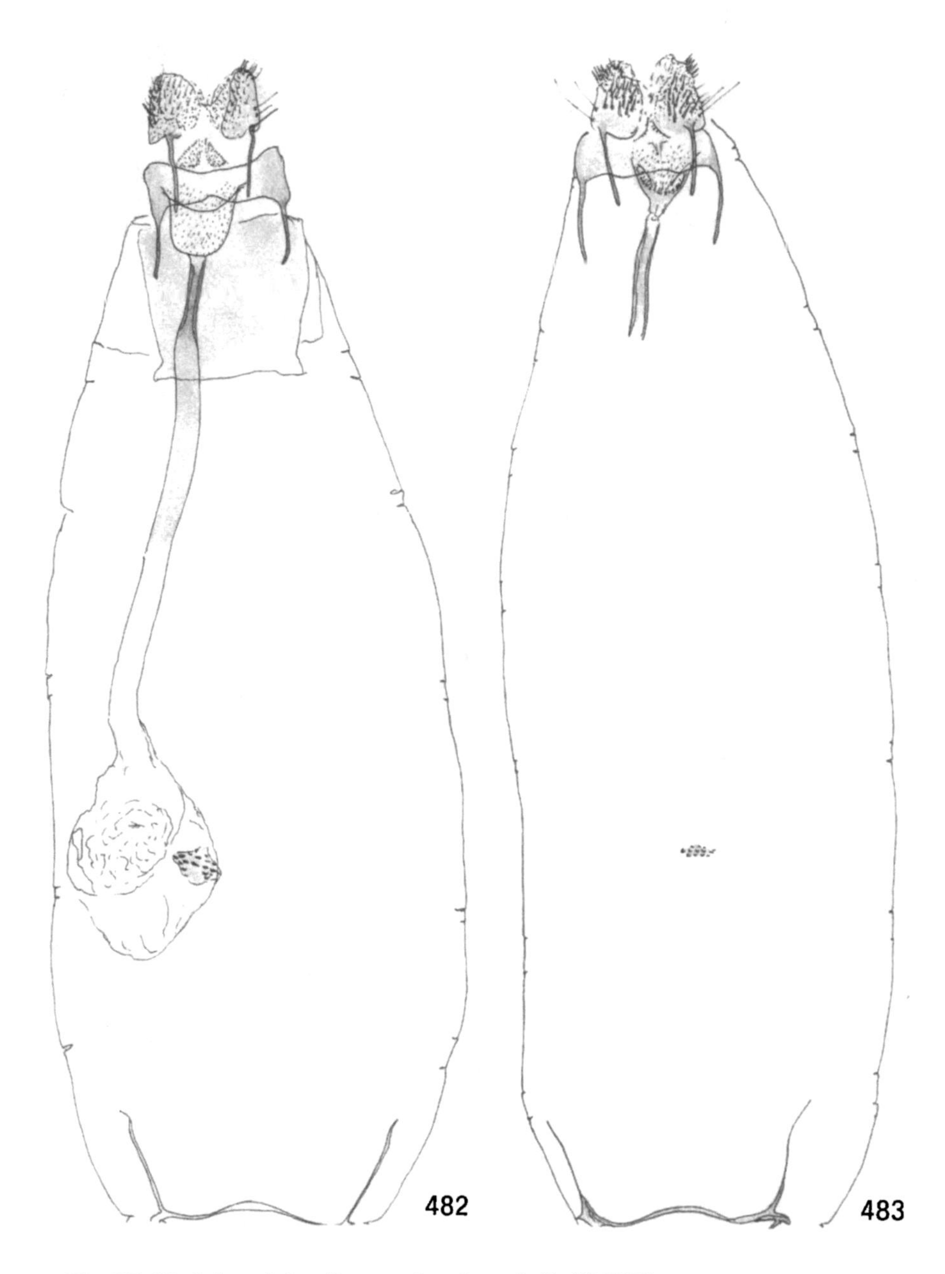

Fig. 482. *Biselachista kebneella* sp. n., female genitalia (IS 5554).
Fig. 483. *Biselachista ornithopodella* (Frey), female genitalia (BMNH 19286).

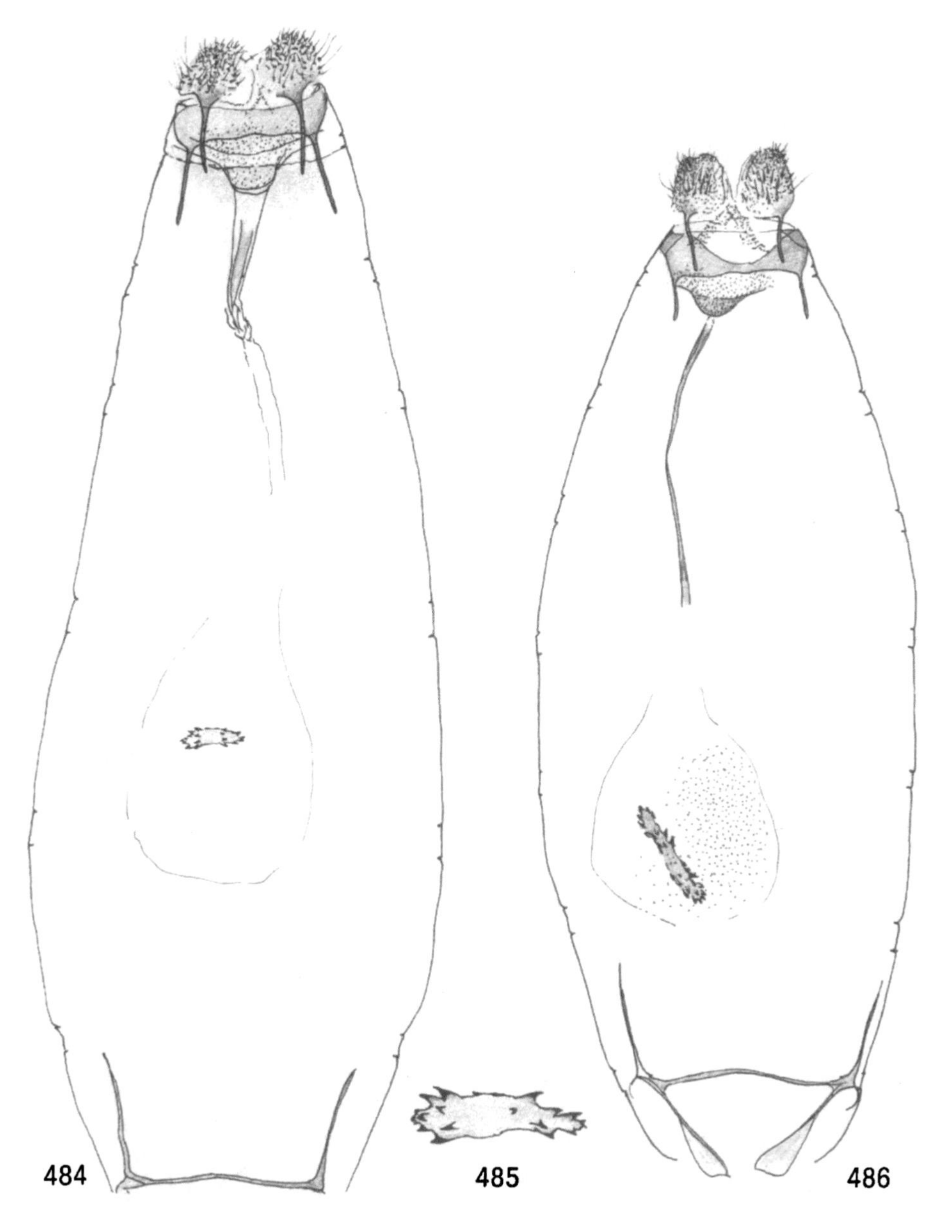

Figs. 484, 485. *Biselachista serricornis* (Stt.). – 484: female genitalia (ETO 2016); 485: signum (ETO 2016).
Fig. 486. *Biselachista freyi* (Stgr.), female genitalia (ETO 5201).

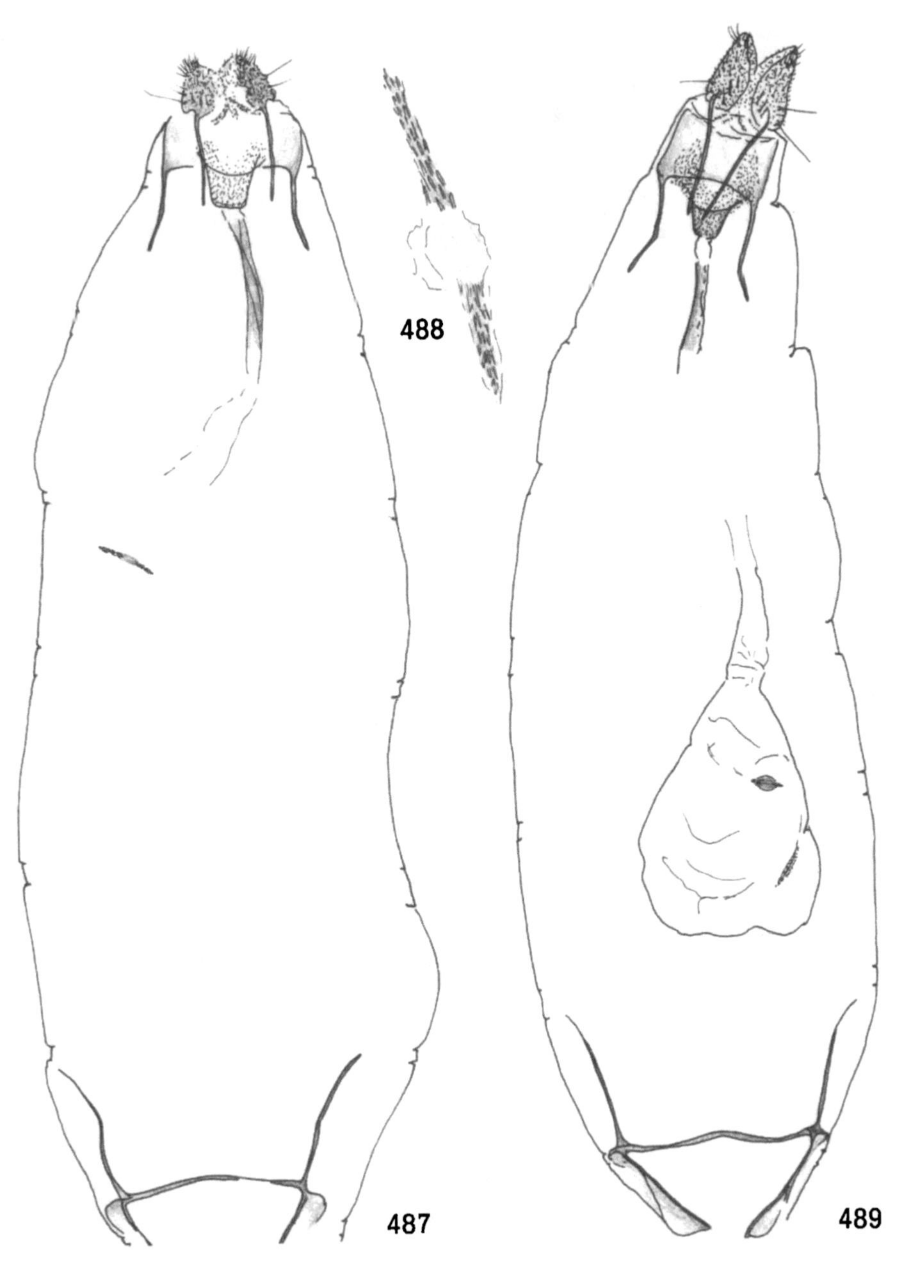

Figs. 487, 488. *Biselachista scirpi* (Stt.). – 487: female genitalia (ETO 1679); 488: signum (ETO 1679).
Fig. 489. *Biselachista eleochariella* (Stt.), female genitalia (ETO 1307).

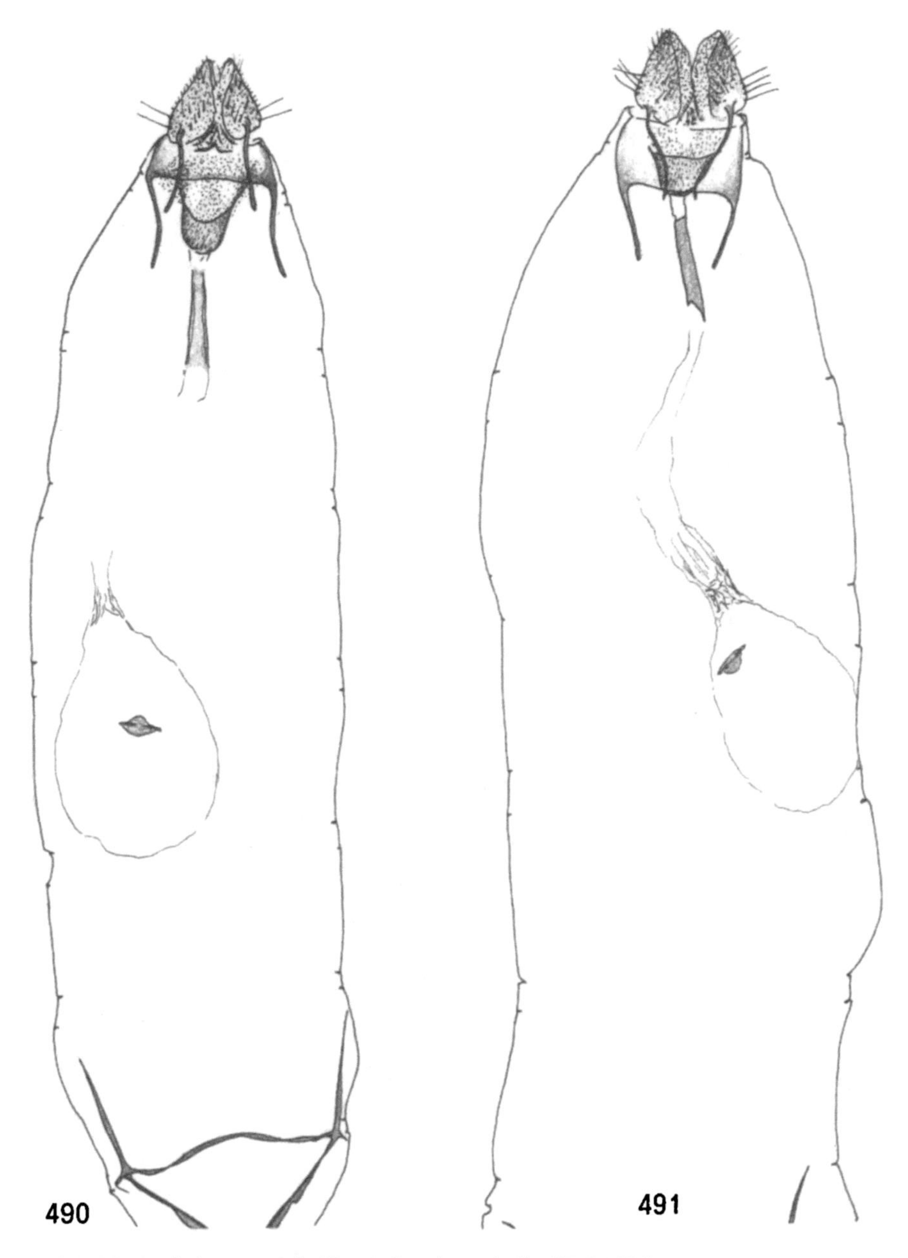

Fig. 490. *Biselachista utonella* (Frey), female genitalia (ETO 1394).
Fig. 491. *Biselachista albidella* (Nyl.), female genitalia (ETO 1300).

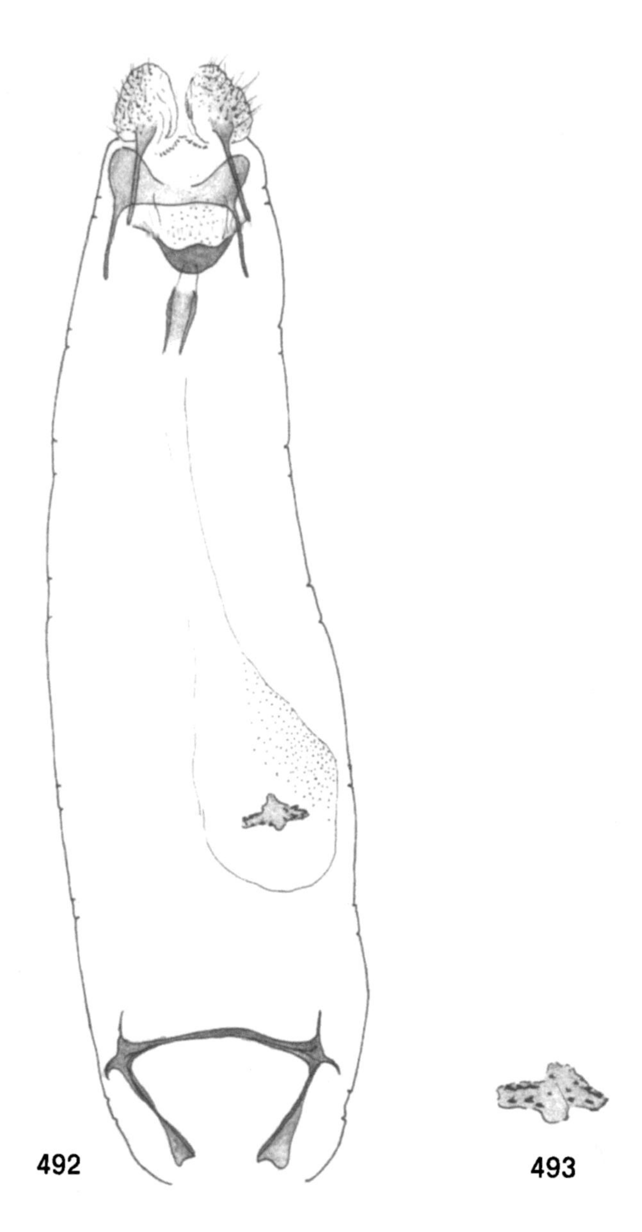

Figs. 492, 493. *Biselachista abiskoella* (Bengts.). - 492: female genitalia (BÅB 547-15046); 493: signum (BÅB 547-15046).

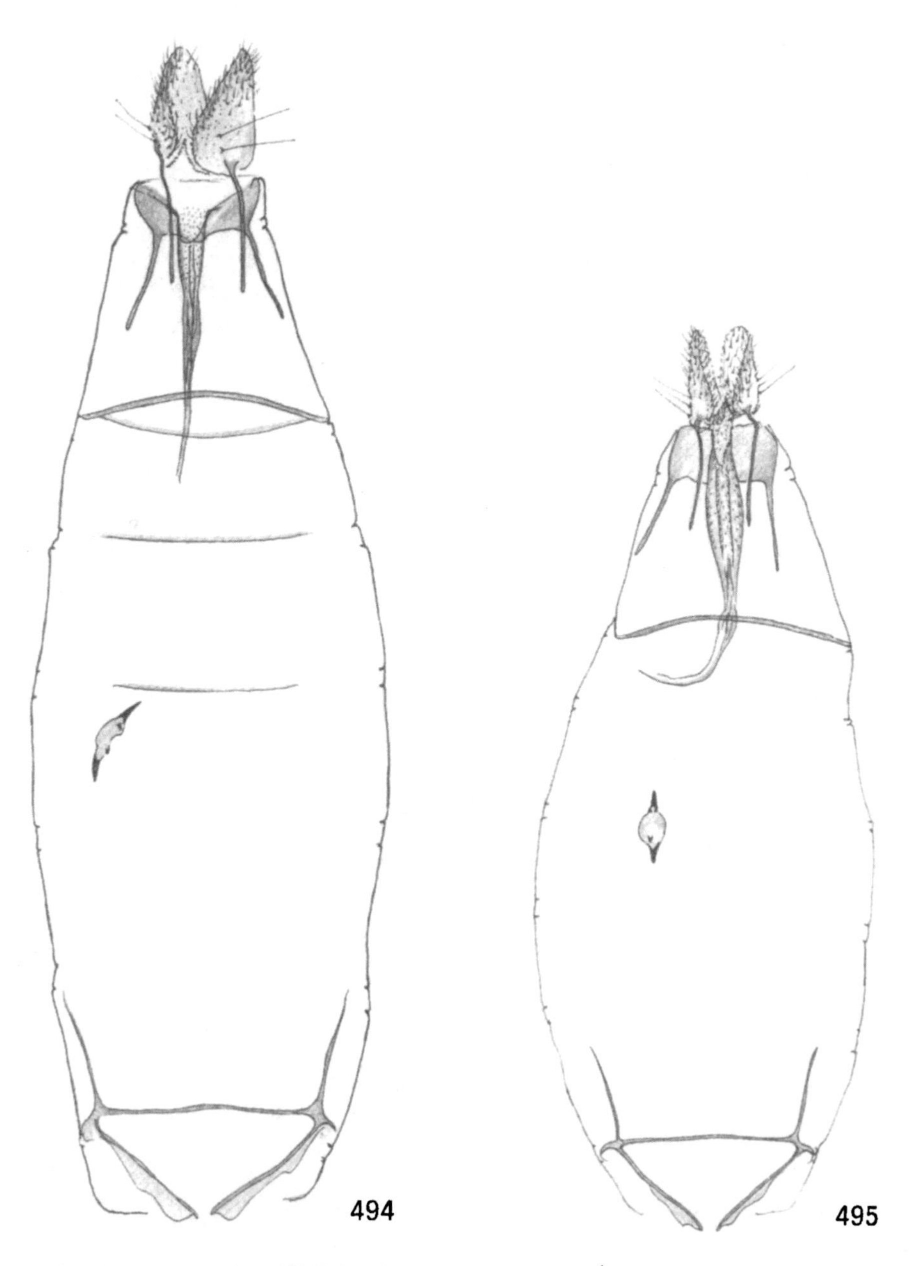

Fig. 494. *Cosmiotes freyerella* (Hb.), female genitalia (ETO 794).
Fig. 495. *Cosmiotes exactella* (HS.), female genitalia (ETO 1942).

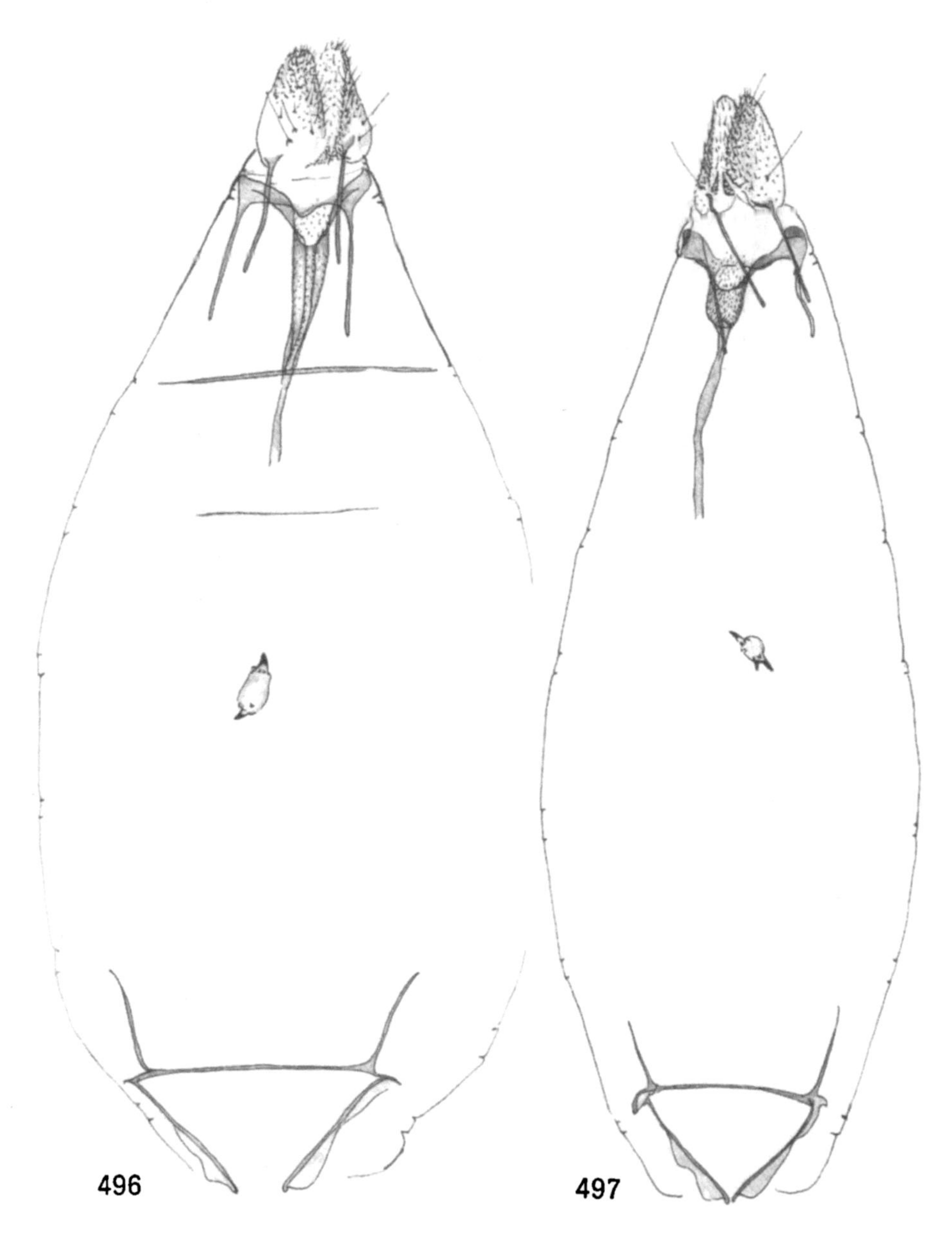

Fig. 496. *Cosmiotes stabilella* (Stt.), female genitalia (ETO 1786).
Fig. 497. *Cosmiotes consortella* (Stt.), female genitalia (ETO 789).

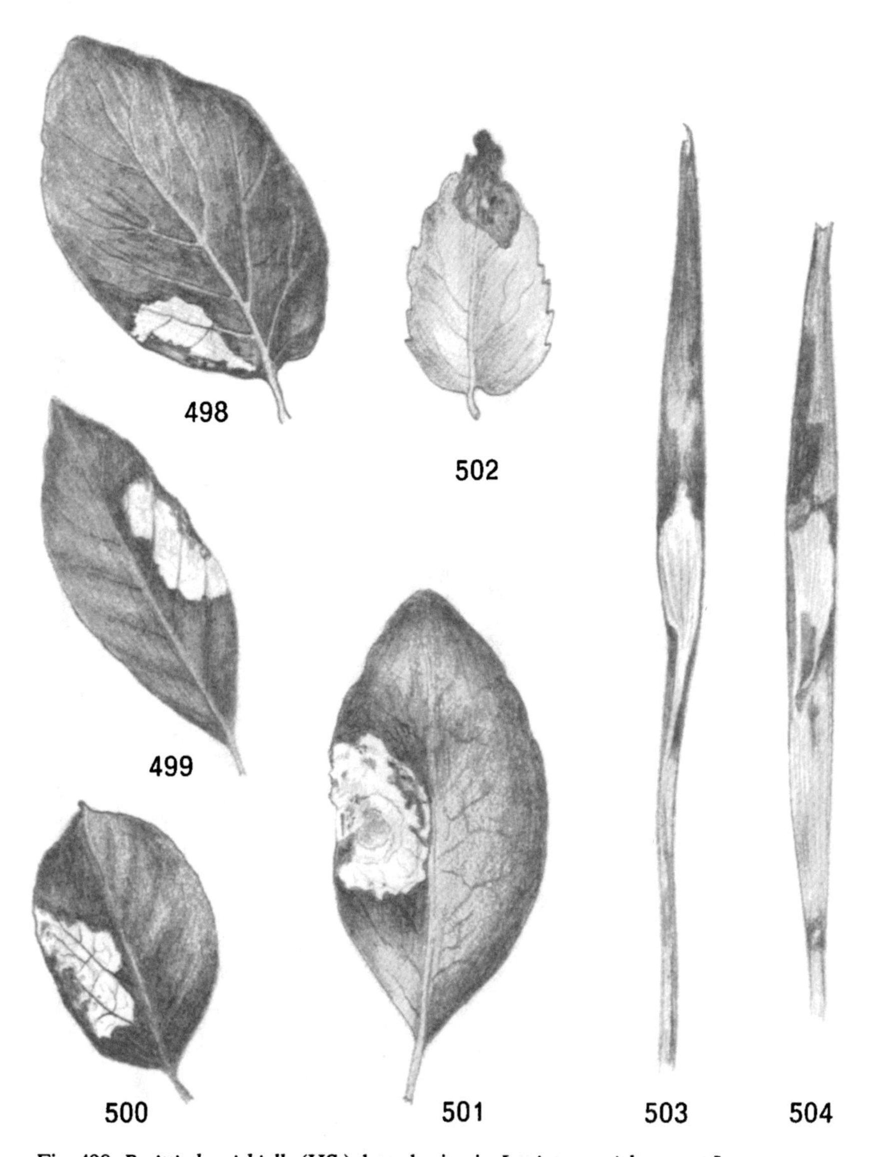

Fig. 498. *Perittia herrichiella* (HS.), larval mine in *Lonicera periclymenum* L.
Figs. 499, 500. *Perittia herrichiella* (HS.), larval mines in *Lonicera xylosteum* L.
Fig. 501. *Perittia obscurepunctella* (Stt.), larval mine in *Lonicera periclymenum* L.
Fig. 502. *Stephensia brunnichella* (L.), larval mine in *Satureja vulgaris* (L.) Fritsch.
Figs. 503. 504. *Elachista regificella* Sirc., larval mines in *Luzula pilosa* (L.) Willd.

Fig. 505. *Elachista gleichenella* (F.), larval mine in *Carex flacca* Schreb.
Fig. 506. *Elachista gleichenella* (F.), larval mine in *Luzula pilosa* (L.) Willd.
Fig. 507. *Elachista poae* Stt., larval mine in *Glyceria maxima* (Hartm.) Holmb.
Fig. 508. *Elachista atricomella* Stt., larval mine in *Melica uniflora* Retz.
Fig. 509. *Elachista elegans* Frey, larval mine in *Calamagrostis arundinacea* (L.) Roth.

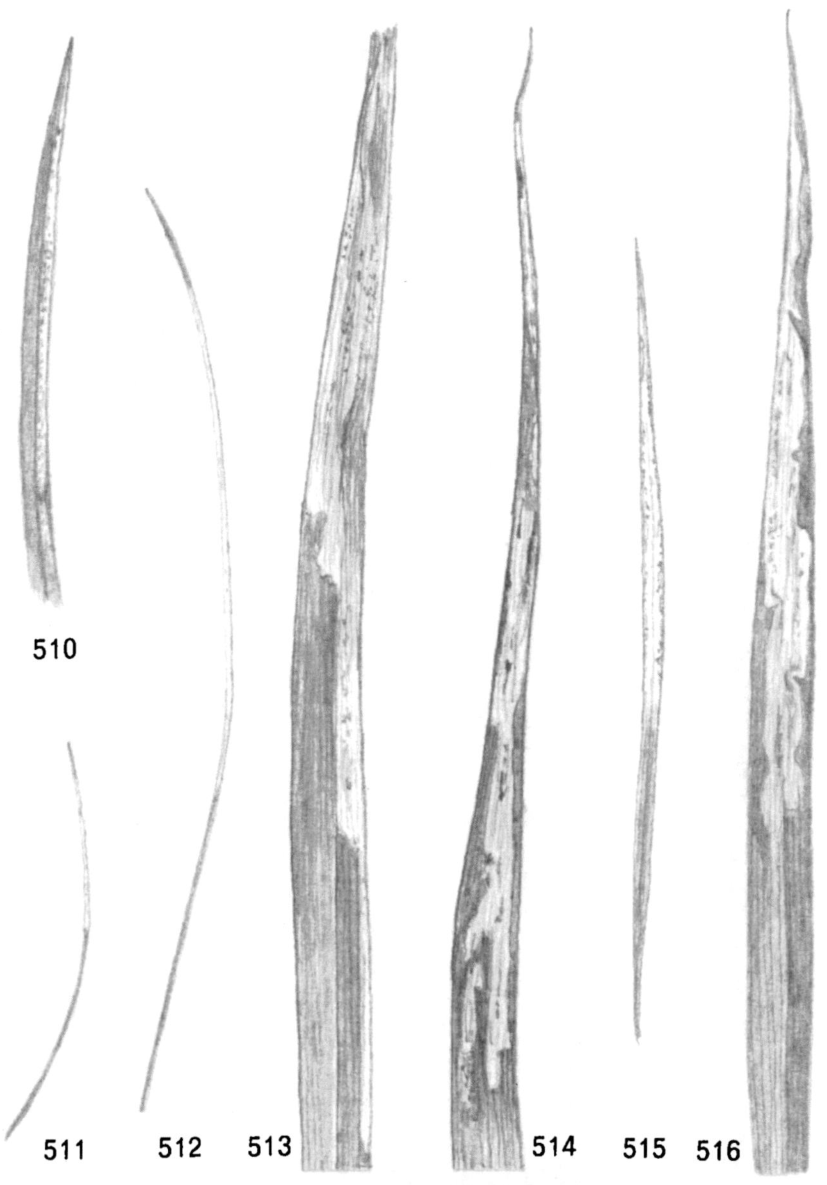

Fig. 510. *Elachista luticomella* Zell., larval mine in *Dactylis glomerata* L.
Figs. 511, 512. *Elachista bifasciella* Tr., larval mines in *Deschampsia flexuosa* (L.) Trin.
Fig. 513. *Elachista apicipunctella* Stt., larval mine in *Festuca altissima* All.
Fig. 514. *Elachista apicipunctella* Stt., larval mine in *Festuca gigantea* (L.) Vill.
Fig. 515. *Elachista lastrella* Chrét., larval mine in *Bromus erectus* Huds.
Fig. 516. *Elachista cerusella* (Hb.), larval mine in *Phalaris arundinacea* L.

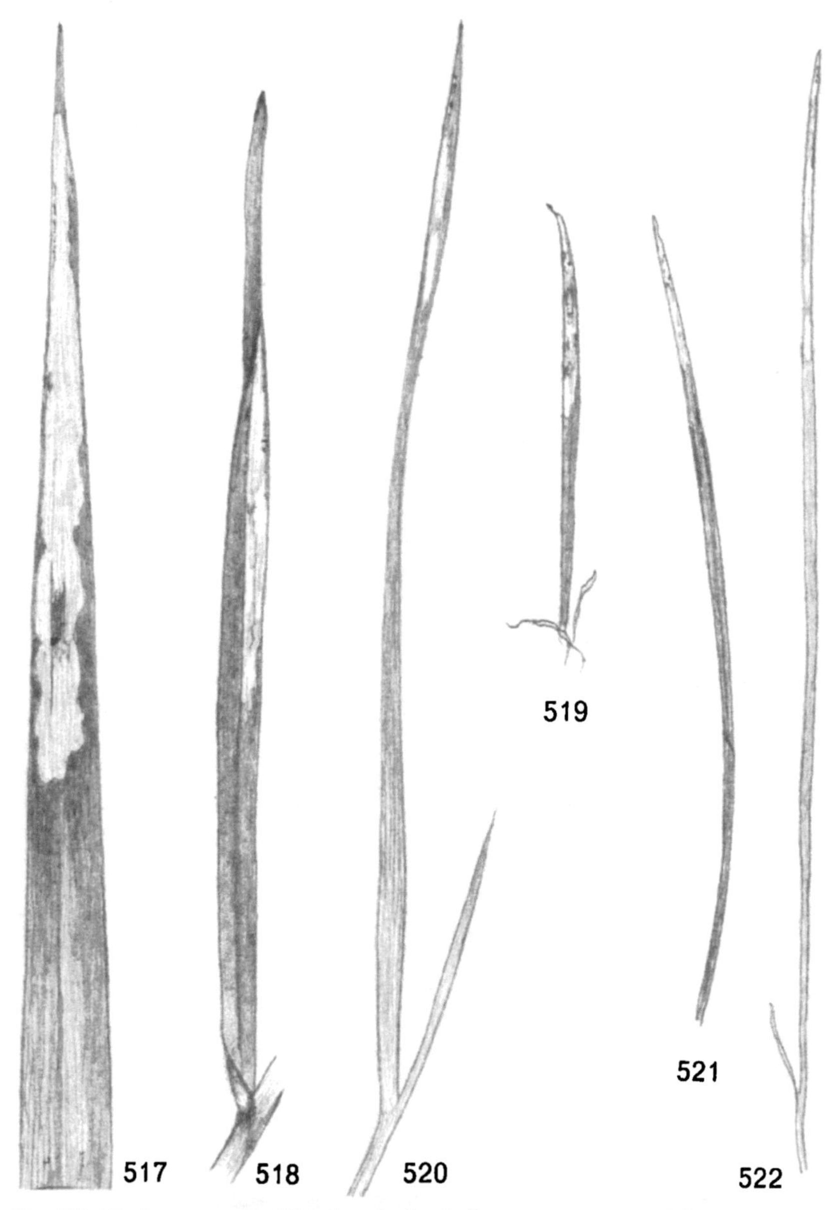

Fig. 517. *Elachista cerusella* (Hb.), larval mine in *Pragmites communis* Trin.
Fig. 518. *Elachista subnigrella* Dougl., larval mine in *Poa chaixii* Vill.
Fig. 519. *Elachista reuttiana* Frey, larval mine in *Koeleria* sp.
Fig. 520. *Elachista humilis* Zell., larval mine in *Deschampsia caespitosa* (L.) PB.
Figs. 521, 522. *Elachista pullicomella* Zell., larval mines in *Festuca ovina* L.

Fig. 523. *Elachista pollinariella* Zell., larval mine in *Festuca ovina* L.
Fig. 524. *Elachista chrysodesmella* Zell., larval mine in *Brachypodium sylvaticum* (Huds.) PB.
Fig. 525. *Elachista unifasciella* (Hw.), larval mine in *Dactylis glomerata* L.
Fig. 526. *Elachista gangabella* Zell., larval mine in *Brachypodium sylvaticum* (Huds.) PB.
Fig. 527. *Elachista subalbidella* Schl., larval mine in *Brachypodium pinnatum* (L.) PB.

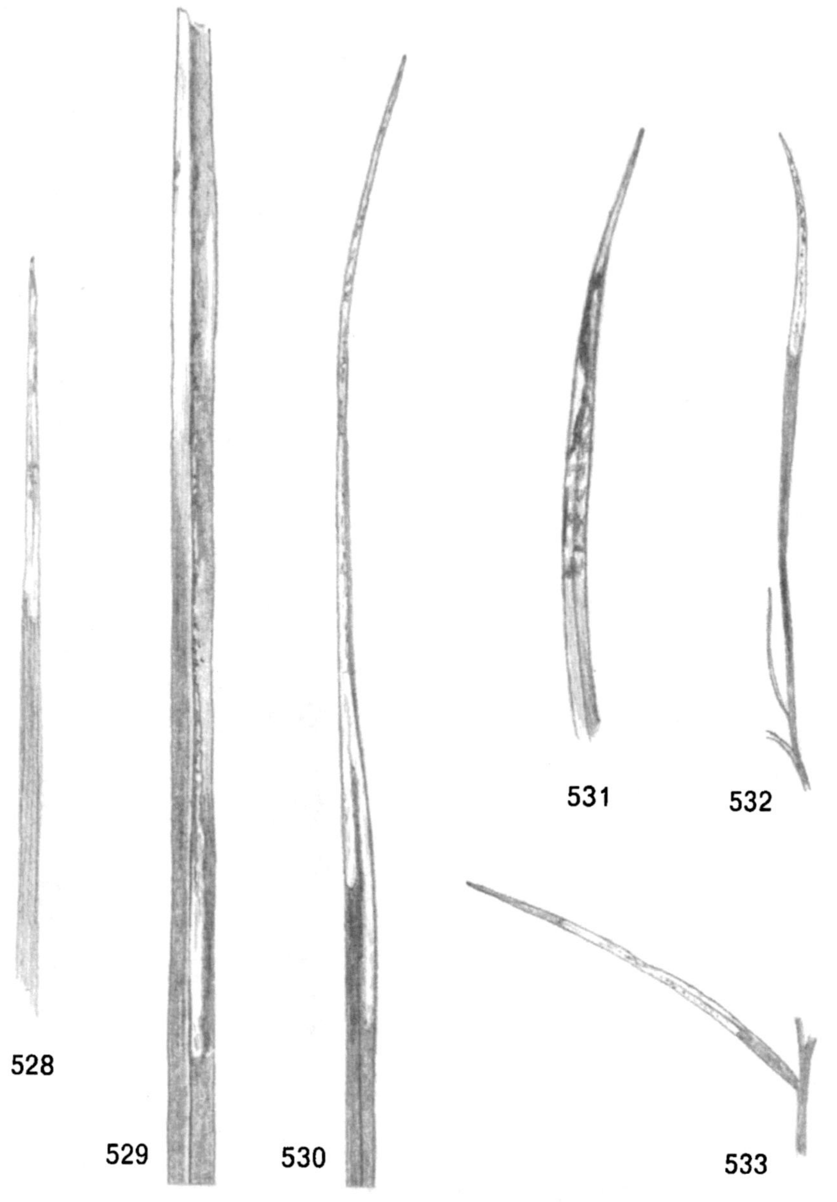

Figs. 528–530. *Elachista revinctella* Zell., larval mines in 528: *Deschampsia caespitosa* (L.) PB.; 529: *Festuca altissima* All. ; 530: *Sesleria caerulea* (L.) Ard.
Fig. 531. *Biselachista scirpi* (Stt.), larval mine in *Scirpus maritimus* L.
Figs. 532, 533. *Cosmiotes freyerella* (Hb.), larval mines in *Poa nemoralis* L.

Fig. 534. *Perittia obscurepunctella* (Stt.). Pupation inside a cork.
Fig. 535. *Elachista poae* Stt. Pupa attached to a leaf of *Glyceria* by a fine web.
Fig. 536. *Elachista albifrontella* (Hb.). Pupa attached to a leaf by a single girdle around abdomen and by the cremaster.

Biology. Larva dull grey-green, with a fine whitish dorsal line; head light brownish, margins darker, prothoracic shield dark, divided, irregular (Stainton, 1858c). It is reported from *Deschampsia caespitosa* (L.) PB., *Brachypodium sylvaticum* (Huds.) Pb., *Calamagostis epigeios* (L.) Roth and *Carex*. It starts mining downwards from the tip, making a rather long, pale brown, almost transparent, not opaque, broad mine; the larva often changes mine, and more than one mine can be present in a single leaf. Pupa light brown, attached by a single girdle and cremaster (Stainton, 1858c). Often reported as bivoltine, larvae mining in April and June and adults flying during May and July–August, but according to Wörz (1957) larvae collected in May can produce adults in August; in Denmark, Sweden and Finland adults are mainly caught from mid-July to late August.

Genus *Biselachista* gen. n.

Type-species: *Elachista freyi* (Staudinger, 1870).

Face (Text-fig. 44) smooth, scales raised on neck. Scales on distal part of antenna often raised, mainly in males. Antenna of ***B. serricornis*** (Stt.) with raised fine cilia (Text-fig. 6). Labial palpi drooping, to recurved, diverging; third segment short or almost as long as middle segment. Pecten made up of a few (6–7) fine hairs.

Pattern on forewing (Text-fig. 38) rather consistent: fold with dark or blackish scales, often forming a short, prominent, black streak in middle, bound by two white spots, of which one or both can be absent, or confluent. A generally distinct whitish mark from middle or beyond middle of costa; a small white streak, often elongate, parallel with termen from costa just before apex; tornal spot present or absent. Cilia line generally distinct.

Venation (Pl.-figs. 215–224) rather uniform. Forewing with R_4 and R_5 stalked but not coalescent, M_1 to stem of $R_4 + R_5$, always branching off before middle of the stalk, CuA_1 closer to M_2 than to CuA_2, only most distal part of CuP present in more generalized species. Hindwing with Rs always simple, M_1 to Rs, cell open, M_2 to CuA_1.

Hind tibia with long hairs above and below; middle spurs one-sixth from proximal end, inner spur twice as long as outer spur; apical spurs just beyond middle, of almost equal length.

Male genitalia. Uncus more or less indented, uncus lobes with rather thick, short, more or less clavate setae. Gnathos always bifurcate. Socii present but rather small. Costa of valva strongly sclerotized, often angular and bulged, and frequently with longitudinal folds. Vinculum triangular, strong to very strong, without saccus or medial sclerotized ridge. Digitate process club-shaped. Juxta lobes often with a small tuft of setae on apical margins. Aedeagus slightly tapering, generally with a longitudinal sclerotized ridge or band in distal part.

Female genitalia. Papillae anales short, stout, rounded to triangular, with coarse, short setae. Ventral margin varying from straight line to a more or less deep anterior curve, anterior part of segment VIII or more anteriorly. Antrum in all species separated from colliculum by a short membranous zone. Antrum more or less bowl-shaped; colliculum tube-shaped. Signum a single dentate plate, often weak and small, or absent. Membranous part of ductus bursae and corpus bursae rarely with spines.

At present, fifteen Palaearctic species previously placed in the genus *Elachista* Treitschke, 1833 are included in *Biselachista* gen.n. Of these, twelve are given full treatment in the present work; the species not treated here are: *Biselachista contaminatella* (Zeller, 1847) **comb.n.** from S. Italy, S. France, Algeria and Tenerife, *B. juliensis* (Frey, 1870), **comb.n.** from C. and SE. Europe and *B. igaloensis* (Amsel, 1951) **comb.n.** from Dalmatia. The generic position of *Elachista totalbella* Chrétien, 1908 is uncertain at present. Further, five species of Nearctic Elachistidae catalogued by Braun (1948) belong to *Biselachista* gen.n.: *Elachista cucullata* Braun, 1948, **comb.n.**, *E. agilis* Braun, 1921, **comb.n.**, *E. leucosticta* Braun, 1948, **comb.n.**, *E. tanyopis* Meyrick, 1932, **comb.n.** and *E. salinaris* Braun, 1925, **comb.n.** Some of these species seem very closely related to Palaearctic species here dealt with, and may be conspecific with them: *agilis* Braun, 1921 with *trapeziella* Stainton, 1849, *leucosticta* Braun, 1948 with *serricornis* Stainton, 1854 and *tanyopis* Meyrick, 1932 with *albidella* Nylander, 1848. No further species of *Elachista* s.lat. known to the authors can at present be included in *Biselachista* gen.n.

Little is yet known about the morphology and biology of the early stages, but apparently all species of *Biselachista* gen.n. feed on Cyperaceae and Juncaceae; a few records from Poaceae cannot be verified. Pupation takes place on a leaf or stem of the food-plant, the pupa attached by a medial girdle and cremaster. None of the species is known to make a cocoon for pupation. The pupa is slender, with rather prominent dorsal and dorso-lateral ridges; between these are often two dark brown lines.

Text-fig. 44. Head of *Biselachista freyi* (Stgr.).

Key to species of *Biselachista* based on external characters.

1		Antenna strongly serrate, distinctly ciliate (Text-fig. 6; Pl.-figs. 131, 132)	58. *serricornis* (Stt.)
–		Antenna not or only slightly serrate, not distinctly ciliate	2
2	(1)	Head blackish brown (Pl.-figs. 129, 130)	57. *ornithopodella* (Frey)
–		Head not blackish brown	3
3	(2)	Head, tegulae and thorax white (Pl.-figs. 141, 142)	63. *albidella* (Nyl.)
–		Head, tegulae and thorax not pure white	4
4	(3)	Forewing with white fascia before middle (Pl.-figs. 143, 144)	64. *abiskoella* (Bengts.)
–		Forewing without white fascia	5
5	(4)	Forewing with two more or less distinct whitish spots before and behind a dark streak in fold	6
–		Inner whitish plical spot missing, or fold more or less whitish	10
6	(5)	Head, thorax, tegulae and forewing ochreous white (Pl.-figs. 137, 138)	61. *eleochariella* (Stt.)
–		Head, thorax, tegulae and forewing mottled greyish brown	7
7	(6)	Forewing strongly mottled, plical spots very indistinct (Pl.-fig. 124)	54. *imatrella* (Schantz)
–		Plical spots white and distinct	8
8	(7)	White inner plical spot smaller than outer (Pl.-figs. 133, 134)	59. *freyi* (Stgr.)
–		White inner plical spot larger than outer	9
9	(8)	Forewing ground colour blackish brown with shining, silky white marks (Pl.-figs. 122, 123)	53. *trapeziella* (Stt.)
–		Forewing ground colour greyish brown, almost plain grey between fold and dorsum, marks not shining (Pl.-figs. 127, 128)	56. *kebneella* sp. n.
10	(5)	White inner plical spot missing	11
–		Fold suffused with whitish, surrounding a distinct blackish streak	12
11	(10)	Whitish costal spot opposite and often confluent with outer plical spot (Pl.-figs. 125, 126)	55. *cinereopunctella* (Hw.)
–		White costal spot beyond and never confluent with outer plical spot (Pl.-figs. 137, 138)	61. *eleochariella* (Stt.)
12	(10)	Apical spot forming a long submarginal streak towards tornal spot (Pl.-figs. 135, 136)	60. *scirpi* (Stt.)
–		Apical spot short, triangular (Pl.-figs. 139, 140)	62. *utonella* (Frey)

Key to species of *Biselachista* based on male genitalia.

1		Uncus slightly indented, uncus lobes short, broad, rounded	2
–		Uncus deeply indented, indentation at least as deep as uncus lobe is wide at base	5
2	(1)	Aedeagus base bulbous, very long and slender, distal end indented (Pl.-figs. 368, 369)	54. *imatrella* (Schantz)
–		Aedeagus base not distinctly bulbous, gradually tapering	3

3 (2) Valva widest distally, aedeagus rather long, straight (Pl.-figs. 370, 371)
55. *cinereopunctella* (Hw.)

– Valva widest at base, aedeagus long and curved or short and straight 4

4 (3) Uncus with U-shaped indentation; aedeagus long and curved (Pl.-figs. 366, 367) 53. *trapeziella* (Stt.)

– Uncus with V-shaped indentation, uncus lobes close; aedeagus short, gradually tapering (Pl.-figs. 372, 373) 56. *kebneella* sp. n.

5 (1) Aedeagus with distinctly serrate sclerotized ridge; uncus lobes three times as long as wide 58. *serricornis* (Stt.)

– Sclerotized ridge or band (if present) in aedeagus not serrate; uncus lobes less than three times as long as wide 6

6 (5) Sacculus with a distinct distal spine below cucullus 7

– Sacculus distally rounded towards cucullus, without a spine 8

7 (6) Uncus lobes broadest distally; costa convex, cucullus a prominent semi-circular dorsal lobe (Pl.-figs. 386, 387) 63. *albidella* (Nyl.)

– Uncus lobes almost conical; proximal two-thirds of valva with almost parallel margins, cucullus narrow, rounded towards costa (Pl.-figs. 388–390)
64. *albiskoella* (Bengts.)

8 (6) Uncus almost as wide as long, set with sparse short, stiff setae; digitate process strong 9

– Uncus lobe distinctly longer than wide, highly setose; digitate process slender even if most distal part is enlarged 11

9 (8) Valva narrowest at base; ventral anellus lobes rounded (Pl.-figs. 380, 381)
60. *scirpi* (Stt.)

– Valva narrowest distally; ventral anellus lobes with short lateral processes 10

10 (9) Uncus lobes almost circular, wide apart; juxta lobes produced apically into two short processes (Pl.-figs. 384, 385) 62. *utonella* (Frey)

– Uncus slightly longer than wide; juxta lobes rounded apically, with short lateral processes (Pl.-figs. 382, 383) 61. *eleochariella* (Stt.)

11 (8) Digitate process long and slender; aedeagus long and slender, gradually tapering (Pl.-figs. 378, 379) 59. *freyi* (Stgr.)

– Digitate process slender, very swollen distally; aedeagus rather short, basal half broad, distal half more narrow, tapering (Pl.-figs. 374, 375)
57. *ornithopodella* (Frey)

Key to species of *Biselachista* based on female genitalia.
(Note: Female of *B. imatrella* (Schantz) is unknown.)

1 Corpus bursae without a signum (Pl.-fig. 480) 53. *trapeziella* (Stt.)

– Corpus bursae with a signum 2

2 (1) Signum a small plate, each end with a fine dentate ridge forming a transverse, medially interrupted line 3

– Signum rather prominent, with multiple teeth in several rows or with scattered teeth 5

3 (2) Ventral margin of antrum deeply anteriorly U-shaped; apophyses slender, apophyses anteriores slightly longer than or of equal length to apophyses posteriores (Pl.-fig. 490) 62. *utonella* (Frey)

– Ventral margin of antrum slightly anteriorly curved or almost straight; apophyses posteriores distinctly longer than apophyses anteriores 4

4 (3) Antrum large, ventral margin almost straight; apophyses posteriores sinuate (Pl.-fig. 491) 63. *albidella* (Nyl.)

– Antrum small, ventral margin anteriorly curved; apophyses posteriores straight (Pl.-fig. 489) 61. *eleochariella* (Stt.)

5 (2) Ventral surface of papillae anales with short, triangular setae; ventral margin of antrum anteriorly U-shaped (Pl.-fig. 483) 57. *ornithopodella* (Frey)

– Setae on ventral surface of papillae anales not triangular; ventral margin of antrum widely V-shaped, straight or slightly anteriorly curved 6

6 (5) Ventral margin of antrum widely V-shaped, coarsely spined on inner side (Pl.-fig. 481) 55. *cinereopunctella* (Hw.)

– Ventral margin of antrum straight or slightly anteriorly curved, with or without fine internal spines 7

7 (6) Colliculum long and narrow, four times as long as apophyses posteriores (Pl.-fig. 486) 59. *freyi* (Stgr.)

– Colliculum short and broad, as long as or slightly longer than apophyses posteriores 8

8 (7) Antrum large, cup-shaped, almost membranous; signum square (Pl.-fig. 482) 56. *kebneella* sp. n.

– Antrum sclerotized, bowl-shaped and anteriorly truncate or cup-shaped with strong lateral rims; signum elongate 9

9 (8) Ventral margin of antrum straight, lateral rims strong; signum oval (Pl.-figs. 484, 485) 58. *serricornis* (Stt.)

– Ventral margin of antrum anteriorly curved; signum long and narrow or oval and strongly swollen medially 10

10 (9) Ventral margin of antrum anteriorly curved, but medially with posteriorly curved extension; signum swollen medially (Pl.-figs. 492, 493) 64. *abiskoella* (Bengts.)

– Ventral margin of antrum anteriorly curved, without medial posterior extension; signum long and very narrow (Pl.-figs. 487, 488) 60. *scirpi* (Stt.)

53. ***Biselachista trapeziella*** (Stainton, 1849) **comb.n.**
Plate-figs. 122, 123, 215, 366, 367, 480.

? *Recurvaria guttifera* Haworth, 1828: 553.
Elachista trapeziella Stainton, 1849: 26.

8–10 mm. (Pl.-fig. 122). Face whitish; neck tufts, tegulae and thorax dark brown. Antenna brown, distal part slightly annulated with greyish and scales slightly raised. Labial palpi drooping, upperside whitish basally, greyish distally, underside brown. Forewing ground colour dark brown with five silky-white marks: one in the fold at one-quarter and a less distinct or sometimes absent one in the fold in the middle of the wing, separated from the former by a dark streak; an elongate, triangular mark from middle of costa, sometimes confluent with outer plical spot; a narrow, triangular sub-apical streak from costa and a more prominent tornal spot. Cilia grey-brown, slightly ochreous beyond sub-apical streak; cilia line very weak, dark brown. Hindwing grey-brown, cilia of same colour. Abdomen dark brown, paler ventrally. Hind tarsus grey, distal end of segments whitish.

Female (Pl.-fig. 123). Very similar to male, but annulation on distal part of antenna more prominent, marks on forewing white, larger and more distinct, mid-costal mark often confluent with distal plical spot, but never extending to dorsum. Abdomen whitish below.

Male genitalia (Pl.-figs. 366, 367). Uncus slightly indented, uncus lobes very short, with short, slightly spatulate setae. Gnathos lobes elongate, elliptical. Costa of valva emarginate beyond middle, widest before middle; transtilla and costa strongly sclerotized. Digitate process prominent, swollen towards apex. Juxta lobes broad, truncate, triangular. Aedeagus long, slender, tapering, distal end chisel-shaped, with sclerotized band in distal third.

Female genitalia (Pl.-fig. 480). Papillae anales with short thick setae. Apophyses of almost equal length. Antrum very short, wide, bowl-shaped, ventral margin a broad anterior curve, finely spined, dorsal wall with fine spines; colliculum very narrow, twice as long as apophyses posteriores, separated from antrum by short, dilated, membranous duct. Corpus bursae without signum.

Distribution. Not in Denmark or Norway; in Sweden from the SE. districts up to Upl.; in Finland from the S. districts and from Ta. – Known from Britain and C. Europe.

Biology. Frey (1859) provides notes on the immature stages. The larva is yellowish white, with two dorso-lateral reddish lines, head blackish brown, prothoracic shield brown, prothoracic legs brown. It mines in the leaves of *Luzula pilosa* (L.) Willd., *L. luzuloides* (Lam.) Dandy & Wilm. and a few other *Luzula* species. The mine starts from the tip of the leaf, and is flat and wide; whitish. The larva starts mining in the autumn and leaves this mine in the following spring in late April – beginning of May. Pupation takes place on a leaf. The adults fly from early June until late July, probably mainly in the first part of July in N. Europe; univoltine. The biotope is shady places in both coniferous and deciduous forests.

Note. This species seems to vary much more in C. Europe, where the white spots can be very large and confluent (Hering, 1891; Groschke, 1937), than in northern Europe, where only little variation has been observed.

54. ***Biselachista imatrella*** (Schantz, 1971) **comb.n.**

Plate-figs. 124, 216, 368, 369.

Elachista imatrella Schantz, 1971: 99.

6–9 mm. Male (Pl.-fig. 124). Face greyish white, with a few scattered brownish scales between antennae; neck tufts brownish, suffused with whitish grey; tegulae and thorax brownish. Antenna greyish brown, distal third with slightly raised scales. Labial palpi porrect, curved; third segment half as long as second segment, upperside whitish, underside brownish white, base of third segment slightly lighter. Forewing ground colour greyish brown, slightly suffused with whitish grey and with scattered blackish brown scales along costa, in middle of fold and long termen; with slightly lighter, very indistinct marks; dark streak in fold inwardly and outwardly bound by slightly lighter spots, which can be whitish; an outwardly oblique streak, most distinct just by costa, from middle of costa, and a narrow whitish streak parallel with termen just before apex; tornal spot very faint; scattered blackish scales inside cilia line, often forming a small dot at apex. Cilia dark grey along termen, ochreous grey at apex and light grey at tornus; cilia line blackish brown, not distinct. Hindwing grey-brown, cilia similar, tinged ochreous-grey. Abdomen greyish, anal tuft light grey. Hind tarsus dull whitish, outside with blackish grey dots.

Female. Unknown.

Male genitalia (Pl.-figs. 368, 369). Uncus lobes slightly indented, with short strong setae. Costa of valva longitudinally folded in proximal bulged part, emarginate before prominent projecting cucullus. Digitate process very narrow at base, widest before apex, blunt. Juxta lobes broad, truncate, with a small group of setae on apical margin. Vinculum elongate, triangular. Aedeagus very long, slender, weakly sclerotized, base bulbous, distal end indented, with sclerotized ridge beyond middle.

Female genitalia. Unknown.

Distribution. Only two records known. Schantz (1971) based his description on three males caught in the vicinity of Imatra, Sa, Finland, but the species is probably now extinct there as the site is built-up. Five further specimens are present in the Grønlien collection at the Zoological Museum, Bergen, bearing a label reading "Norway"; there is no doubt that these specimens were caught in Norway, but as no locality is stated, they cannot with certainty be referred to a definite district, as Grønlien travelled and collected in many parts of Norway (A. Løken, *in litt.*).

Biology. Unknown, but Schantz (1971) reports that the specimens from Imatra were collected flying or running on *Eriophorum vaginatum* L., which therefore could be the food-plant. Adults were collected in Finland in late June.

55. ***Biselachista cinereopunctella*** (Haworth, 1828) **comb.n.**

Plate-figs. 125, 126, 217, 370, 371, 481.

Tinea cinereopunctella Haworth, 1828: 581.

7–9 mm. Male (Pl.-fig. 125). Face whitish; neck tufts mottled greyish; tegulae and thorax grey-brown. Antenna grey-brown, scape cream-white on underside, scales slightly raised on the distal third. Labial palpi drooping, slightly curved; upperside dull white, third segment with narrow brownish ring just before tip, underside greyish white on third segment, grey-brown on second segment. Forewing ground colour dark grey-brown, mottled with whitish, and with scattered blackish scales in fold and distal part; with four greyish white marks: one triangular, outwardly oblique, elongate from middle of costa, nearly reaching fold; beyond this a small, indistinct spot in fold, sometimes confluent with costal mark; from costa just before apex a short, oblique streak; tornal spot just opposite the middle of the two costal spots, apical spot faintly ochreous-tinged. Cilia grey; cilia line distinct, blackish. Hindwing grey-brown, cilia similar. Abdomen grey, anal tuft ochreous grey. Hind tarsus whitish, with narrow greyish rings.

Female. (Pl.-fig. 126). Forewing ground colour more uniform greyish brown, with more scattered blackish scales beyond middle; marks on forewing pure white, distinct. Inner costal spot and small spot in fold confluent; apical mark preceded by blackish scales.

Male genitalia (Pl.-figs. 370, 371). Uncus slightly indented, setae short, stout. Gnathos rather short, tapering. Costa of valva slightly emarginate before broad, truncate cucullus. Digitate process rather short, narrow at base, broader distally. Juxta lobes broad, apical margins laterally produced into short, blunt, setose processes. Vinculum triangular. Aedeagus straight, slightly tapering, with a longitudinal, short, sclerotized band from beyond middle.

Female genitalia (Pl.-fig. 481). Papillae anales with short setae. Apophyses posteriores a little longer than apophyses anteriores. Antrum very short, bowl-shaped, ventral margin broadly V-shaped, with coarse spines on inner side, dorsal wall with patch of minute spines; colliculum narrow, slightly longer than apophyses posteriores. Signum elongate, weak, finely dentate.

Distribution. Not in Denmark (the specimen reported by Larsen, 1927: 141 is *E. compsa* Tr.-0.), Norway or Finland; in Sweden only from Gtl. – Otherwise reported from Britain, C. Europe, Belgium, France and Italy.

Biology. According to Stainton (1858c) the larva has yellowish sides and whitish back; all segments, except prothorax, with red dorso-lateral spots; between these spots a greenish dorsal line; head and prothoracic shield dark brown. It mines in the leaves of *Carex flacca* Schreb. from late summer to early autumn, hibernating in the mine, and is mature in April-early May the following spring. The mine, which starts from the tip, can be short and as wide as the blade, or long and slender, between the midrib and one edge. More than one mine of the slender type can be present on the same leaf. Excrement within the mine in a rather broad brownish track. Apparently no change of leaf takes place after hibernation. Hering (1957), Wörz (1957) and Lhomme (1951) further report the following food-plants: *Carex digitata* L., *C. ornithopoda* Willd., *Sesleria caerulea* (L.) Ard., *Melica* and *Deschampsia*. Pupation takes place in the angle of a leaf, the pupa is yellowish brown with reddish dorsal lines and is held by a girdle. The adults fly from late May to early July; univoltine in N. Europe, but specimens reported from August in S. Europe may indicate a second brood there. The biotope is sandy or calcareous areas with shady places.

56. ***Biselachista kebneella*** **sp.n.**

Plate-figs. 127, 128, 218, 372, 373, 482.

8–9 mm. Male (Pl.-fig. 127). Head dark grey; thorax and tegulae dark grey, tinged with light grey. Antenna dark greyish brown, with slightly raised scales on distal half. Labial palpi porrect and slightly curved, upperside pale greyish, underside dark grey-brown; tip of second segment with light grey ring, basal part of terminal segment blackish, tip dark grey. Forewing ground colour grey-brown, scattered with dark brown scales; with five whitish marks: a rather distinct, elongate, square spot in fold near base, preceding a black streak in the fold and succeeded by a small, less distinct spot, an elongate, oblique, indistinct streak not reaching to fold from middle of costa, a small, less distinct, whitish spot at apex and a similar spot at tornus. Blackish scales between outer plical spot and tornal spot; below the fold the ground colour is plainer greyish, without blackish scales. Cilia grey, cilia line blackish, not reaching tornus from apex. Hindwing grey-brown, cilia of same colour. Abdomen dark-brown. Hind tarsus with broad blackish rings and narrow white rings.

Female (Pl.-fig. 128). Head ochreous-grey, labial palpi less contrasted. Marks on forewing more distinct, white, costal spot round, apical and tornal spots more prominent, without brown scales; dorsum of forewing, hindwing and abdomen slightly lighter than in male.

Male genitalia (Pl.-figs 372, 373). Uncus indented, incision short and narrow, V-shaped, uncus lobes rounded, with short, distally widening, stiff setae. Transtilla and costa of valva strongly sclerotized; costa emarginate before rounded cucullus which is not as wide as base of valva. Digitate process narrow at base, rather slender, widest in middle, tapering. Juxta lobes broad, apical margin laterally produced into short, blunt processes. Vinculum triangular. Aedeagus short, straight, tapering, distal end chisel-shaped, outwards with sclerotized band not reaching distal end from one-half.

Female genitalia (Pl.-fig. 482). Papillae anales with short stiff setae. Apophyses short, of almost equal length. Antrum large, cup-shaped, membranous, ventral margin slightly anteriorly curved, dorsal wall and inner side of antrum with fine spines; colliculum narrow, as long as apophyses posteriores. Signum weakly sclerotized, square, finely dentate.

Distribution. Only known from the mountain area east of Kebnekaise in T. Lpm., N. Sweden and Ks and Li in Finland.

Type-material. Holotype: 1 o, labelled: "HOLOTYPE" (round red), "Suecia, T.Lpm., Kåkittjårro, UTM 34 W DA 1329, 9.VII.1974, Ingvar Svensson", "Genital præparat C. 9.11.76 ♂, E. Traugott-Olsen". Paratypes: 2 ♂, 1 ♀, labelled "PARATYPE" (round blue) with same locality label as holotype and "Genitalpreparat 5554 ♀, Ingvar Svensson"; 1 ♂, 1 ♀, Suecia, T.Lpm., Nikkaluokta, UTM 34 W DA 1831, 2.VII.1974, Ingvar Svensson", "Genitalpreparat 5542 ♀, Ingvar Svensson" and same locality label, "Højre vingepar præp. 3.2.76 by E. Traugott-Olsen", "Genitalpreparat 5541 ♂, Ingvar Svensson". All in coll.I. Svensson, Österslöv, Sweden. 1 ♀, labelled "Fennia, Ks. Kuusamo, 15.7.1963, Max von Schantz", "Gen. slide 1317 ♀, E.S. Nielsen"; 1 ♀, labelled "Fennia, Karigasniemi, Li 12.7.[19]59, Nybom", "Genital præparat nr. A. 3.12.73 ♀, E. Traugott-Olsen; 1 ♀, labelled "Fennia, Karigasniemi, Li 9.7.[19]57, Nybom", "Gen. slide 1325 ♀, E.S. Nielsen" in coll. O. Nybom and M. v. Schantz, Helsinki, Finland.

Biology. Immature stages unknown. Adults from Sweden were collected in the first half of July in the alpine region just above the timber line by sweeping low scattered stands of *Salix lapponum* L. in the shelter of which grew a *Luzula* sp. It seems reasonable to suppose the latter to be the food-plant.

Note. The species is named after the highest peak in Sweden, Kebnekaise, in the vicinity of which the species was caught by I. Svensson.

57. ***Biselachista ornithopodella*** (Frey, 1859) **comb.n.**
Plate-figs. 129, 130, 374, 375, 483.

Elachista ornithopodella Frey, 1859: 194.

7–8 mm. Male (Pl.-fig. 129). Head, tegulae and thorax blackish brown. Antenna blackish brown. Labial palpi blackish. Ground colour of forewing blackish, tinged dark brown, with four pure white marks: one in the fold close to the base, extending towards dorsum, but here suffused with scales of ground colour, preceded by black scales in the fold; a distinct, inwards oblique streak from middle of costa to or close to fold; a narrow, inwards oblique streak from costa just before apex and a small elongate spot at tornus. Cilia pale brownish grey; cilia line distinct, blackish. Hindwing grey-brown, tinged with ochreous, cilia ochreous grey, base shining brownish. Abdomen dorsally blackish, ventral side greyish. Hind tarsus black/grey ringed.

Female (Pl.-fig. 130). Head with greyish scales on face. Marks on forewing more prominent and distinct, slightly more shining white; plical spot more rounded, always to dorsum. Hindwing greyish, cilia brownish grey.

Male genitalia (Pl.-figs. 374, 375). Uncus deeply indented, proximal part of incision strongly sclerotized, uncus lobes widest in middle, densely setose. Costa of valva almost straight, only slightly emarginate before somewhat prominent, slightly angled cucullus. Digitate process long, base very narrow, tip markedly swollen. Juxta lobes elongate, tapering, bent outwards. Vinculum triangular. Aedeagus rather short, basal half wide, distal half narrower, tapering, with slightly sclerotized ridge in medial region.

Female genitalia (Pl.-fig. 483). Papillae anales with short, triangular setae on ventral side. Apophyses of almost equal length. Antrum short, bowl-shaped, weakly sclerotized, ventral margin wide, U-shaped, with fine spines, dorsal wall with fine spines; colliculum narrow, longer than apophyses posteriores. Signum oval, with three rows of fine spines.

Distribution. Apparently a very rare species. Not in Denmark, Norway or Sweden; in Finland three males from Ab, Lojo and Sa, Imatra. – Further reported from Leningrad, USSR (Spuler, 1910), Zürich, Switzerland (Frey, 1859), Venezia and Bolzano, Italy (Hartig, 1956; 1964).

Biology. According to Frey (1859) the larva is yellowish, head brown, and with two wide, red dorso-lateral lines. It mines downwards from the tip of the leaves of *Carex ornithopoda* Willd. during May; the mine is flat, whitish. Pupation on a leaf. Adults have been collected in June. The biotope is reported to be shady places in both coniferous and deciduous woods.

One of the specimens from Finland was caught in an open deciduous forest with much *Corylus avellana* L. and a rich herb layer; *Carex ornithopoda* is not present in this area, but *C. digitata* L. is common.

58. ***Biselachista serricornis*** (Stainton, 1854) **comb.n.**

Text-fig. 6; plate-figs. 131, 132, 219, 376, 377, 484, 485.

Elachista serricornis Stainton, 1854a: 260.
Elachista serricornella Morris, 1870: 227.
Elachista mitterbergeri Rebel, 1906: 643. **Syn.n.**
Elachista preisseckeri Krone, 1911: 40.

7–8 mm. Male (Pl.-fig. 131). Head and thorax dark grey-brown, tips of scales on posterior part of thorax and tegulae slightly lighter grey. Antenna pale greyish brown, with fine whitish cilia, scales on distal part of each segment raised, giving the antenna a strongly serrate appearance. Ground colour of forewing brownish grey, suffused with ochreous or reddish brown and blackish brown scales, especially at base of costa, in fold and along termen; a black dot in fold beyond middle and one between apex and black plical dot; an irregular, outwardly oblique, triangular streak from three-fourths of costa, ochreous white along costa and white towards termen, inwardly bound by a reddish brown and dark brown area; tornal spot inconspicuous, greyish. Cilia grey, suffused with blackish, cilia line very distinct, blackish, especially around apex. Hindwing dark grey-brown, often slightly tinged reddish brown; cilia light grey-brown. Abdomen dark grey-brown, with slightly lighter anal tuft. Hind tarsus light grey, only slightly ringed.

Female (Pl.-fig. 132). Scales on antenna less raised. Ground colour of forewing lighter than in male, with greyish fascia across the wing before the middle.

It should be noted that in rather worn specimens the reddish brown scales beyond the middle of the forewing are nearly always present.

Male genitalia (Pl.-figs. 376, 377). Uncus deeply indented, uncus lobes long, slender, sides almost parallel, tips setose, setae clavate. Valva rather wide, transtilla and costa strongly sclerotized, cucullus dorsally acute, rounded towards sacculus, sacculus well developed at base. Digitate process rather long, slender, widest before apex. Juxta lobes broad, truncate, tapering dorso-laterally. Vinculum triangular, broad, rounded distally. Aedeagus long, slender, almost straight, only slightly tapering, distal end rounded, distally with one long, prominent, sclerotized ridge with 5–7 teeth in the distal end.

Female genitalia (Pl.-figs. 484, 485). Papillae anales with slender setae. Apophyses short, of almost equal length. Antrum cup-shaped, with broad lateral rims, weakly sclerotized, ventral margin almost straight, not sclerotized, dorsal wall and antrum with fine spines; colliculum narrow, slightly longer than apophyses posteriores. Posterior portion of membranous ductus bursae with many longitudinal folds. Signum broad, with coarse and widely separated teeth.

Distribution. Scattered and local in Denmark; in Sweden from Sk., Hall., Sdm. and Nb.; not in Norway; from several districts in southern and central Finland. – Apparently widespread in Europe, from NW. USSR to Britain.

Biology. The larva is grey-greenish, dorsal line whitish; head and prothoracic shield light brown, meso – and metathorax with opposed lateral marks, connected by a whitish streak (Hering, 1891; Meyrick, 1928). It is known from *Carex ericetorum* Poll. and especially *C. sylvatica* Huds. The larva hibernates in a long, narrow, red-brown mine, but changes to a new leaf in the spring; the new mine starts from the base of the blade, is about 3–5 cm. long, rather wide, with irregular sides, yellowish; excrement blackish, deposited inside mine in distinct sections of the mine. Martini (1912a) gives a detailed description of the spring mine in *C. sylvatica:* the mine is made up of from one to six areas between the parallel ribs. The larva starts mining in a single area, but after having mined here for 6 mm it includes three more areas for 4 mm, and then only mines two areas for 3 mm, hereafter two more areas for 5 mm, and ends up with six areas; apparently never crossing the mid-rib. The mine can be separated from that of *E. gleichenella* (F.), which can be present on the same plant during the spring, in that this species start mining from the tip of the blade. Probably univoltine; adults fly from early June to late July. Mainly in boggy areas with *Carex* and *Eriophorum* and shaded, humid places in forests.

Note. The nomenclature of this species is much confused. *B. serricornis* Stt. has currently been termed *mitterbergeri* Rbl. in Scandinavia. Previous records of *serricornis* from Sweden and Finland refer to *freyi* Stgr.

Gaedike (1975: 345, figs. 20-22) figured the male genitalia of *mitterbergeri,* and stated that no cornutus is present; this was an error – the distal part of aedeagus of the figured specimen was broken (Parenti, *in litt.*).

59. ***Biselachista freyi*** (Staudinger, 1870) **comb. n.**
Text-figs. 38, 44; plate-figs. 133, 134, 220, 378, 379, 486.

Elachista freyi Staudinger, 1870: 322.
Elachista serricornis Stainton *sensu* Spuler, 1910: 429.

7–8 mm. Male (Pl.-fig. 133). Face white; neck tufts whitish with blackish scale tips; tegulae and thorax dark grey-brown, suffused with whitish. Antenna dark brown, distal third slightly serrate. Labial palpi descending and only slightly curved, upperside whitish, underside of second segment brownish, but tip whitish, third segment whitish on underside with brown ring at base and one just before tip. Forewing ground colour dark grey-brown, mottled by whitish basal parts of scales; fold with blackish streak bound inwardly by very small, indistinct, whitish spot and outwardly by larger, distinct, white spot in the middle of the wing; an outwards oblique, triangular white streak with scattered dark brown scales from two-thirds of costa; a small greyish white spot before apex, and a rounded white spot on tornus beyond the costal spot; subterminal dark-tipped scales in rows. Cilia greyish; cilia line distinct, blackish. Hindwing grey, cilia of same colour. Abdomen greyish. Hind tarsus dark brown, with narrow white rings.

Female (Pl.-fig. 134) Marks on forewing more distinct; basal parts mottled with more greyish white, distal half of wing almost uniformly dark brown. Cilia line less distinct.

Male genitalia. (Pl.-figs. 378, 379). Uncus deeply indented, medial margins of uncus lobes and base of incision strongly sclerotized, sides of lobes almost parallel, tips densely setose, setae slightly clavate. Gnathos lobes elongate, rounded. Valva slightly bent, costa almost straight, base of valva wider than cucullus, cucullus more rounded towards sacculus than towards costa. Digitate process long and slender, apex pointed. Juxta lobes short, broad, square. Vinculum short, triangular. Aedeagus long, straight, slender, distally chisel-shaped with a long straight sclerotized ridge present in distal half, often with two to three additional teeth.

Female genitalia (Pl.-fig. 486). Papillae anales with rather short, curved setae. Apophyses short, of almost equal length. Antrum short, triangularly bowl-shaped, weakly sclerotized, ventral margin slightly curved anteriorly, dorsal wall and inner side of antrum spined; colliculum narrow, about four times as long as apophyses posteriores. Signum elongate, distally dentate, middle part enlarged, not dentate.

Distribution. Not in Denmark; in Sweden from Baltic islands, Sm. and Vg.;

Norway: two specimens in the Grønlien collection at the Zoological Museum, Bergen bearing locality – labels inscribed only "Norway"; in Finland from the southern districts. – Widespread on the Continent from NW. USSR to Italy, but only few records from the western part; not in Britain.

Biology. According to Hofmann (1896) the larva is greenish grey, prothoracic segment yellowish and broad. It mines upwards in the leaves of *Carex humilis* Leyss and other *Carex* species from autumn to late May, making a mine very similar to that of *E. martinii* Hofm. At first the mine of *freyi* occupies half the width of the blade, but later it widens to the whole width; this takes place especially in narrow leaves. Pupa brown, dorsal and lateral ridges dark brown, not very distinct. Univoltine; adults from mid-June to late July. The biotope is sunny and dry places.

Note. From the type series of *E. freyi* Staudinger in the Humboldt Museum, Berlin, which consists of three specimens, a lectotype has been designated by the present authors, bearing the following labels "29/7", "Macugnago m.", "Origin", "freyi Stdgr.", "Genital præparat 1984♂, E. Traugott-Olsen, Lectotype".

60. ***Biselachista scirpi*** (Stainton, 1887) **comb.n.**
Plate-figs. 135, 136, 221, 380, 381, 487, 488, 531.

Elachista scirpi Stainton, 1887: 253.

10–12 mm. (Pl.-fig. 135). Head cream-white; tegulae and thorax ochreous white. Antenna beige, except for scape and pedicel, which are of the same colour as head. Labial palpi recurved, upperside and underside of third segment cream-white, second segment ochreous brown on underside. Forewing ground colour varying from almost plain ochreous to plain grey, but generally with whitish or white-blurred pattern in central area; the fold and central area are whitish, surrounded by ochreous; costa more brownish and dorsum with two poorly defined brownish blotches, one near the base and another to beyond middle; a short, distinct black dash in fold just above the latter, often beyond this an inconspicuous outward-angled band of ground colour from brown area at costa to outer brown blotch on dorsum; a whitish spot opposite an often indistinct tornal spot at four-fifths of costa; a whitish apical streak along termen, almost to tornus, inwardly bound by ground colour. Cilia pale ochreous grey, with distinct brown cilia line from apex to tornus. Hindwing pale beige, with similar cilia. Abdomen ochreous brown. Hind tarsus ochreous white.

Female (Pl.-fig. 136). Very similar to male; thorax and tegulae of same colour as head; pattern on forewing even more diffuse.

Male genitalia (Pl.-figs. 380, 381). Uncus deeply indented, uncus lobes rather short, rounded, wide apart, with slender, curved setae, margin between lobes rather strongly sclerotized. Valva widest distally; cucullus prominent, rounded towards the longitudinally folded, strong costa, more right-angled towards sacculus. Digitate process short, very stout. Juxta lobes short, distally rounded, vinculum very short and strong (difficult to mount correctly!). Aedeagus almost straight, gradually tapering, often dilated just before apex, distal end with weakly sclerotized fold or ridge.

Female genitalia (Pl.-figs. 487, 488). Papillae anales with short setae. Apophyses slender, posteriores longer than anteriores. Antrum large, bowl-shaped, anteriorly truncate, ventral margin curved anteriorly, dorsal wall and inner side of antrum set with spines, colliculum broad, about length of apophyses posteriores. Signum long and very narrow, middle part without teeth, distal portions strongly dentate.

Distribution. In Denmark from west coast of Jutland (WJ) and from SZ, LFM and NEZ; in Sweden from many localities along the Kattegat and the Baltic Sea up to Boh. and Sm.; not in Norway and from only a single locality in Ab in Finland. – Well known from Britain, but from the Continent otherwise only from Belgium (Janmoulle, 1949a).

Biology. The larva is pale greenish yellow, with an indistinct dorsal line; head pale yellow, mouth part darker. It mines in the upper part of the leaves of *Scirpus maritimus* L., *Juncus gerardii* Lois. (Fletcher, 1887; Gudmann, 1930) and *J. compressus* Jacq. (Meyrick, 1928) during the spring. The mine is short and broad (Pl.-fig. 531); the excrement is deposited as a long dark mass in the middle of the mine and in the end opposite to that in which the larva feeds, or can be let out through a hole in the basal part of the mine. The mine is greenish white, and easy to see. Pupation takes place on the mid-rib of a leaf of the food-plant in a fine silk web. The larva can be obtained from early March to late May, the pupa from mid-May to mid-June and adults fly from mid-June to early August; univoltine. The biotope is salt-marshes, where the food-plants grow; the adults fly around the food-plants in the sunset and come to light. This species will probably be found to be more widespread and common if searched for in these localities.

61. ***Biselachista eleochariella*** (Stainton, 1851) **comb.n.**
Plate-figs. 137, 138, 222, 382, 383, 489.

Elachista eleochariella Stainton, 1851: 10.

7–8 mm. Male (Pl.-fig. 137). Head grey, often tinged beige; thorax and tegulae greyish. Antenna grey-brown. Labial palpi drooping to porrect, almost plain greyish, upperside a little paler than underside. Forewing rather narrow, ground colour varying from greyish to beige-brown, costa and dorsum mainly grey, the fold almost plain beige-brown, never with a whitish streak, but with a distinct black dash in the middle, inwardly bound by a tiny or missing white dot and outwardly by a small, often distinct whitish spot; a whitish, outwards oblique, triangular streak from three-quarters of costa, inwardly bound by dark brown, just opposite a small similar coloured tornal spot; apical area often with scattered dark brown scales beyond a very tiny apical white streak. Cilia greyish; cilia line distinct from apex to middle of termen. Hindwing greyish brown, cilia similar. Abdomen grey. Hind tarsus ringed light grey/grey.

Female (Pl.-fig. 138). Head whitish grey, thorax and tegulae ochreous-white. Antenna grey, proximal half suffused with blackish brown, anterior part of scape blackish brown, posterior part ochreous white. Ground colour of forewing varying as in male, but generally paler ochreous; pattern similar to male, but costal, tornal and plical spots more distinctly white; dark scales before costal spot more prominent. Base of cilia white. Cilia on hindwing lighter than ground colour. Hind tarsus plain yellowish grey.

Male genitalia (Pl.-figs. 382, 383). Uncus deeply indented, uncus lobes close, distally rounded, with short, stiff setae, basal part of incision strongly sclerotized. Socii rather distinct. Valva slightly bent, widest before middle, costa slightly angled before middle, slightly emarginate beyond, cucullus small, rounded towards costa and sacculus. Digitate process prominent, club-shaped, distal end very broad. Juxta lobes distally rounded, lateral processes short. Vinculum triangular, elongate, very heavily sclerotized. Aedeagus rather long, slightly bent, distal end with two sclerotized ridges.

Female genitalia (Pl.-fig. 489). Papillae anales with short setae. Apophyses slender, posteriores longer than apophyses anteriores, anteriores bent at one-third from distal end. Antrum rather small, bowl-shaped, rounded anteriorly, ventral margin curved anteriorly, dorsal wall and inner side of antrum with fine spines; colliculum narrow, shorter than apophyses posteriores. Signum very small, with a fine dentate distal ridge at each end.

Distribution. Known from several districts in Denmark and Sweden; not from Norway; from several districts in the S. part of Finland and from Ob. – Widespread in Europe, from USSR to Italy, France and Britain.

Biology. Apparently few reliable records of the biology are available. The larva is reported to mine in May in the leaves of *Eriophorum angustifolium* Honck, *Scirpus palustris* L. (Hering, 1891; Wörz, 1957) and *Carex* sp. (Hering, 1957). The mine is not described. Adults fly from late June through July, in wet and moist places.

62. ***Biselachista utonella*** (Frey, 1856) **comb.n.**
Plate-figs. 139, 140, 223, 384, 385, 490.

Elachista utonella Frey, 1856: 300.
Elachista caricis Stainton, [1858] 1859: 155.
Elachista paludum Frey, 1859: 283.
Elachista palustrella Morris, 1870: 225.
Elachista carinisella Morris, 1870: 225.

8–9 mm. Male (Fig. 139). Head greyish to cream-white, or with greyish neck and whitish face, thorax and tegulae greyish. Antenna grey-brown, scales slightly raised on distal third. Labial palpi descending and only slightly curved, upperside silk-white, underside dark greyish. Forewing ground colour varying from warm orange-brown to whitish, but most often grey-brown with white marks; fold with a white or whitish streak from base almost to tornus, interrupted in middle by a short, blackish streak; base of costa dark brown; distinct whitish spot, right-angled outwards to costa, from three-quarters of costa; a small, white, elongate streak just before apex and a small tornal spot a little beyond inner costal spot; whitish fold and white costal and tornal marks always separated by transverse band of ground colour which is darkest towards termen; often a few small dark scales inside apical spot. Hindwing grey-brown, cilia of same colour. Abdomen grey with paler anal tuft. Hind tarsus plain grey-brown.

Female (Fig. 140). A little larger than male. Head cream-white; tegulae and thorax pale beige. Inner side and underside of labial palpi whitish. Forewing pattern generally more distinct. Hind tarsus with a narrow white ring on the distal end of each segment.

Male genitalia (Pl.-figs. 384, 385). Uncus lobes short, rounded, almost circular, with setae. Tegumen widening proximally. Gnathos short, triangular. Basal half of valva widest, costa strong, straight with many longitudinal folds, cucullus small, rounded. Digitate process very broad, club-shaped. Juxta lobes short, truncate, apical margin tapering into pointed lateral process, not setose. Vinculum short, strong, triangular. Aedeagus cylindrical, slightly tapering; without sclerotized ridges.

Female genitalia (Pl.-fig. 490). Papillae anales with short, coarse setae. Apophyses slender, posteriores shorter than anteriores. Antrum large, bowl-shaped, rounded anteriorly, ventral margin deeply U-shaped, spined, dorsal wall and inner side of antrum with spines; colliculum with longitudinal folds, as long as apophyses posteriores. Signum small, with a finely dentate distal ridge at each end.

Distribution. From many districts in Denmark and S. Sweden up to Upl.; not in Norway, but known from four districts in SW. Finland. – Apparently known from almost all parts of Europe.

Biology. Larva yellowish green, head black, prothoracic shield light brown. The larva has been reported to mine in several species of *Carex, C. acutiformis* Gnrh., *C. disticha* Hllds., *C. flacca* Schreb., *C. paniculata* L., *C. vesicaria* L., *C. brizoides* Jusl. and probably other species too (Hauder, 1918; Lhomme, 1951). Hering (1891) states that it is a somewhat polyphagous species, also mining in *Festuca*. Larvae are reported from March, April and May, but also from August (Wörz, 1957), but it is not known if the larva generally hibernates. The mine is flat and rather narrow, as it only occupies half the width of the blade; the mine is yellowish white or pale greenish; excrement grey or greenish. Pupation takes place in an angle of a leaf, the pupa attached by a single girdle. Adults fly from the beginning of June to early August; generally considered to be univoltine in N. Europe; Lhomme (1951) records two broods. The biotope is damp meadows and moors.

63. ***Biselachista albidella*** (Nylander, 1848) **comb.n.**
Plate-figs. 141, 142, 224, 386, 387, 491.

Elachista albidella Nylander *in* Tengström, [1848] 1847: 150.
Aphelosetia rhynchosporella Stainton, 1848: 2165.
Poeciloptilia uliginosella Herrich-Schäffer, 1855: 310.

9–10 mm. Male (Pl.-fig. 141). Head white; thorax and tegulae ochreous white. Antenna grey-brown, with scales on distal end of segments slightly raised. Labial palpi descending and only slightly curved, whitish, second segment grey below. Forewing ground colour white; basal two-thirds of costa grey-brown; fold and central area of wing white, fold with a distinct, short black streak in the middle and with indistinct basal patch of ochreous-brown scales, often a similar patch below fold; ochreous brown irregular band beyond middle, acutely produced medially towards apex and tornus, giving it an H-shaped appearance; white elongate streak below termen beyond post-medial band, distally suffused with ochreous brown or blackish brown scales, surrounding white costal and tornal spots. Cilia ochreous white; cilia line distinct near apex, dark brown. Hindwing dark grey, cilia similar. Abdomen grey. Hind tarsus plain, whitish.

Female (Pl.-fig. 142). Antenna scales less raised; forewing band less distinct. Abdomen ochreous grey, with ochreous-tinged anal tuft.

Male genitalia (Pl.-figs. 386, 387). Uncus lobes wide apart, widest distally, with long, slender, slightly clavate setae. Gnathos small. Valva with strongly angled costa, proximal part with many longitudinal folds, cucullus a prominent semicircular lobe, projecting dorsally, below which sacculus projects into a distinct spine; valva narrowest beyond middle before cucullus. Digitate process club-shaped, middle part narrow. Juxta lobes elongate, slender, tapering apically. Vinculum strong, short, triangular. Aedeagus rather long, tapering, without sclerotized ridges.

Female genitalia (Pl.-fig. 491). Papillae anales with short, coarse ventral and long lateral setae. Apophyses posteriores sinuate, longer than apophyses anteriores. Antrum wide, triangularly bowl-shaped, rounded anteriorly, ventral margin almost straight, dorsal wall and inner side of antrum with fine spines; colliculum without longitudinal folds, as long as apophyses posteriores. Signum small, with a fine dentate distal ridge at each end.

Distribution. Rather common and known from many districts in Denmark and Sweden; in Norway from SW. part; from many districts in Finland. – Reported from almost all parts of the Continent and from Britain.

Biology. The larva is rather large, 7–10 mm long, slender, pale green or greyish green, head shining greyish black, prothoracic shield greyish brown. It is reported from *Scirpus caespitosus* L. (Frey, 1859), *Carex riparia* Curt. and *C. acuta* L. (Hering, 1891), *Scirpus palustris* L. and *Eriophorum angustifolium* Honck (Lhomme, 1951), mining in the leaves during late April–May, often together with the larva of *B. utonella* (Frey), which generally is full grown a fortnight before and which looks much more blackish in the mine than *B. albidella*. The larva starts mining from the tip of the leaf, always mining downwards, making a 4–15 cm long mine, which first is whitish, but later turns yellowish grey and more transparent. Pupation in late May; pupa 5–7 mm long, slender, reddish brown, with a prominent dorsal ridge, attached to a leaf or stem by a single girdle. Adults fly from mid-June throughout July to the first part of August; apparently univoltine. On wet meadows and moors.

64. ***Biselachista abiskoella*** (Bengtsson, 1977) **comb.n.**
Plate-figs. 143, 144, 388–390, 492, 493.

Elachista abiskoella Bengtsson, 1977: 55.

8–9 mm. Male (Pl.-fig. 143). Face whitish, faintly brownish-tinged; vertex grey; neck tufts light brownish grey; tegulae and thorax mottled dark grey, light posteriorly, scales with whitish grey bases and dark grey tips. Antenna distally serrate, dark grey, scape and pedicel speckled with whitish posteriorly. Labial palpi long and slender, third segment almost as long as second segment, whitish grey above, dark grey beneath. Forewing ground colour dark grey, strongly mottled, scales as on thorax, tips almost black; a silky, whitish-blurred fascia before middle, bent on fold; a large silky whitish costal spot, inner margin of which is irregular, beyond a smaller tornal spot, spots almost confluent, forming an irregular whitish fascia; only a few dark tipped scales beyond spots, apical spot indistinct. Cilia whitish, faintly tinged with brownish; cilia line very weak. Hindwing lighter grey, cilia similar. Abdomen dark grey. Hind tarsus dark grey, distal part of segments whitish.

Female (Pl.-fig. 144). Antenna more strongly annulated, finely ciliate. Forewing ground colour lighter from base to fascia than beyond fascia, costal spot square, tornal spot triangular, connected by fine outward-angled fascia.

Male genitalia (Pl.-figs. 388–390). Uncus lobes wide apart, small, conical, distally blunt, with setae of varying lengths. Proximal two-thirds of valva wide, with almost parallel margins, cucullus distinctly more narrow, rounded towards costa, sacculus with small distal spine below cucullus. Digitate process slender, club-shaped. Juxta lobes very prominent, distally curved outwards, pointed, apical margins with scattered setae. Vinculum strong, triangular, broad distally, blunt. Aedeagus bulbous basally, gradually tapering from middle to triangular distal end, no sclerotized ridge present.

Female genitalia (Pl.-figs. 492, 493). Apophyses posteriores enlarged at base, slightly longer than apophyses anteriores. Posterior margin of tergite VIII emarginate. Antrum very strongly sclerotized, short bowl-shaped, ventral margin anteriorly curved, medially with posteriorly curved extension; dorsal wall with minute spines; colliculum almost as long as apophyses posteriores. Corpus bursae with minute internal spines. Signum elongate, middle portion enlarged, distally dentate.

Distribution. Only two males and a female known, all from the Abisko area, T.Lpm., Sweden.

Biology. Nothing is known about immature stages; known from moors in the subalpine zone. Adults were caught in early July.

Genus *Cosmiotes* Clemens

Cosmiotes Clemens, 1860: 8.

Type-species: *Cosmiotes illectella* Clemens, 1860.

Head (Text-fig. 45) smooth-scaled, neck tuft raised; pecten of antenna consisting of only few fine hairs. Labial palpi very long and slender, strongly diverging, drooping to porrect, only slightly recurved; third segment much shorter than second.

Wing pattern (Text-fig. 39) uniform; a more or less distinct light fascia before middle, two opposite spots beyond middle, separated by an often narrow streak of scales of ground colour; a characteristic small whitish spot at apex inside cilia line, termed cilial spot. Sexual dimorphism rather pronounced, males generally more unicolourous, pattern with less contrast than in females.

Venation (Text-fig. 9, pl.-figs. 225–28). Forewing R_4, R_5 and M_1 stalked, R_4 forks before fork R_5–M_1, M_2 absent, CuA_1 and CuA_2 distant, CuP reduced, 1A + 2A simple. Hindwing half as wide as forewing, acute, Sc + R_1 remote from Rs, free branch to costa only seen in *C. consortella* (Stt.), M_1 often very reduced, M_2 absent, CuA_1 and CuA_2 close or almost fused on common stalk, cell always open between M_2 and CuA_2.

Mid tibia with two rather small distal spurs, inner slightly longer than outer. Hind tibia with rather long hairs above on distal half, shorter and fewer hairs on underside;

middle spurs just before middle of the segment, inner spur longer than outer; apical spurs near distal end, of almost equal length.

Male genitalia. Uncus deeply indented, uncus lobes widely separated, claw-like, apices incurved, setose on distal part of inner surface, basal and distal parts with very short setae or bare. Socii two small inconspicuous papillae. Gnathos small, simple, rounded or elongate. Valva elongate, rather narrow near base, cucullus enlarged, truncate or rounded, costa rather strong, with few longitudinal folds, base produced medially into transtilla with distinct labides, sacculus produced into an acute spine below cucullus, cucullus and sacculus setose. Digitate process very long, slender, apex enlarged, and setose on inner surface. Juxta lobes short, conical, tapering, more or less setose, incision deeply V-shaped, narrow, ventral part of juxta prominent, duck-beak shaped. Vinculum prominent, produced into a distinct, often long saccus. Aedeagus long, slender, often curved, without cornuti or with a single cornutus.

Female genitalia. Papillae anales with short setae, sometimes with a few long basal setae on lateral and ventral sides. Segment VIII narrow. Ventral margin of antrum is an anterior sinus from posterior margin of sternite VIII, antrum with internal spines from ostium to colliculum; inception of ductus seminalis near to ductus bursae. Signum a circular or oval disc with 1–3 prominent teeth at either end, and often with a few small additional teeth. Anterior margin of sternite VII strongly reinforced, except in *C.consortella* (Stt.).

No larval material is accessible to the authors, but according to Braun (1948: 90) larvae are very similar to those of *Elachista*. Dugdale (1971: 80) gives a short diagnosis based on some South Pacific species. All species known are miners in leaves of grasses, apparently more or less polyphagous. The pupa (Text-fig. 35) is similar to that of the more specialized *Elachista* in structure, but can be distinguished

Text-fig. 45. Head of *Cosmiotes freyerella* (Hb.).

by a pair of backwardly directed spines on the upper part of the vertex. Pupation takes place in a flat silk cocoon on the leaf-surface.

The genus *Cosmiotes* Clem. is apparently widespread, known from Europe, N. Africa, N. America (Braun, 1948), New Guinea (Diakonoff, 1955), Australia (Common, 1970), New Zealand (Dugdale, 1971) and Rennell Island (Bradley, 1956). The genera *Dicranoctetes* Braun, 1918, from N. America and Cuba, and *Eupneusta* Bradley, 1974 from New Guinea are closely related to *Cosmiotes* Clem.

The W. Palaearctic species of this genus show a high degree of similarity which has led to much nomenclatural confusion. Of the five species recognized from this region by the authors, all except *C. ksarella* Chrét. are given full treatment. All other *Cosmiotes* species examined by the authors are placed in synonym with the treated species as here defined.

Key to species of *Cosmiotes* based on external characters.

1		Head and neck tufts cream-white (Pl.-figs. 149, 150)	67. *stabilella* (Stt.)
–		Head and neck tufts not wholly cream-white	2
2	(1)	Frons sordid white, neck tufts suffused with brownish grey. Wing expanse generally over 7,5 mm (Pl.-figs. 145, 146)	65. *freyerella* (Hb.)
–		Frons not sordid white. Wing expanse generally less than 7 mm	3
3	(2)	Face almost unicolourous pale beige, shining (Pl.-figs. 151, 152)	68. *consortella* (Stt.)
–		Face grey (Pl.-figs. 147, 148)	66. *exactella* (HS.)

Key to species of *Cosmiotes* based on male genitalia.

1		Spine at distal end of sacculus appressed to cucullus; saccus short and stout; aedeagus almost straight (Pl.-figs. 394, 395)	68. *consortella* (Stt.)
–		Spine at distal end of sacculus distinctly free; saccus slender; aedeagus not straight	2
2	(1)	Saccus long and slender, gradually tapering to distal end; aedeagus bent at one-quarter from distal end (Pl.-fig. 392)	66. *exactella* (HS.)
–		Saccus slender, distally extended; aedeagus evenly curved from base to distal end	3
3	(2)	Valva of almost equal width throughout its length; spine on sacculus large, very distinct; saccus rather short and broad (Pl.-fig. 393)	67. *stabilella* (Stt.)
–		Valva broadening distally; spine shorter; saccus slender and long (Pl.-fig. 391)	65. *freyerella* (Hb.)

Key to species of *Cosmiotes* based on female genitalia.

1		Anterior edge of sternite VII not reinforced, antrum ovoid, distinctly constricted towards colliculum, which is twice as long as apophyses posteriores (Pl.-fig. 497)	68. *consortella* (Stt.)

– Anterior edge of sternite VII reinforced, antrum more or less funnel-shaped, gradually tapering into colliculum which is shorter than or as long as apophyses posteriores 2

2 (1) Antrum longer than apophyses anteriores, widest in the middle, colliculum weakly sclerotized (Pl.-fig. 495) 66. *exactella* (HS.)

– Antrum not longer than apophyses posteriores, antrum gradually tapering anteriorly, colliculum sclerotized 3

3 (2) Antrum rather narrow, with almost parallel sides in posterior two-thirds, then tapering into colliculum, which has several longitudinal folds (Pl.-fig. 494) 65. *freyerella* (Hb.)

– Antrum widest at ostium bursae, gradually tapering toward colliculum, which has 2–3 longitudinal folds (Pl.-fig. 496) 67. *stabilella* (Stt.)

65. ***Cosmiotes freyerella*** (Hübner, 1825)
Text-figs. 39, 45; plate-figs. 145, 146, 225, 391, 494, 532, 533.

Tinea nigrella Hübner, [1805], pl. 41, fig. 285, *nec* Fabricius, 1775.
Antispila freyerella Hübner, [1825]: 419.
Elachista tristictella Nylander *in* Tengström, [1848] 1847: 148.
Elachista bistictella Tengström, [1848] 1847: 149. **Syn.n.**
Elachista aridella Heinemann, 1854: 7.
Elachista gregsoni Stainton, 1855: 48.
Elachista incertella Frey, 1859: 231.
Poeciloptilia grisescens Wocke, 1862: 248.
Elachista gregsonella Morris, 1870: 216.
Elachista sublimis Frey, 1866: 140.
Elachista baltica Hering, 1891: 207. **Syn.n.**

7–8 mm. Male (Pl.-fig. 145). Face whitish; neck tufts suffused with light grey; tegulae and thorax dark greyish brown. Antenna dark grey-brown, distal part finely serrate. Labial palpi drooping, upperside with a whitish grey streak, underside light brown. Forewing ground colour grey-brown to dark brown, often suffused with darker scales in fold; an almost straight, outwardly oblique, inconspicuous fascia, generally only indicated by slightly lighter scales, from two-thirds of costa; costal spot slightly beyond tornal spot, spots faintly indicated by slightly lighter scales; small cilial spot often distinct, whitish, but often lost in worn specimens, remaining cilia dark grey; cilia line weak. Hindwing greyish brown. Abdomen lead-grey, anal tuft lighter.

Female (Pl.-fig. 146). Face more yellowish or whitish grey; neck tufts, tegulae and thorax coke-grey. Ground colour of forewing to more distinct whitish fascia light grey with scattered blackish brown scales, beyond fascia blackish brown; costal and tornal spots distinct, whitish, separated by a narrow streak of ground colour; cilial spot distinct; cilia line around apex distinct.

Male genitalia (Pl.-fig. 391). Acute spine from sacculus below cucullus distinct. Digitate process long, swollen before apex, with fine setae on distal part. Juxta lobes short, conical, with fine setae. Vinculum short, triangular, with convex margins before the long, narrow, blunt saccus with slightly swollen tip. Aedeagus long and slender, evenly curved from base to distal end.

Female genitalia (Pl.-fig. 494). Apophyses posteriores longer than apophyses anteriores. Antrum long, as long as apophyses posteriores, with almost parallel sides, tapering anteriorly into colliculum, with several longitudinal folds, ventral margin U-shaped, dorsal wall and inner side of antrum set with fine spines; colliculum tapering anteriorly, slightly longer than apophyses posteriores. Signum with one prominent tooth at each end and with additional small triangular teeth.

Distribution. Known from many districts in Denmark, Sweden and Norway; in Finland from all districts in the southern part and from ObS and Li. – Apparently from all parts of Europe, including Britain.

Biology. The larva is yellowish grey, head and prothoracic shield black. It mines in March to May and July in the leaves of *Poa trivialis* L., *P. annua* L., *P. nemoralis* L. and probably other species of meadow-grass (*Poa* sp.). It is also reported known from *Festuca arundinacea* Schreber (Hering, 1891) and *Koeleria cristata* (L.) Pers. (Herms, 1888; Hering, 1891); many other food-plants are reported by Wörz (1957), and the species seems to be a polyphage, but some of the records could be erroneous, owing to identification problems. At present, records from *Poa, Festuca, Bromus, Triticum* and *Koeleria* are verified. The larva mines from the distal end of the leaf downwards, making a long and rather broad, flat, white blotch (Pl.-figs. 532, 533), in which the excrement is irregularly scattered. A single larva may mine several leaves. Pupation in a dense web on a leaf or a stem on the food-plant (Stainton, 1878); pupa sometimes fixed with long threads from the ends as in *Lyonetia*. The adults occur from late May to late June and again in August. They fly around the food-plant during the sunset and in the dusk and are easy to sweep. It is known from many biotopes: wet meadows, forest clearings and edges, from dunes and sea shores.

Notes *Elachista tristictella* Nyl. is considered conspecific with *C. freyerella* (Hb.) in accordance with current usage, as no authentic specimens are present (see introduction).

Elachista bistictella Tgstr. The type series at the Zoological Museum, Helsinki, consists of three males, all without abdomen and in worn condition, apparently representing three different species; one specimen is *E. pulchella* (Hw.) or *E. humilis* Zell., the second is probably a *E. humilis* Zell., and the third no doubt conspecific with *C. freyerella* (Hb.); the latter is hereby designated as lectotype and bears the following labels "H:fors", "Tengström", "bistictella", "3847, Wlsm. 1903", "Elachista bistictella Tgstr, Homo-type Tgstr (?)".

Elachista baltica E. Her. The type series in the Institut Zoologii, PAN, Warsaw, consists of five apparently conspecific specimens. Three are reared from *Koeleria* and two from *Festuca;* a lectotype has been designated by the present authors, labelled "27/7 18 89, Fest.ar., Misdroy" slide E. Traugott-Olsen A. 29.9.76, ♀, labelled accordingly.

66. ***Cosmiotes exactella*** (Herrich-Schäffer, 1855).
Plate-figs. 147, 148, 226, 392, 495.

Poeciloptilia exactella Herrich-Schäffer, 1855: 300, 304.
Poeciloptilia parvulella Herrich-Schäffer, 1855: 300, 304.

Elachista infuscata Frey, 1882: 373.
Elachista spectrella Frey, 1885: 107.

6–7 mm Male (Pl.-fig. 147). Head mottled greyish, frons paler; neck tufts, tegulae and thorax grey-brown. Antenna blackish brown, finely serrate distally. Labial palpi drooping to porrect, upperside whitish, underside of second segment and middle part of third segment brownish grey, joint whitish grey. Forewing ground colour mottled greyish brown with dark scales in fold, expecially beyond fascia, which is indistinct and slightly curved outwards; costal and tornal spots more distinct, paler, opposite and narrow; cilial spot distinct. Cilia grey, cilia line distinct, dark brown. Hindwing grey, cilia similar. Abdomen grey-brown, anal tuft ochreous. Hind tarsus grey, with blackish lateral spots.

Female (Pl.-fig. 148). Face more yellowish or whitish grey; neck tufts, tegulae and thorax mottled. Antenna slightly annulated light grey/blackish grey. Forewing light grey to fascia, sometimes slightly ochreous-tinged, plain blackish brown beyond fascia; fascia wide, whitish; costal and tornal spots opposite and distinct, white; cilial spot white, often blurred. Cilia line distinct, black.

Male genitalia (Pl.-fig. 392). Acute spine from sacculus distinct. Distal part of digitate process wide, setose. Juxta lobes conical, rounded distally, proximal part of incision strongly sclerotized. Vinculum short, triangular with almost straight margins, produced into a long, narrow, tapering, almost pointed saccus. Aedeagus long and slender, characteristically bent one-quarter from distal end.

Female genitalia (Pl.-fig. 495). Apophyses of almost equal length. Antrum longer than apophyses posteriores, with few longitudinal folds, widest in middle, ventral margin V-shaped; dorsal wall and inner side of antrum strongly spined; colliculum tapering anteriorly, shorter than apophyses posteriores. Signum with two opposed prominent teeth and small additional ones.

Distribution. Known from many districts in all four Scandinavian countries. – Not reported from Britain. Otherwise known from almost all parts of the Continent.

Biology. Larva not described; as the notes given by Frey (1859) and adopted by Hering (1957) probably concern *C. freyerella* (Hb.). The same applies to records of *Poa nemoralis* L. and *Festuca ovina* L. as food-plants; the larva is, apparently, only definitely known from *Deschampsia flexuosa* (L.) Trin., mining the leaves from early spring to late autumn in several generations. According to Hering (1891: 206) the species hibernates as a pupa, but according to Wörz (1957: 287) as a larva; the correct alternative is uncertain. This species is almost always present where the food-plant grows, on dry heath and under conifers on raw, humus-containing ground.

Notes. The identity of *P. exactella* Herrich-Schäffer has been established by examination of authentic specimens from the Staudinger collection in the Humboldt Museum, Berlin. All four syntypes are mounted on a common piece of polyporus. From these R. Gaedike has made genitalia dissections of two, without indicating the respective specimens; therefore a third genitalia dissection has been made from a specimen hereby designated as lectotype. The common pin bears the following labels "H-Sch.", "Origin" (pink), "exactella H.S.", "Gen.præp.Gaed.Nr.1415", "Gen.præp.Gaed.Nr.1416", "Genital præparat 5201, ♀, E. Traugott-Olsen, the front spec. to the left".

Assumption of identity of *Poeciloptilia parvulella* HS. follows Rebel (1901).

67. ***Cosmiotes stabilella*** (Stainton, 1858)
Plate-figs. 149, 150, 227, 393, 496.

Elachista stabilella Stainton, 1858b: 303.

7–8 mm. Male (Pl.-fig. 149). Head and neck tufts white; tegulae and thorax grey, posterior edges suffused with greyish white. Antenna grey, distal half faintly serrate. Labial palpi porrect, upperside dull white, underside suffused light brownish. Forewing narrow, ground colour blackish grey; distinct, white fascia, outwards oblique to fold, right-angled to dorsum, outer margin irregular from one-third of costa; costal and tornal spots distinct, triangular, white; tornal area often tinged ochreous; cilial spot blurred, white. Cilia grey, with fine whitish line inside distinct black cilia line. Hindwing grey. Hind tarsus dark grey, with prominent whitish rings.

Female (Pl.-fig. 150). Head and neck tufts pure white. Forewing pale grey to fascia, blackish grey beyond fascia, marks more distinct, fascia curved outwards, widest towards dorsum. Hindwing whitish grey, cilia similar.

Male genitalia (Pl.-fig. 393). Acute spine from sacculus very distinct. Digitate process distally broad, setose. Juxta lobes narrow, tapering, setose on lateral margins. Vinculum broad, with long, wide saccus, which is widest just before blunt tip. Aedeagus long and slender, evenly curved from base to distal end.

Female genitalia (Pl.-fig. 496). Apophyses of almost equal length, or posteriores slightly shorter than anteriores. Antrum funnel-shaped, longer than apophyses posteriores, with 2–3 longitudinal folds, ventral margin U-shaped, lateral rims very strong and narrow, dorsal wall and inner side of antrum set with spines; colliculum weakly sclerotized, less than one-third length of apophyses posteriores. Signum with two opposed, almost triangular, teeth, and a few small additional ones.

Distribution. From a few districts in Denmark; in SE. Sweden up to Upl.; not in Norway; in Finland only from Al. – Known from almost all parts of the Continent and from Britain.

Biology. Larva yellowish, head and prothoracic shield pale brownish. It mines in the leaves of *Avena, Agrostis, Deschampsia caespitosa* (L.) Beauv. and *Milium effusum* L. (Sønderup, 1949; Hering, 1957) from early spring to late May and in July, probably hibernating as a small larva. The mine is begun from the tip of the blade, is narrow and rather long, yellowish. Often more than one larva in the mine (Warren, 1878; Sønderup, 1949). Pupation in a dense web on a leaf. Apparently bivoltine; adults from mid-June to first part of July and in August.

Note. Until now the specific name *stabilella* has been erroneously attributed to Frey, even though Frey (1859: 237) himself cites Stainton as author.

68. ***Cosmiotes consortella*** (Stainton, 1851)
Plate-figs. 151, 152, 228, 394, 395, 497.

Elachista consortella Stainton, 1851: 9.

Elachista exiguella Frey, 1885: 105.
Elachista roesslerella Wocke *in* Heinemann, [1876]: 489. **Syn.n.**
Elachista tyrrhaenica Amsel, 1951: 143. **Syn.n.**

7–8 mm. Male (Pl.-fig. 151). Head with frons almost white, vertex mottled grey; neck tufts, tegulae and thorax grey-brown, tegulae lighter posteriorly. Antenna grey with darker annulation, distal half slightly serrate. Labial palpi drooping, upperside whitish, proximal two thirds of second segment light brown on underside, third segment whitish with two blackish spots, one at base and one in the middle. Forewing ground colour brownish grey, with scattered dark grey scales in fold, especially beyond fascia and between costal and tornal spots; fascia curved slightly outwards, indicated by slightly lighter scales, narrow towards costa; costal and tornal spots blurred, opposed, almost confluent, pointing towards cilial spot which is whitish and slightly ochreous-tinged. Cilia dark grey; cilia line fine, blackish brown. Hindwing dark, cilia similar. Abdomen grey. Hind tarsus light grey, with blackish distal spots.

Female (Pl.-fig. 152). Antenna distinctly annulated greyish/blackish, not or only slightly serrate. Distal spot on third segment of labial palpi elongate. Forewing ground colour light grey to fascia, blackish beyond fascia; fascia wide, distinct, almost straight; costal and tornal spots distinct, whitish, almost square, opposed, separated by dark streak; cilial spot prominent, often blurred. Cilia grey to dark grey, whitish beyond cilial spot. Hindwing blackish grey.

Male genitalia (Pl.-figs. 394, 395). Valva enlarged distally, acute spine from distal end of sacculus small, appressed along cucullus. Digitate process gradually thickened distally, setose. Juxta lobes short, pointed, setose. Vinculum broad, margins almost parallel, abruptly tapering into a stout, blunt, short saccus. Aedeagus long and slender, straight, gradually tapering.

Female genitalia (Pl.-fig. 497). Apophyses posteriores longer than apophyses anteriores. Antrum short, ovoid, distinctly constricted towards colliculum, ventral margin U-shaped, dorsal wall and inner side of antrum set with spines; colliculum narrow, twice as long as apophyses posteriores. Signum with one small and two opposed, large teeth, often with a few additional ones.

Distribution. From a few localities in Denmark; in Sweden from several districts in the southern part up to Boh.; not in Norway or Finland. – Only a few reliable records from Europe, but it may be discovered to be widespread; known from Britain, Germany, Switzerland, S. Italy and S. Spain.

Biology. Due to identification problems no reliable records of the immature stages and biology are known. The habitat is open grassland, often on calcareous ground. The adult has been collected in July and August.

Notes. The holotype of *E. roesslerella* Wocke, a male in fair condition but without abdomen, labelled "Kollektion Dr. A. Rössler", "TYPE", "Elachista roesslerella Hein. Wocke", "Museum Wiesbaden", "Holotypus, Elachista roeslerella Wck.-Hein., teste U. Parenti, 1976" is conspecific with *C. consortella* Stt.

There is no doubt *C. tyrrhaenica* Amsel is also conspecific with *C. consortella* (Stt.); in the original description it was only separated from *freyerella* and *exactella* and not compared with other *Cosmiotes* species, but it has a small appressed distal spine on the sacculus and a short, stout saccus.

Catalogue

				DENMARK												
		Germany	G. Britain	SJ	EJ	WJ	NWJ	NEJ	F	LFM	SZ	NWZ	NEZ	B	Sk.	Bl.
Mendesia farinella (Thnbg.)	1		●				●	●					●		●	●
Perittia herrichiella (HS.)	2	●									●		●	●	●	
P. obscurepunctella (Stt.)	3	●		●	●		●	●		●	●		●		●	
Stephensia brunnichella (L.)	4	●	●							●	●				●	
Elachista regificella Sirc.	5	●	●		●	●		●	●		●		●	●	●	●
E. gleichenella (F.)	6	●	●		●	●		●	●	●	●		●		●	●
E. pigerella (HS.)		●														
E. quadripunctella (Hb.)	7	●														
E. tetragonella (HS.)	8	●														
E. biatomella (Stt.)	9	●	●			●	●	●					●			
E. martinii Hofm.		●														
E. poae Stt.	10	●	●								●		●	●		
E. atricomella Stt.	11	●	●		●	●	●	●	●	●	●	●	●	●	●	
E. kilmunella Stt.	12	●	●												●	
E. parasella Tr.-O.	13															
E. alpinella Stt.	14	●	●	●		●	●	●	●	●	●		●		●	●
E. diederichsiella Her.	15	●								●			●		●	
E. compsa Tr.-O.	16	●			●					●	●		●		●	●
E. elegans Frey	17	●													●	
E. luticomella Zell.	18	●	●		●	●	●	●	●	●	●	●	●	●	●	●
E. albifrontella (Hb.)	19	●	●		●	●	●	●	●	●	●		●	●	●	●
E. bifasciella Tr.	20	●		●	●	●		●	●	●						
E. nobilella Zell.	21	●			●	●	●	●	●	●	●		●	●	●	●
E. apicipunctella Stt.	22	●	●					●	●	●	●	●	●	●	●	
E. subnigrella Dougl.	23	●	●								●		●		●	
E. reuttiana Frey		●														
E. orstadii Palm	24		●	●	●			●					●		●	
E. ingvarella Tr.-O.	25															
E. krogeri Svens.	26															
E. nielswolffi Svens.	27															
E. pomerana Frey	28	●	●						●	●	●		●		●	
E. humilis Zell.	29	●	●		●	●		●	●	●	●		●	●	●	●
E. vonschantzi Svens.	30															
E. pulchella (Hw.)	31	●	●		●	●	●	●	●	●	●	●	●	●	●	●
E. anserinella Zell.	32	●											●		●	●
E. rufocinerea (Hw.)	33	●	●	●	●	●			●			●				
E. lastrella Chrét.		●														
E. cerusella (Hb.)	34	●	●	●	●	●	●	●	●	●	●	●	●	●	●	●

SWEDEN

	Hall.	Sm.	Öl.	Gtl.	G. Sand.	Ög.	Vg.	Boh.	Dlsl.	Nrk.	Sdm.	Upl.	Vstm.	Vrm.	Dlr.	Gstr.	Hls.	Med.	Hrj.	Jmt.	Ång.	Vb.	Nb.	Ås. Lpm.	Ly. Lpm.	P. Lpm.	Lu. Lpm.	T. Lpm.
1		●	●	●		●	●	●			●	●			●													
2			●	●		●	●			●	●	●																
3																												
4				●																								
5	●	●	●	●	●		●	●	●		●	●			●			●										
6	●	●	●	●	●	●	●	●	●		●	●																
7							●																					
8						●	●		●																			
9	●																											
10		●	●	●																								
11		●	●	●			●	●				●																
12	●	●					●	●		●	●	●	●	●	●		●		●	●	●	●	●	●		●	●	●
13																				●						●		
14	●	●	●			●	●	●			●	●			●					●	●	●	●			●		●
15																												●
16		●		●			●		●			●																
17				●		●				●	●	●		●														
18	●	●	●	●		●	●	●		●	●	●		●														
19	●	●	●	●	●	●	●	●	●	●	●	●	●	●	●		●			●	●	●	●					
20																												
21	●	●	●	●	●		●	●			●	●																
22		●	●	●			●			●	●	●	●		●			●	●	●	●		●					
23			●	●			●				●	●																
24		●	●				●		●	●	●	●																
25																			●	●			●			●		●
26																							●					●
27																				●								
28						●					●												●					
29	●	●	●	●			●	●	●	●	●	●		●	●		●		●		●	●	●					
30																					●		●					
31	●	●	●			●	●	●	●	●	●	●	●	●	●		●		●	●		●	●	●		●		
32		●	●	●		●	●	●			●	●																
33																												
34	●		●				●	●																				

		NORWAY														
		Ø + AK	HE (s + n)	O (s + n)	B (ø + v)	VE	TE (y + i)	AA (y + i)	VA (y + i)	R (y + i)	HO (y + i)	SF (y + i)	MR (y + i)	ST (y + i)	NT (y + i)	Ns (y + i)
Mendesia farinella (Thnbg.)	1															
Perittia herrichiella (HS.)	2															
P. obscurepunctella (Stt.)	3															
Stephensia brunnichella (L.)	4															
Elachista regificella Sirc.	5	◗				●										
E. gleichenella (F.)	6	◖				●			◖		◗					
E. pigerella (HS.)																
E. quadripunctella (Hb.)	7															
E. tetragonella (HS.)	8															
E. biatomella (Stt.)	9															
E. martinii Hofm.																
E. poae Stt.	10															
E. atricomella Stt.	11															
E. kilmunella Stt.	12	◖									◗			◗	◗	
E. parasella Tr.-O.	13													◗		
E. alpinella Stt.	14	●														◗
E. diederichsiella Her.	15										◗					
E. compsa Tr.-O.	16															
E. elegans Frey	17															
E. luticomella Zell.	18					N	O	R	W	A	Y					
E. albifrontella (Hb.)	19	●				●			◖		●	◗	◗			
E. bifasciella Tr.	20															
E. nobilella Zell.	21	◗									●			◗	◗	
E. apicipunctella Stt.	22										●	◖			◗	
E. subnigrella Dougl.	23			●			◖									
E. reuttiana Frey																
E. orstadii Palm	24															
E. ingvarella Tr.-O.	25											◗				
E. krogeri Svens.	26															
E. nielswolffi Svens.	27															
E. pomerana Frey	28	◖														
E. humilis Zell.	29															
E. vonschantzi Svens.	30															
E. pulchella (Hw.)	31	◖		◖	◖	●	◖		◗		●					
E. anserinella Zell.	32	◗														
E. rufocinerea (Hw.)	33															
E. lastrella Chrét.																
E. cerusella (Hb.)	34															

					FINLAND																				USSR		
	Nn (ø + v)	TR (y + i)	F (v + i)	F (n + ø)	Al	Ab	N	Ka	St	Ta	Sa	Öa	Tb	Sb	Kb	Om	Ok	Ob S	Ob N	Ks	LkW	LkE	Le	Li	Vib	Kr	Lr
1					●						●														●	●	
2					●	●	●			●	●																
3						●	●			●	●															●	
4																											
5					●	●	●		●	●	●														●	●	
6					●	●	●	●	●	●	●														●		
7																											
8																											
9																											
10						●				●																	
11																											
12	◖	●	●	◖		●	●	●	●	●	●	●	●		●	●	●	●	●	●	●	●	●		●	●	
13		◗		◖														●	●	●	●		●	●			
14				●	●	●	●	●		●	●	●	●		●			●		●					●	●	
15	◖	●	◗			●																		●			
16					●	●	●			●	●																
17						●	●		●	●																	
18					●																						
19					●	●	●	●	●	●	●	●	●	●	●	●	●	●	●						●	●	
20																											
21	◖				●	●	●	●	●		●															●	
22					●	●	●	●	●	●	●	●	●	●	●		●	●	●	●				●	●	●	●
23																											
24																											
25																							●				
26																							●				
27		◗																									
28	◖					●	●	●	●	●	●							●		●							
29	◖				●	●	●	●	●	●	●		●		●	●	●	●	●	●	●				●	●	
30																											
31	◖		●		●	●	●	●	●	●	●	●	●	●	●	●	●	●		●				●	●	●	
32																											
33																											
34					●		●																				

				DENMARK												
		Germany	G. Britain	SJ	EJ	WJ	NWJ	NEJ	F	LFM	SZ	NWZ	NEZ	B	Sk.	Bl.
Elachista argentella (Cl.)	35	●	●		●	●	●	●	●	●	●	●	●	●	●	●
E. pollinariella Zell.	36	●								●						
E. triatomea (Hw.)	37	●	●		●	●	●	●	●	●	●	●	●	●	●	●
E. collitella (Dup.)		●														
E. subocellea (Stph.)	38	●	●													
E. festucicolella Zell.	39	●	●													
E. nitidulella (HS.)		●														
E. dispilella Zell.	40	●			●	●	●	●		●			●	●	●	●
E. triseriatella Stt.	41		●													
E. dispunctella (Dup.)		●	●													
E. rudectella Stt.		●														
E. squamosella (HS.)		●														
E. bedellella (Sirc.)	42	●	●		●	●		●		●		●	●	●	●	●
E. pullicomella Zell.	43	●								●	●		●	●	●	●
E. littoricola Le March.	44									●						
E. chrysodesmella Zell.	45	●														
E. megerlella (Hb.)	46	●	●							●	●					
E. cingillella (HS.)	47	●	●													
E. unifasciella (Hw.)	48	●	●		●				●	●	●		●		●	
E. gangabella Zell.	49	●	●						●	●	●		●	●	●	
E. subalbidella Schl.	50	●	●		●	●	●	●	●	●	●	●	●	●	●	●
E. revinctella Zell.	51	●	●	●	●	●		●	●	●	●		●	●	●	●
E. bisulcella (Dup.)	52	●	●						●	●	●		●		●	●
Biselachista trapeziella (Stt.)	53	●	●													
B. imatrella (Schantz)	54															
B. cinereopunctella (Hw.)	55	●	●													
B. kebneella sp. n.	56															
B. ornithopodella (Frey)	57															
B. serricornis (Stt.)	58				●	●			●	●	●		●		●	
B. freyi (Stgr.)	59	●	●													
B. scirpi (Stt.)	60		●			●				●	●		●		●	●
B. eleochariella (Stt.)	61	●	●		●		●			●	●	●	●		●	●
B. utonella (Frey)	62	●	●		●		●	●	●	●	●		●	●	●	●
B. albidella (Nyl.)	63	●	●		●	●	●	●	●	●	●		●	●	●	●
B. abiskoella (Bengts.)	64															
Cosmiotes freyerella (Hb.)	65	●	●				●	●	●	●	●	●	●	●	●	●
C. exactella (HS.)	66	●			●	●	●	●	●	●	●	●	●	●	●	●
C. stabilella (Stt.)	67	●	●		●			●			●		●		●	
C. consortella (Stt.)	68		●						●	●			●	●	●	●

SWEDEN

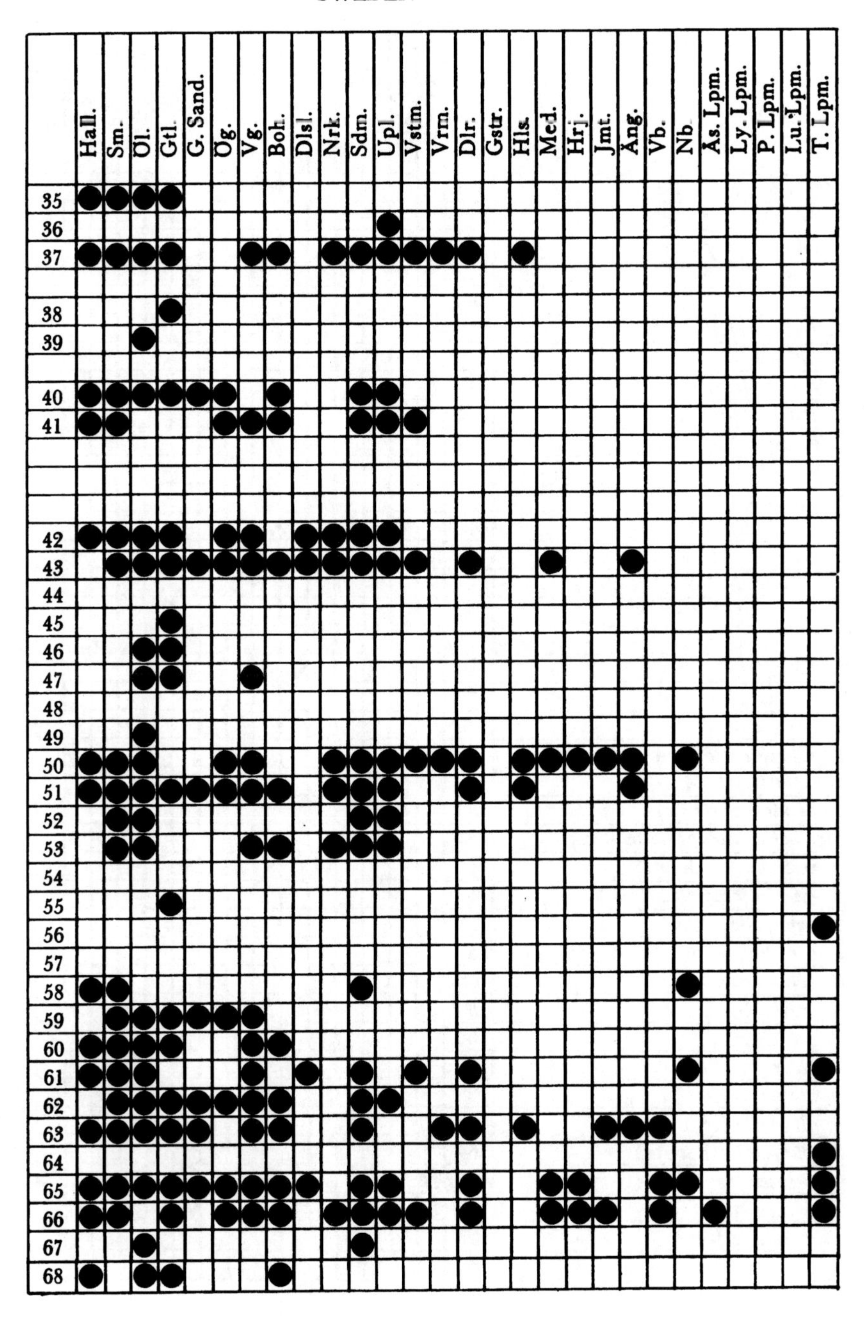

	Hall.	Sm.	Öl.	Gtl.	G. Sand.	Ög.	Vg.	Boh.	Dlsl.	Nrk.	Sdm.	Upl.	Vstm.	Vrm.	Dlr.	Gstr.	Hls.	Med.	Hrj.	Jmt.	Ång.	Vb.	Nb.	Ås. Lpm.	Ly. Lpm.	P. Lpm.	Lu. Lpm.	T. Lpm.
35	●	●	●	●																								
36												●																
37	●	●	●	●			●	●		●	●	●	●	●	●		●											
38				●																								
39			●																									
40	●	●	●	●	●	●		●			●	●																
41	●	●				●	●	●			●	●	●															
42	●	●	●	●		●	●		●	●	●	●																
43		●	●	●	●	●	●	●	●	●	●	●	●		●			●			●							
44																												
45				●																								
46			●	●																								
47			●	●			●																					
48																												
49			●																									
50	●	●	●			●	●			●	●	●	●	●	●		●	●	●	●	●		●					
51	●	●	●	●	●	●	●	●		●	●	●			●		●				●							
52		●	●								●	●																
53		●	●				●	●		●	●	●																
54																												
55				●																								
56																												●
57																												
58	●	●									●												●					
59		●	●	●	●	●	●																					
60	●	●	●	●			●	●																				
61	●	●	●				●		●		●		●		●								●					●
62		●	●	●	●	●	●	●			●	●																
63	●	●	●	●	●		●	●			●			●	●		●			●	●	●						
64																												●
65	●	●	●	●	●	●	●	●	●		●	●			●			●	●			●	●					●
66	●	●		●		●	●	●		●	●	●	●		●			●	●	●		●		●				●
67			●								●																	
68	●		●	●				●																				

		NORWAY														
		Ø+AK	HE (s+n)	O (s+n)	B (ø+v)	VE	TE (y+i)	AA (y+i)	VA (y+i)	R (y+i)	HO (y+i)	SF (y+i)	MR (y+i)	ST (y+i)	NT (y+i)	Ns (y+i)
Elachista argentella (Cl.)	35															
E. pollinariella Zell.	36															
E. triatomea (Hw.)	37															
E. collitella (Dup.)																
E. subocellea (Stph.)	38															
E. festucicolella Zell.	39															
E. nitidulella (HS.)																
E. dispilella Zell.	40	◖														
E. triseriatella Stt.	41															
E. dispunctella (Dup.)																
E. rudectella Stt.																
E. squamosella (HS.)																
E. bedellella (Sirc.)	42															
E. pullicomella Zell.	43	◖														
E. littoricola Le March.	44															
E. chrysodesmella Zell.	45															
E. megerlella (Hb.)	46										◖			◗		
E. cingillella (HS.)	47						◖									
E. unifasciella (Hw.)	48															
E. gangabella Zell.	49															
E. subalbidella Schl.	50	●									◗				◗	
E. revinctella Zell.	51	●		◗							◗					
E. bisulcella (Dup.)	52															
Biselachista trapeziella (Stt.)	53															
B. imatrella (Schantz)	54					N	O	R	W	A	Y					
B. cinereopunctella (Hw.)	55															
B. kebneella sp. n.	56															
B. ornithopodella (Frey)	57															
B. serricornis (Stt.)	58															
B. freyi (Stgr.)	59					N	O	R	W	A	Y					
B. scirpi (Stt.)	60															
B. eleochariella (Stt.)	61															
B. utonella (Frey)	62															
B. albidella (Nyl.)	63								◖	◗	◗	◗	◖			
B. abiskoella (Bengts.)	64															
Cosmiotes freyerella (Hb.)	65	●			◖			◖			●				◗	◗
C. exactella (HS.)	66	◖			◖						●				●	
C. stabilella (Stt.)	67															
C. consortella (Stt.)	68															

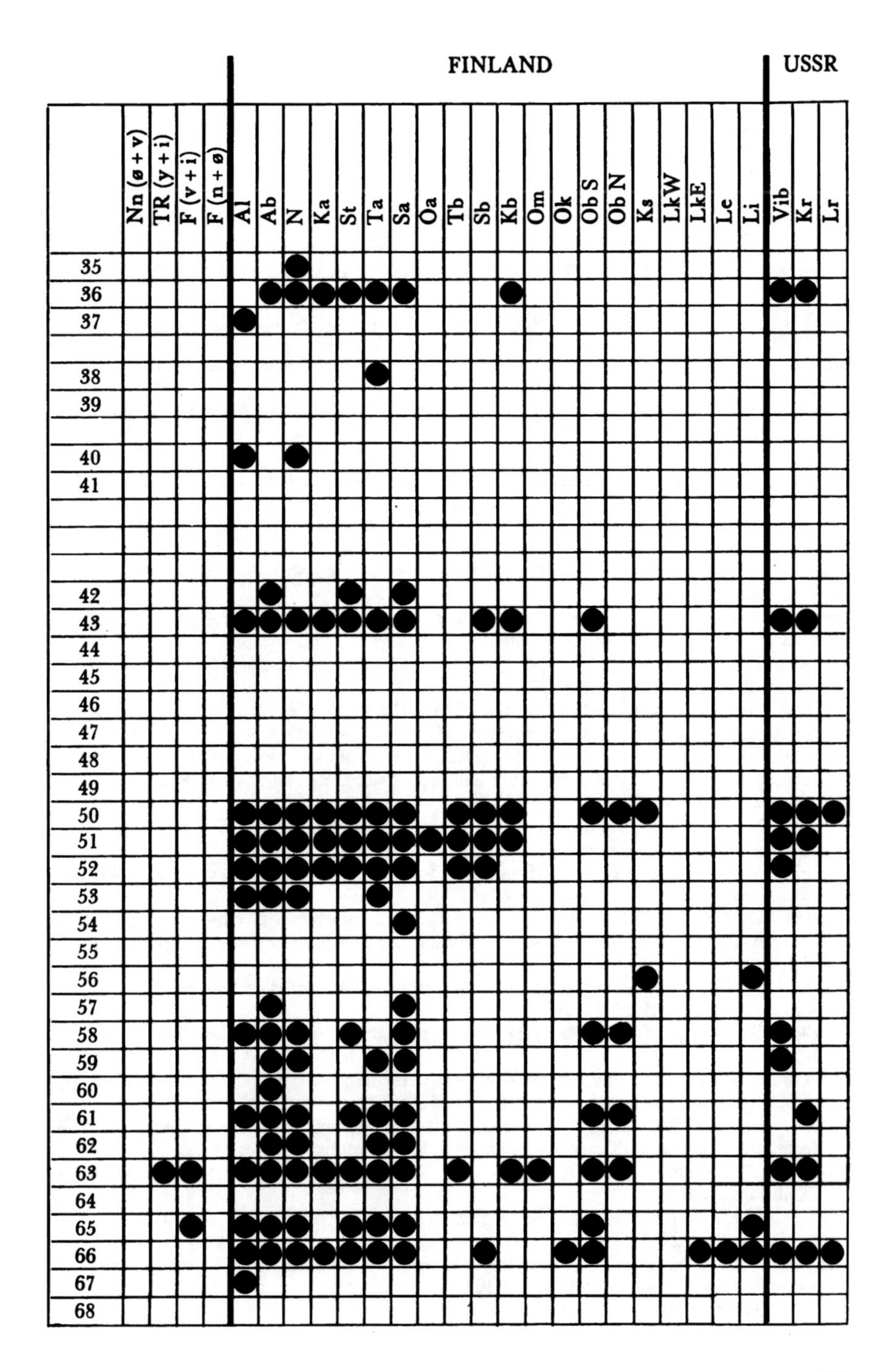

					FINLAND																				USSR		
	Nn (ø + v)	TR (y + i)	F (v + i)	F (n + ø)	Al	Ab	N	Ka	St	Ta	Sa	Öa	Tb	Sb	Kb	Om	Ok	Ob S	Ob N	Ks	LkW	LkE	Le	Li	Vib	Kr	Lr
35							●																				
36						●	●	●	●	●	●				●										●	●	
37					●																						
38										●																	
39																											
40					●		●																				
41																											
42						●			●		●																
43					●	●	●	●	●	●	●			●	●			●							●	●	
44																											
45																											
46																											
47																											
48																											
49																											
50					●	●	●	●	●	●	●		●	●	●			●	●	●					●	●	●
51					●	●	●	●	●	●	●	●	●	●	●										●	●	
52					●	●	●	●	●	●	●		●	●											●		
53					●	●	●			●																	
54											●																
55																											
56																				●				●			
57						●					●																
58					●	●	●		●		●							●	●						●		
59						●	●			●	●														●		
60						●																					
61					●	●	●		●	●	●							●	●							●	
62						●	●			●	●																
63		●	●		●	●	●	●	●	●	●		●		●	●		●	●						●	●	
64																											
65			●		●	●	●		●	●	●							●						●			
66					●	●	●	●	●	●	●			●			●	●				●	●	●	●	●	●
67					●																						
68																											

Literature

Amsel, H.G., 1935: Neue Palästinensische Lepidoptera. – Mitt. zool. Mus. Berl. 20: 271-319, pls. 9-18.
– 1951: Lepidoptera Sardinica. Parte III. Descrizioni di specie nuove ed osservazioni sistematiche di carratere generale. – Fragm. ent. 1: 101-144.
Benander, P., 1941: Zwei durch einander gemischte Schmetterlingsarten der Familie Elachistidae. – Opusc. ent. 6: 44-50.
– 1946: Förteckning över Sveriges småfjärilar. Catalogus Insectorum Sueciae VI. Microlepidoptera. – Ibid. 11: 1-82.
– 1957: Notiser om småfjärilar (Lep.). – Ibid. 22: 53-56.
– 1961: Die Microlepidoptera in Thunbergs "Insecta Suecica" 1784-1794. – Ibid. 26: 243-247.
Bengtsson, B.Å., 1977: Two new species of Microlepidoptera from northern Sweden (Lepidoptera: Elachistidae, Scythrididae). – Ent. scand. 8: 55-58.
Bentinck, G.A.G. & Diakonoff, A., 1968: De Nederlandse Bladrollers (Tortricidae). – Monogr. Ned. Entomol. Ver. 3: 1-201, 98 pls.
Boisduval, J.B.A.D. de, 1836: Historie naturelle des Insectes. Spécies général des Lépidoptères. 1, xii + 690 pp., 24 pls. + 6 pp. – Paris.
Bradley, J.D., 1950: *Mendesia farinella* (Thunberg) (Lepidoptera Elachistidae): A new species to Britain. – Microscope 7: 299-303.
– 1952: Microlepidoptera collected in the Burren, Co. Clare, Ireland in 1951, including two species new to the British list. – Entomologist's Gaz. 3: 185-192, 1 pl.
– 1956: Records and descriptions of Microlepidoptera from Lord Howe Island and Norfolk Island collected by the British Museum (Natural History) Rennell Island Expedition, 1953. – Bull. Br. Mus. nat. Hist. (Ent.) 4: 145-164.
– 1963: A review of the nomenclature of certain species in the genus *Elachista* Treitschke (Lep., Elachistidae). – Entomologist's Gaz. 14: 150-161.
– 1974: A new species of Elachistid moth (Lepidoptera, Elachistidae) reared on sugar-cane in Papua New Guinea. – Bull. ent. Res. 64: 73-79.
Bradley, J.D., Fletcher, D.S. & Whalley, P.E.S., 1972: Lepidoptera. *In* Kloet, G.S. & Hincks, W.D.: A Check List of British Insects. 2nd edn. Part 2. – Handbk Ident. Br. Insects 11 (2): i-viii, 1-153.
Braun, A.F., 1933: Pupal tracheation and imaginal venation in Microlepidoptera. – Trans. Am. ent. Soc. 59: 229-268.
– 1948: Elachistidae of North America (Microlepidoptera). – Mem. Am. ent. Soc. 13: 1-110, 26 pls., i-ii.
Brock, J.P., 1971: A contribution towards an understanding of the morphology and phylogeny of the Ditrysian Lepidoptera. – Jnl. nat. Hist. 5: 29-102.
Bruand, M.T., 1850: Catalogue systématique et synonymique des Lépidoptères du département du Doubs. Tinéides. – Mém. Soc. Emul. Doubs (1) **3** (3): 23-68.
Brues, C.T., Melander, A.L. & Carpenter, F.M., 1954: Classification of Insects. v + 917 pp. – Cambridge, Mass.
Buhr, H., 1942: Einiges über bekannte und unbekannte europäische Schmetterlings-Minen. – Z. wien. EntVer. 27: 72-78.
Busck, A., 1909: Notes on Microlepidoptera, with descriptions of new North American species. – Proc. ent. soc. Wash. 11: 87-103.
– 1914: On the classification of the Microlepidoptera. – Ibid. 16: 46-54.

– 1934: Microlepidoptera of Cuba. – Entomologica am. 13: 151-203.
Büttner, F.O., 1880: Die Pommerschen, inbesondere die Stettiner Microlepidoptern. – Stettin. ent. Ztg 41: 383-473.
Buxton, P.A., 1914: On *Elachista poae*. – Entomologist's Rec. J. Var. 26: 184-185.
Byers, J.R. & Hinks, C.F., 1973: The surface sculpturing of the integument of lepidopterous larvae and its adaptive significance. – Can. J. Zool. 51: 1171-1179.
Caradja, A., 1920: Beitrag zur Kenntnis der geographischen Verbreitung der Mikrolepidopteren des palaearktischen Faunen-gebietes nebst Beschreibung neuer Formen. – Dt. ent. Z. Iris 34: 75-179.
Chrétien,P., 1896: Description de Microlépidoptères nouveaux. – Bull. Soc. ent. Fr. 1896: 190-193.
Chambers, V.T., 1875: On Tineina from Texas, with description of new species. – Can. ent. 7: 105-108.
– 1880: Illustrations of the Neuration of the Wings of American Tineina. – J. Cincinn. Soc. nat. Hist. 2: 194-204, 58 figs.
Clarke, J.F. Gates, 1941: The preparation of slides of the Genitalia of Lepidoptera. – Bull. Brooklyn ent. Soc. 36: 149-161.
Clemens, B., 1860: Contributions to American Lepidopterology. – Proc. Acad. nat. Sci. Philad. 1860: 4-15.
Clerck, C., 1759-64: Icones Insectorum rariorum. [xii] + [iii] pp., 55 pls. – Holmiae.
Common, I.F.B., 1970: Lepidoptera (Moths and Butterflies) pp. 765-866. *In* Mackerras, I.M.: The Insects of Australia. xiii + 1029 pp., 8 pls. – Canberra.
Costa, O.G., [1836] – 1850: Fauna del Regno di Napoli . . . Lepidotteri. xi + [434] pp., 41 pls. – Napoli.
Dattin, E., 1932: Note sur une Elachistidae nouvelle. *Mendesia subargentella* n. sp. – Amat. Papillons 6: 161-168.
[Denis, M. & Schiffermüller, I.], 1775: Ankündung eines systematischen Werkes von den Schmetterlinge der Wienergegend. 323 pp., 3 pls. – Wien.
Dethier, V.G., 1941: The antennae of lepidopterous larvae. – Bull. Mus. comp. Zool. Harv. 87: 455-507.
Diakonoff, A., 1955: Microlepidoptera of New Guinea. Results of the Third Archbold Expedition (American-Netherlands Indian Expedition 1938-39). Part V. – Verh. K. ned. Akad. Wet. (Afd. Natuurk.) (2) 50 no. 3: 1-210, 142 pls.
Doubleday, H., 1859: The Zoologist's synonymic list of British Butterflies and Moths. 40 pp. – London.
Douglas, J.W., 1848: Description of a new British Moth. – Trans. ent. Soc. Lond. 5: 21, 1 pl.
– 1850: Descriptions of a new species of *Elachista* and *Grapholitha*. – Proc. ent. Soc. Lond. 1850: 7-8.
– 1852-54: Contributions towards the Natural History of British Micro-Lepidoptera. – Trans. ent. Soc. Lond. 2: 52-54, 75-81, 119-124, 207-208, 209-212, 6 pls.
Dufrane, A., 1943: Microlepidopteres de la Fauna Belge. (3^{e} note). – Bull. Mus. r. Hist. nat. Belg. 19 (31): 1-8.
– 1957: Microlepidopteres de la Fauna Belge (Huitieme note). – Bull. Inst. r. Sci. nat. Belg. 33 (32): 1-16.
Dugdale, J.S., 1971: Entomology of the Aucklands and other Islands south of New Zealand: Lepidoptera, excluding non-crambine Pyralidae. – Pacif. Insects 27: 55-172.
Duponchel, P.-A.-J., 1838-[40]. *In* Godart, J.-B.: Histoire naturelle des Lépidoptères ou Papillons de France (Nocturnes 8). 11, 720 pp., pls. 287-314. – Paris.
– 1842-[44]. *In* Godart, J.-B.: Historie naturelle des Lépidoptères ou Papillions de France. (Nocturnes, Supplement aux Tomes quatrième et suivants). Supplement 4, 534 pp., pls. 51-90. – Paris.
– 1844-[45]. *In* D'Orbigny, C.: Dictionnaire universel d'Histoire Naturelle. 5, 768 pp. Paris.

Dyar, H.G., 1902: A List of the North American Lepidoptera and key to the Literature of this order of Insects. – Bull. U.S. natn. Mus. 52: i-xix, 1-723.
Fabricius, J.C., 1781: Species Insectorum exhibentes eorum Differentias specificas, Synonyma Auctorum, Loca natalia, Metamorphosin adjectis Observationibus, Descriptionibus 2, 517 pp. – Hamburgi et Kilonii.
– 1794: Entomologia systematica emendata et aucta 3 (2), 349 pp. – Hafniae.
Fischer von Röslerstamm, J.E., 1834-42: Abbildungen zur Berichtigung und Ergänzung der Schmetterlingkunde, besonderes der Microlepidopterologie als Supplement zu Treitschke's und Hübner's europaeischen Schmetterlingen, mit erlaeuterndem Text. ii + 304 pp., 100 pls. Leipzig.
Fletcher, W.H.B., 1887: On the Life-History of *Elachista scirpi*. – Entomologist's mon. Mag. 23: 254-255.
Forbes, W.T.M., 1923: The Lepidoptera of New York and neighboring States. – Mem. Cornell Univ. agric. Exp. Stn 68: 1-729.
Fracker, S.B., 1915: The classification of lepidopterous larvae. – Illinois biol. Monogr. 2: 1-169.
Frey, H., 1856: Die Tineen und Pterophoren der Schweiz. xii + 430 pp. – Leipzig.
– 1859: Das Tineen-Genus *Elachista*. Ein Versuch. – Linn. ent. 13: 172-314.
– 1866: Die schweizerischen Microlepidopteren. – Mitt. schweiz. ent. Ges. 2: 136–146.
1870: Ein Beitrag zur Kenntnis der Microlepidopteren. – Ibid. 3: 277–296.
– 1882: Zweiter Nachtrag zur Lepidopteren-Fauna der Schweiz. – Ibid. 6: 349-375.
– 1885: Zur Kenntnis des Tineen- Genus *Elachista*. – Stettin. ent. Ztg 46: 97-108.
– 1886: Vierter Nachtrag zur Lepidopteren-Fauna der Schweiz. – Mitt. schweiz. ent. Ges. 7: 196-262.
Gaedike, R., 1975: Zum Status der von Rebel, Krone und Gozmány beschriebenen *Elachista*-Arten (Lepidoptera, Elachistidae). – Annls. hist.-nat. Mus. natn. hung. 65: 239-248.
Groschke, F., 1939: Die Kleinschmetterlinge der Grafschaft Glatz. – Mitt. münch. ent. Ges. 29: 643-734, pls. xvi-xix.
Gudmann, F., 1924: Interessante Tilpasningsevne. – Flora Fauna 1924: 29.
– 1930: Nogle biologiske Iagttagelser vedrørende Smaasommerfugle, samt nogle praktiske Vink med Hensyn til Klækningen af disse. – Ibid. 1930: 8-24.
Hannemann, H. J., 1953: Natürliche Gruppierung der Europäischen Arten der Gattung *Depressaria* s.l. (Lep. Oecoph.). – Mitt. zool. Mus. Berl. 29: 269-373.
Haanshus, K., 1933: Fortegnelse over Norges Lepidoptera. – Norsk ent. Tidsskr. 3: 165-216.
Hampson, G.F., 1918: Some small families of the Lepidoptera which are not included in the key to the families in the catalogue of Lepidoptera Phalaenae, a list of the families and subfamilies of the Lepidoptera with their types and a key to the families. – Nov. Zool. 25: 366-395.
Handlirsch, A., 1924: Ordnung: Lepidoptera (Schmetterlinge), pp. 852-941. *In* Schröder, C. Handbuch der Entomologie. 3, viii + 1202 pp. – Jena.
Hartig, F., 1956: Prodromus dei Microlepidotteri della Venezia Tridentina a delle regioni adiacenti. – Studi trent. Sci. nat. 33: 89-148.
– 1964: Microlepidotteri della Venezia Tridentina e delle regioni adiacenti. Parte III. (Fam. Gelechiidae-Micropterygidae). – Ibid. 41: 1-292.
Hauder, F., 1918: *Elachista paludum* Frey. – Z. öst. EntVer. 3: 21.
Haworth, A.H., 1803-28: Lepidoptera Britannica. 1-4, xxxvi + 610 pp. – Londini.
Heinemann, H. v., 1854: Zehn neue Microlepidopteren. – Z. ent. 8: 1–7.
Hering, E., 1889: Beiträge zur Mitteleuropäischen Micro-Lepidopterenfauna. – Stettin. ent. Ztg 40: 290-320.
– 1891: Ergänzungen und Berichtigungen zu F.O. Büttner's Pommerschen Mikrolepidopteren (Stett. ent. Ztg. 1880 pag. 383-473). – Ibid. 52: 135-227.

– 1893: Zuträge und Bemerkungen zur Pommerschen Microlepidopteren-Fauna. (cfr. Stett. ent. Zeitung 1880 p. 383-473 und 1891 p. 135-227). – Ibid. 54: 80-120.
Hering, E.M., 1924: Beitrag zur Kenntnis der Microlepidopteren-Fauna Finlands. – Notul. ent. 4: 75-84.
– 1925: Minenstudien V. – Z. wiss. InsektBiol. 20: 125-136.
– 1926: Synopsis der Minen an Caprifoliaceen. Minenstudien VII. – Z. Morph. Ökol. Tiere. 5: 447-488.
– 1932: Die Schmetterlinge nach ihre Arten dargestellt. *In* Brohmer, P., Ehrmann, P. & Ulmer, G.: Die Tierwelt Mitteleuropas. Ergänzungband 1, 545 pp. – Leipzig.
– 1936: 24. Ordnung: Schmetterlinge, Lepidóptera. Abt. 8, 94 pp. *In* Brohmer, P., Ehrmann, P. & Ulmer, G.: Die Tierwelt Mitteleuropas. 6. Insekten, 3. Teil. – Leipzig.
– 1951: Biology of the Leaf Miners. iv + 420 pp. – Dr. W. Junk, 's-Gravenhage.
– 1957: Bestimmungstabellen der Blattminen von Europa. 1-2, 1185 pp; 3, 221 pp. – Dr. W. Junk, 's-Gravenhage.
– 1963: Neue Blattminen-Studien III (Dipt., Lep.). – Dt. ent. Z. (N.F.) 10: 221-250.
Herms, H., 1888: Lepidopterologische Beobachtungen. – Stettin. ent. Ztg 49: 81-83.
Herrich-Schäffer, G.A.W., 1847-55: Systematische Bearbeitung der Schmetterlinge von Europa 5, 394 pp., pls. 1-124 (Tineide), 1-7 (Pterophorides), 1 (Micropteryges). – Regensburg.
– 1857: Kritischer Anzeiger des zoologisch-mineralogischen Vereines in Regensburg. – KorrBl. zool.-min. Ver. Reg. 11: 33-72.
Hodges, R.W., in press: Walshiidae & Cosmopterigidae. – The Moths of America North of Mexico 6 (1).
Hodgkinson, J.B., 1879: Two new Micro-Lepidoptera. – Entomologist 12: 55-57.
Hoffmann, O., 1893: Beiträge zur Naturgeschichte der Tineinen. – Stettin. ent. Ztg 54: 307-311.
Hofmann, O., 1896: *Elachista martinii* nov. sp. – Dt. ent. Z. Iris 11: 143-146.
Hübner, J., 1796-[1836]: Sammlung europäischer Schmetterlinge 8, 78 pp. (1796), 71 pls. (1796-[1836]). – Augsburg.
– 1816-[25]: Verzeichniss bekannter Schmetterlinge. 431 pp. – Augsburg.
Janmoulle, E., 1945: Une Espèce méconnue: *Mendesia subargentella* Dattin. – Lambillionea 45: 3-4.
– 1947: Une question de nomenclature. – Ibid. 47: 64-72.
– 1949a: Espèces nouvelles pour la fauna belge. – Ibid. 49: 90-91.
– 1949b: *Elachista pulchella* Hw., bona species. – Ibid. 49: 119-123, pl. 9.
– 1962: Notes sur les Microlépidoptères de Belgique. VII. A propos d'*Elachista postremella* Dufrane. – Bull. Inst. r. Sci. nat. Belg. 38 (14): 1-3.
Joannis, J. de, 1902: Note sur un Microlèpidoptére nouveau de Portugal. – Bull. Soc. ent. Fr. 1902: 230-232.
– 1915: Étude synonymique des Espèces de Microlépidoptères décrites comme nouvelles par Duponchel. – Annls soc. ent. Fr. 84: 62-164.
– 1931: *Atachia pigerella* H.S. – Bull. Soc. ent. Fr. 1931: 223–227.
Karsholt, O. & Nielsen, E. Schmidt, 1976: Systematisk fortegnelse over Danmarks sommerfugle (A Catalogue of the Lepidoptera of Denmark). 128 pp. – Scandinavian Science Press, Klampenborg.
Klimesch, J., 1939: Zur Kenntnis der Biologie der *Scirtopoda myosotivora* M.-R. (Lep., Elachistidae). – Z. öst. EntVer. 24: 65-69.
– 1961: Lepidoptera I. Teil. pp. 481-789. *In* Franz, H.: Die Nordost-Alpen im Spiegel ihrer Landtierwelt. 2, 792 pp. – Innsbruck.
Klots, A.B., 1956: Lepidoptera. pp. 97-111, figs. 121-132. *In* Tuxen, S.L.: Taxonomist's Glossary of Genitalia in Insects. 284 pp. – Munksgaard, Copenhagen.
Krogerus, H., Opheim, M., Schantz, M. von, Svensson, I. & Wolff, N. L., 1971: Catalogus

Lepidopterorum Fenniae et Scandinaviae. Microlepidoptera. 40 pp. – Helsingin Hyönteisvaihtoyhdistys, Helsinki.
Krone, W., 1909: Drei neue Microlepidopteren. – Jber. wien. ent. Ver. 19: 129–133.
– 1911: Neubeschribungen einiger Arten und Varietäten von Microlepidopteren. – Ibid. 21: 39–42.
Larsen, C. S., 1916: Fortegnelse over Danmarks Microlepidoptera. – Ent. Meddr 11: 28–319.
– 1927: Tillæg til Fortegnelse over Danmarks Microlepidoptera. – Ibid. 17: 7–212.
Le Marchand, S., 1938: *Elachista littoricola,* n. sp. – Revue fr. Lépidopt. 9: 95–98.
– 1952: Tineina – Les Elachistidae. – Ibid. 13: 167–171.
Lhomme, L., 1935 [63]: Catalogue des Lépidoptères de France et de Belgique. 2, 1253 pp. – Douelle (Lot).
Lid, J., 1974: Norsk og Svensk Flora. 808 pp. – Det norske Samlaget, Oslo.
Linnaeus, C., 1761: Fauna Svecica. [43] + 578 pp., 2 pls. – Stockholmiae.
– 1767: Systema Naturae. Editio 12. 1 (2): 533–1328 + [36] pp. – Holmiae.
Machin, W., 1880: *Elachistra cerussella.* – Entomologist 13: 244.
MacKay, M. R., 1972: Larval sketches of some Microlepidoptera, chiefly North American. – Mem. ent. Soc. Can. 88: 1–83.
Martini, W., 1912a: Beiträge zur Kenntnis der *Elachista*-Raupen. – Dt. ent. Z. Iris 26: 92–95.
– 1912b: Über die Heinemann'sche *Elachista*-Gruppe D. – Ibid. 26: 185–188.
– 1912c: Beiträge zur Kenntnis von Arten der Gattung *Elachista* – Ibid. 26: 208–211.
– 1916: Verzeichnis Thüringer Kleinfalter aus den Familien Pyralidae – Micropterygidae. – Ibid. 30: 153–186.
Mehta, D. R., 1933: Comparative morphology of the male genitalia in Lepidoptera. – Rec. Indian Mus. 35: 197–266.
Mendes, C., 1909: Altera nova *Mendesia* ex Lusitania. – Broteria, Ser. Zool. 8: 65–67, pl. x.
Mesnil, L., 1965: Larvaevorinae (Tachininae). *In* Lindner, E.: Die Fliegen der Palaearktischen Region. 10 (2): 555–879, 27 pls.
Matsuda, R., 1976: Morphology and Evolution of the Insect Abdomen with special reference to developmental patterns and their bearings upon systematics. viii + 534 pp. – Pergamon Press, Oxford.
Meyrick, E., 1915: Descriptions of New Zealand Lepidoptera. – Trans. N.Z. Inst. 47: 201–244.
– 1919: Descriptions of New Zealand Lepidoptera. – Ibid. 51: 349–354.
– 1928: A revised handbook of British Lepidoptera. vi + 914 pp. – Watkins and Doncaster, London.
Morris, F. O., 1870: A Natural History of British Moths. 4, viii + 304 pp., index, pls. 97–132. – London.
Mosher, E., 1916: A classification of the Lepidoptera based on characters of the pupa. – Bull. Ill. St. Lab. nat. Hist. 12: 17–159.
Nielsen, E. Schmidt & Traugott-Olsen, E., 1977: On the status of the genera *Irenicodes* Meyrick, 1919 and *Euproteodes* Viette, 1954 (Lepidoptera, Elachistidae). – Ent. scand. 8.
– in press: Elachistidae (Lepidoptera) described by O. Staudinger, J. Mann and C. Mendes. – Entomologist's Gaz.
Palm, N.-B., 1943: Two new Swedish species of Tineina. – Opusc. ent. 8: 25–28.
Parenti, U., 1972: Revisione degli Elachistidi (Lepidoptera, Elachistidae) paleartici. I.-I Tipi di Elachistidi del Museo di Storia naturale di Parigi. – Boll. Mus. Zool. Univ. Torino 2: 29–56.
– 1973: Revisione degli Elachistidi (Lepidoptera, Elachistidae) paleartici. III. – Le specie di Elachistidi descritte da H. G. Amsel e F. Hartig. – Ibid. 3: 41–58.
– 1977: Revisione degli Elachistidi (Lepidoptera, Elachistidae) paleartici. IV. Le specie di Elachistidi descritte da H. Frey e P. C. Zeller. – Ibid. 3: 19–50.
Pierce, F. N. & Metcalfe, J. W., 1935: The Genitalia of the Tineid Families of the Lepidoptera of the British Islands. xxii + 116 pp., 68 pls. – Oundle, Northants.

Ragonot, E. L., 1889: Description de diverses espèces nouvelles de Microlépidoptères de France et d'Algérie: *Cochylis, Nemotois, Depressaria, Oecophora, Elachista* et *Eriocephala*. – Bull. Soc. ent. Fr. 1889: 105–107.
Razowski, J., 1973: Lepidoptera of Poland I – General Part. (Motyle (Lepidoptera) Polski Część I – Ogólna). 157 pp. (Translation of Monographie Fauny Polski 2), 1976 – Washington, Warsaw.
Rebel, H., 1899: Ueber einige heimische Arten der Gattung *Elachista* Tr. – Verh. zool.-bot. Ges. Wien 49: 523–526.
– 1901. *In* Staudinger, O. & Rebel, H.: Catalog der Lepidopteren des palaearctischen Faunengebietes. II. Theil. Famil. Pyralidae – Micropterygidae. 368 pp. – Friedländer & Sohn, Berlin.
–. 1906: Beschreibung eines neuen heimischen Kleinschmetterlings. – Verh. zool.-bot. Ges. Wien 56: 643.
– 1930: Neue Lepidoptera aus Bulgarien. – Ibid. 80: 12–15.
– 1936: Neue Mikrolepidopteren von Sardinia. – Dt. ent. Z. Iris 50: 92–100.
Reid, J., 1976: Techniques. pp. 117–134. *In* Heath, J.: The Moths and Butterflies of Great Britain and Ireland. 1. Micropterigidae – Heliozelidae. 343 pp., 13 pls. – Blackwell Scientific Publications, Oxford.
Robinson, G. S., 1976: The preparation of slides of Lepidoptera genitalia with special reference to the Microlepidoptera. – Entomologist's Gaz. 27: 127–132.
Rössler, A., 1864–67: Verzeichniss der Schmetterlinge des Herzogthums Nassau, mit besonderer Berücksichtigung der biologischen Verhältnisse und Entwicklungsgeschichte. – Jb. nassau. Ver. Naturk. 19–20: 99–442.
– 1880–81: Die Schuppenflügler (Lepidopteren) des Kgl. Regierungsbezirks Wiesbaden und ihre Entwicklungsgeschichte. – Ibid. 33–34: 1–393.
Sattler, K., 1973: A Catalogue of the family-group and genus-group names of the Gelechiidae, Holcopogonidae, Lecithoceridae and Symmocidae (Lepidoptera). – Bull. Br. Mus. nat. Hist. (Ent.) 28: 155-282.
Schantz, M. von, 1971: Zwei neue Kleinschmetterlinge (Microlepidoptera) aus Ostfennoskandien. – Notul. ent. 51: 99-104.
Schläger, F., 1847: Entdeckungen, Ergänzungen, Berichtigungen und sonstige Bemerkungen. – Ber. lepidopt. Tausch-Ver. 1847: 223–244.
Schmid, A., 1886: Die Lepidopteren-Fauna der Regensburger Umgebung mit Kehlheim und Wörth. II. Microlepidopteren. – KorrespBl. naturw. Ver. Regensb. 40: 19-224.
Schrank, F. v. P., 1802: Fauna Boica 2(2), 412 pp. – Ingolstadt.
Schütze, K.T., 1931: Die Biologie der Kleinschmetterlinge unter besonderer Berücksichtigung ihrer Nährpflanzen und Erscheinungszeiten. 235 pp. – Frankfurt am Main.
Sherborn, C.D., 1940: Where is the ______Collection? 148 pp. – Cambridge University Press, Cambridge.
Sircom, J., 1848: Description of *Lophoptilus Staintoni* and *Microsetia Bedellella*, two new British Moths of the family Tineidae. – Zoologist 6: 2037.
– 1849: Description of three new British Tineidae. – Zoologist 7, App.: 42.
Spuler, A., 1903-10: Die Schmetterlinge Europas. 2, 523 pp. (Elachistidae species treated by A. Meess, genera by A. Spuler). – Stuttgart.
Stainton, H.T., 1848: Description of three undescribed species of the genus *Aphelosetia*. – Zoologist 4 (7): 2164-2165.
– 1849: An attempt at a Systematic Catalogue of the British Tineidae & Pterophoridae. 32 pp. – London.
– 1851: A Supplementary Catalogue of the British Tineidae & Pterophoridae. Appendix. A Catalogue of the Tineidae obtained from Herr Joseph Mann, of Vienna, in 1849. 28 pp. – London.
– 1854a: Insecta Brittanica. Lepidoptera: Tineina. viii + 313 pp., 10 pls. – London.

– 1854b: On some of the difficulties of Entomological students, as exemplified by recent experience in the genus *Elachista*. – Trans. ent. Soc. Lond. (2) 3: 84-89.
– 1855: The new British Species in 1854. – Entomologist's Ann. 1855: 40-50.
– 1856: New British Species in 1855. – Ibid. 1856: 26–45.
– 1857: New British Species in 1856. – Ibid. 1857: 97-112.
– 1858a: On the aberrant species hitherto placed in the genus *Elachista*. – Trans. ent. Soc. Lond. 4: 261-271.
– 1858b: Synopsis of the genus *Elachista*. – Trans. ent. Soc. Lond. 4: 292-338.
– 1858c: The Natural History of the Tineina. 3, Elachista. Part 1. Tischeria. Part 1. vii + 269 pp., 8 pls. London, Paris, Berlin.
– 1859: New British Species (Lepidoptera) in 1858. – Entomologist's Ann. 1859: 145-157.
– 1878: Pupation of *Elachista gregsoni*. – Entomologist's mon. Mag. 14: 279.
– 1879: On *Elachista kilmunella* and some closely allied species. – Ibid. 15: 174-179.
– 1887: Description of a new species of *Elachista* allied to *rhynchosporella*, Stn. – Entomologist's mon. Mag. 23: 253-254.
Staudinger, O., 1870: Beschreibung neuer Lepidopteren des Europäischen Faunengebietes. – Berl. ent. Z. 14: 273-336.
Stephens, J.F., 1834-35: Illustrations of British Entomology. Haustellata 4, 436 pp., pls. 33-40. – London.
Steuer, H., 1973: Beiträge zur Kenntnis der Elachistiden (Lepidoptera). Teil I. – Dt. ent. Z. (N.F.) 20: 153-169.
– 1976a: Beiträge zur Kenntnis der Elachistiden (Lepidoptera). Teil II. – Ibid. 23: 165-179.
– 1976b: Zucht von Elachistidae. – Ent. Ber., Berl. 1976: 35-36.
Strand, E., 1919: Neue Kleinschmetterlinge aus Ranen in Nordland. – Nytt. Mag. 56: 121-127.
– 1925: Neubenennungen palaearktischer Lepidopteren und Apidae. – Arch. Naturgesch. A, 12: 281-283.
Svensson, I., 1966: New and confused species of Microlepidoptera. – Opusc. ent. 31: 183-200, 4 pls.
– 1974: Catalogus Insectorum Sveciae. VI. Microlepidoptera (1946). Additamenta II. – Ent. Tidskr. 95: 151-171.
– 1976a: Six new species of Microlepidoptera from northern Europe. – Ent. scand. 7: 195-206.
– 1976b: Anmärkningsvärda fynd av Microlepidoptera i Sverige 1975. – Ent. Tidskr. 97: 124-134.
Swezey, O.H., 1950: Notes on the occurrence and life history of the Honeysuckle leafminer, *Swezeyula lonicerae*, in Hawaii. – Proc. Hawaii. ent. Soc. 14: 197-200.
Swinhoe, C. & Cotes, E.C., 1889: A Catalogue of the Moths of India. Part VI. – Crambites, Tortrices, and addenda. pp. 671-777. – Calcutta.
Sønderup, H.P.S., 1924: Fra Smaasommerfugleomraadet. – Flora Fauna 1924: 70-75.
– 1949: Fortegnelse over de Danske Miner (Hyponomer). – Spolia zool. Mus. haun. 10: 1-256.
Tengström, J.M.J., 1848: Bidrag till Finlands Fjäril-Fauna. – Notis. Sällsk. Faun. Fl. fenn. Förh. 1: 69-164.
Thompson, W.R., 1944-58: A Catalogue of the parasites and predators of insect pests. 14 vols. – Commonzealth Institute of Biological Control, Belleville & Ottawa.
Thunberg, C., 1794: D.D. Dissertatio entomologica sistens Insecta svecica 7: 83-98, 1 pl. – Upsaliae.
Traugott-Olsen, E., 1973: Noter om europæiske arter af *Elachista* Tr. (Lep., Elachistidae). – Ent. Meddr 41: 77-82.
– 1974: Description of three new *Elachista* species, and nomenclatural remarks on other species of the genus (Lep., Elachistidae). – Entomologist's Gaz. 25: 259-268.

Treitschke, F., 1833: Die Schmetterlinge von Europa. 9 (2), 294 pp. – Leipzig.
Toll, S., 1936: Untersuchung der Genitalien bei *Pyrausta purpuralis* L. und *P. ostrinalis* Hb., nebst Beschreibung 11 neuer Microlepidopteren-Arten. – Annls Mus. zool. pol. 11: 403-413, 3 pls.
Viette, E.L., 1954: Une nouvelle espèce de Lépidoptère brachytère de l'île Campbell. – Ent. Meddr 27: 19-22.
Wallengren, H.D.J., 1852: Nya Svenska Lepidoptera. – Övers. K. VetenskAkad. Förh. 9:214–220.
– 1875: Species Tortricum et Tinearum Scandinaviae. – Bih. K. svenske VetenskAkad. Handl. 3 (5): 1-90.
– 1881: Genera nova Tinearum. – Ent. Tidskr. 2: 94-97.
Walsingham, Lord, 1907: Microlepidoptera of Tenerife. – Proc. zool. Soc. Lond. 1907: 911-1028, 3 pls.
– 1908: Spanish and Moorish Microlepidoptera. – Entomologist's mon. Mag. 44: 52-55, 226-229.
– 1909: Descriptions of Micro-Lepidoptera from Bolivia and Peru. – Trans. R. ent. Soc. Lond. 1909: 13-43.
Warren, W., 1878: *Elachista stabilella* bred. – Entomologist's mon. Mag. 15: 16.
Wocke, M.F., 1862: Reise nach Finmarken. II. Microlepidopteren. – Stettin. ent. Ztg 23: 233-257.
– 1876. *In* Heinemann, H.v.: Die Schmetterlinge Deutschlands und der Schweiz. 2. Abtheilung, Kleinschmetterlinge. 2, Die Motten und Federmotten. vi + 825 + 102. – Braunschweig.
Wörz, A., 1957: Die Lepidopterenfauna von Württemberg. II. Microlepidopteren. Kleinschmetterlinge. (8. Fortsetzung). – Jh. Ver. vaterl. Naturk. Württ. 112: 282-313.
Wolff, N.L., 1970: Notes on the synonymy of two *Elachista* species (Lep., Elachistidae). – Ent. Meddr 38: 173-176.
Zeller, P.C., 1839: Versuch einer naturgemässen Eintheilung der Schaben. – Isis, Leipzig 1839: 167-220.
– 1847: Bemerkungen über die auf Reise nach Italien und Sicilien gesammelten Schmetterlingsarten. – Ibid. 1847: 881-914.
– 1850: Verzeichniss der von Herrn Jos. Mann beobachteten Toscanischen Microlepidoptera. – Stettin. ent. Ztg 11: 195-212.
– 1853a: Caroli Clerck Icones insectorum rariorum ... critisch bestimmt. – Stettin. ent. Ztg 14: 199-214, 239-254, 271-294.
– 1853b: Lepidopterologisches. – Ibid. 14: 408-416.
– 1868: Beitrag zur Kenntniss der Lepidopteren – Fauna der Umgegend von Raibl in Oberkärnthen und Preth im angrenzenden Küstengebiet. – Verh. zool.-bot. Ges. Wien 18: 563-628.
Zetterstedt, J.W., 1838-40: Insecta Lapponica. vi + 1140 pp. – Lipsiae.
Zimmerman, E.C. & Bradley, J.D., 1950: A new species of Elachistidae mining Lonicera leaves in Hawaii (Lepidoptera). – Proc. Hawaii. ent. Soc. 14: 191-196.

Index to host-plants

Only generic names are listed, and families where actually mentioned in the text.

Index to entomological names

Synonyms are given in italics. The number in bold refers to the main treatment of the taxon.

Authors' addresses:

E. Traugott-Olsen
Calle de Munguia 8
El Mirador
Marbella
Spain

E. Schmidt Nielsen
Zoological Museum
Universitetsparken 15
DK-2100 Copenhagen Ø
Denmark

SWEDEN

Sk.	Skåne	Vrm.	Värmland
Bl.	Blekinge	Dlr.	Dalarna
Hall.	Halland	Gstr.	Gästrikland
Sm.	Småland	Hls.	Hälsingland
Öl.	Öland	Med.	Medelpad
Gtl.	Gotland	Hrj.	Härjedalen
G. Sand.	Gotska Sandön	Jmt.	Jämtland
Ög.	Östergötland	Ång.	Ångermanland
Vg.	Västergötland	Vb.	Västerbotten
Boh.	Bohuslän	Nb.	Norrbotten
Dlsl.	Dalsland	Ås. Lpm.	Åsele Lappmark
Nrk.	Närke	Ly. Lpm.	Lycksele Lappmark
Sdm.	Södermanland	P. Lpm.	Pite Lappmark
Upl.	Uppland	Lu. Lpm.	Lule Lappmark
Vstm.	Västmanland	T. Lpm.	Torne Lappmark

NORWAY

Ø	Østfold	HO	Hordaland
AK	Akershus	SF	Sogn og Fjordane
HE	Hedmark	MR	Møre og Romsdal
O	Opland	ST	Sør-Trøndelag
B	Buskerud	NT	Nord-Trøndelag
VE	Vestfold	Ns	southern Nordland
TE	Telemark	Nn	northern Nordland
AA	Aust-Agder	TR	Troms
VA	Vest-Agder	F	Finnmark
R	Rogaland		

n northern s southern ø eastern v western y outer i inner

FINLAND

Al	Alandia	Kb	Karelia borealis
Ab	Regio aboensis	Om	Ostrobottnia media
N	Nylandia	Ok	Ostrobottnia kajanensis
Ka	Karelia australis	ObS	Ostrobottnia borealis, S part
St	Satakunta	ObN	Ostrobottnia borealis, N part
Ta	Tavastia australis	Ks	Kuusamo
Sa	Savonia australis	LkW	Lapponia kemensis, W part
Oa	Ostrobottnia australis	LkE	Lapponia kemensis, E part
Tb	Tavastia borealis	Li	Lapponia inarensis
Sb	Savonia borealis	Le	Lapponia enontekiensis

USSR

Vib Regio Viburgensis Kr Karelia rossica Lr Lapponia rossica